A121X4

ADVANCED DYNAMICS FOR ENGINEERS

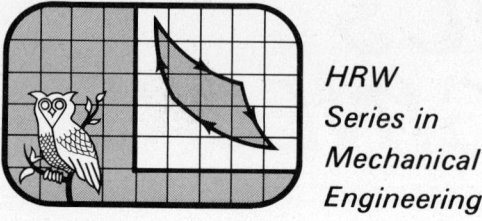

HRW Series in Mechanical Engineering

L. S. Fletcher, Series Editor

F. L. Stasa   APPLIED FINITE ELEMENT ANALYSIS FOR ENGINEERS
B. J. Torby   ADVANCED DYNAMICS FOR ENGINEERS

ADVANCED DYNAMICS FOR ENGINEERS

BRUCE J. TORBY

Department of Mechanical Engineering
California State University, Long Beach

Holt, Rinehart and Winston
New York   Chicago   San Francisco   Philadelphia
Montreal   Toronto   London   Sydney   Tokyo
Mexico City   Rio de Janeiro   Madrid

Copyright © 1984 CBS College Publishing
All rights reserved.
Address correspondence to:
383 Madison Avenue, New York NY 10017

Library of Congress Cataloging in Publication Data

Torby, Bruce J.
  Advanced dynamics for engineers.
  (HRW series in mechanical engineering)
  Includes index.
  1. Dynamics.   I. Title.   II. Series.
TA352.T67   1984   531'.11   84–688

ISBN 0-03-063366-4

Printed in the United States of America
Published simultaneously in Canada

    5 6 7   038   9 8 7 6 5 4 3 2

CBS COLLEGE PUBLISHING
Holt, Rinehart and Winston
The Dryden Press
Saunders College Publishing

TO MY WIFE BIRGITTA AND DAUGHTER LINDA

CONTENTS

PREFACE — xi

Part One: Review of Newtonian Mechanics — 1

1 INTRODUCTORY REMARKS — 3

- 1.1 Newton's Laws — 3
- 1.2 The Law of Gravitation — 4
- 1.3 Units — 4
- 1.4 Inertial Reference — 7

2 PARTICLE DYNAMICS: KINEMATICS — 10

- 2.1 Definitions — 10
- 2.2 Cartesian Coordinates — 11
- 2.3 Normal and Tangential Coordinates — 25
- 2.4 Radial and Transverse Components (Polar Coordinates) — 27
- 2.5 Cylindrical Coordinates — 29
- 2.6 Spherical Coordinates — 30
- 2.7 Relative Motion — 31
- 2.8 Numerical Solutions to Ordinary Differential Equations with Initial Conditions — 33
- Problems — 47

3 PARTICLE DYNAMICS: KINETICS — 53

- 3.1 Equations of Motion — 53
- 3.2 Integrals of the Motion — 55
- Problems — 70

4 SYSTEMS OF PARTICLES — 76

- 4.1 The Center of Mass — 76
- 4.2 The Equations of Motion for a System of Particles — 78
- 4.3 Integrals of the Equations of Motion — 81
- Problems — 94

5 RIGID BODY DYNAMICS: KINEMATICS — 100

- 5.1 Degree of Freedom — 100
- 5.2 Translation — 101

viii CONTENTS

5.3	Rotation about a Fixed Point	101
5.4	General Motion	108
5.5	General Plane Motion	110
5.6	Time Derivative of a Vector Referenced to a Rotating Frame	121
5.7	Motion Relative to the Rotating Earth	143
	Problems	147

6 RIGID BODY PLANE MOTION — 157

6.1	Equations of Motion	157
6.2	The Mass Moment of Inertia Matrix: Transformations	163
6.3	Eigenvalues and Eigenvectors of the Inertia Matrix	167
6.4	The Moment Equation about an Arbitrary Point	181
6.5	Principle of Work and Kinetic Energy for a Rigid Body	184
6.6	Principle of Impulse and Momentum for a Rigid Body	189
	Problems	193

Part Two: Analytical Mechanics — 203

7 RIGID BODY DYNAMICS — 205

7.1	Introduction	205
7.2	General Rigid Body Motion	205
7.3	Modified Euler Equations	216
7.4	Eulerian Angles	218
7.5	General Force-Free Motion of a Rigid Body: Geometric Solution	221
7.6	Force-Free Motion of Axially Symmetric Shapes	224
7.7	The Motion of a Symmetrical Top	230
7.8	Gyroscopic Instruments	238
	Problems	248

8 ENERGY METHODS — 258

8.1	Introduction	258
8.2	Independent Generalized Coordinates and Constraint Equations	259
8.3	Virtual Work and Generalized Forces	263
8.4	Lagrange's Equations	268
8.5	Lagrange Multipliers	282
8.6	The Routhian	284
8.7	Hamilton's Principle	295
8.8	The Hamiltonian	303
8.9	Hamilton's Equations	307
	Problems	310

9 THEORY OF SMALL OSCILLATIONS — 320

9.1	Introduction	320
9.2	Normal Modes of Vibration	322

9.3	Repeated Roots	327
9.4	Normal Coordinates	327
9.5	Forced Vibrations with Damping Present	333
9.6	Numerical Methods	337
9.7	Small Oscillations about Steady Motion	347
	Problems	355

10 DIGITAL COMPUTER SIMULATION   *367*

10.1	Introduction	367
10.2	The Analog Computer	367
10.3	Digital Computer Simulation	373
10.4	CSSL, Version 4	374
	Problems	399

Appendix A   Short Summary of Elementary Dynamics   401

Appendix B   Mass Moments of Inertia of Common Shapes   403

Answers to Even-Numbered Problems   406

List of Symbols   416

Index   421

PREFACE

Much has happened recently in the practice of engineering. The introduction of advanced semiconductor technology has brought down the cost of computers to such a great extent that they are nearly universally available. As a result, the engineering problems of only two decades ago that were too difficult to solve without elaborate and expensive computer facilities have now become accessible for analysis in the classroom. Textbooks are no longer forced to linearize equations to affect solutions, nor must students wait until they graduate to deal with more concrete problems.

In response to these changes, this text on advanced dynamics introduces the use of computer analysis and numerical methods in the solution of dynamics problems. I believe there is much advantage to this approach. Too often the engineering student takes an isolated service area course covering the fundamentals of programming and numerical analysis and then is never required to use this knowledge in his or her major area of study. The student therefore never integrates what has been learned and rarely experiences how a computer is put to use in his or her own chosen field. In addition to the benefits provided by the extra training and exposure to computers, it is also my belief that the learning of mechanics is enhanced in this manner. Computer solutions provide results for problems whose system equations are so complex and nonlinear that dynamic behavior is impossible to visualize. When motions can be accurately described, cause-and-effect relationships can better be understood and grasped by the student.

The introduction of numerical methods, however, must be kept secondary to the subject matter of dynamics. I decided, when planning this book, that the best way to keep this balance was to have a rigorous standard text in advanced dynamics as the main body of the work while letting the sample problems provide the extra material. For example, sample dynamic problems requiring computer solutions are introduced. They are then solved, and the source listing and results are shown. The numerical methods that assisted in the solution are discussed briefly. At the end of key chapters there are additional problems that require computer analysis as well. The sample problem format also allowed more specialized applications to be introduced. For instance, I believe that a discussion of rocket propulsion, satellite motion (the two-body problem), the gyrocompass, etc., should not be placed in the main structure of a book dealing with general dynamics theory. They are important enough not to be neglected; they are included as lengthy sample problems.

The numerical methods that are introduced in the book are limited to those that I have found most useful in solving advanced dynamics problems. They obviously can be supplemented. In the text there are programs illustrating Euler's method, the modified Euler method, the Runge-Kutta algorithm, matrix inversion, Newton-Raphson method, Jacobi's method, and matrix iteration with modal sweeping to attain natural frequencies.

The last chapter of the book, Chapter 10, is devoted solely to the introduction and study of a typical continuous system simulation language. The language demonstrated here, CSSL, version 4, is a higher-level language that includes a command that accomplishes integration of a function with respect to time, INTEG. The CSSL translator writes the Fortran program for the user and allows the user to select from several optional integration routines.

The book is divided into two parts. Part One contains an extensive review of elementary dynamics. I believe that the advantage of such a lengthy written review outweights its omission. It has been my experience that students entering an advanced dynamics class have various backgrounds and preparation levels. Some have not studied dynamics for several years; some have come from different universities. Including in one reference all the necessary resource material that a student will need, written by the same author, offers the reader a uniformity of approach and nomenclature. It also allowed the building of a larger, more thorough, pedagogical framework, with the assurance that all the primary building blocks are in place. To help keep reader interest, the material is presented on a much higher level than what is found in elementary textbooks. It is suggested that for a one-semester course in intermediate or graduate dynamics that the time scheduled for Part One be kept brief.

The second part of the book covers the topics of general rigid body motion, Euler's equations, the routhian, Hamilton's principle, the hamiltonian, Hamilton's equations, the theory of small oscillation about an equilibrium point or about steady motion, and finally, digital computer simulation.

Chapter 1 reviews fundamentals, Newton's laws, the law of gravitation, units, and the use of an inertial reference frame.

Chapter 2 reviews the kinematics of particles, introducing normal and tangential, cylindrical, and spherical coordinates. The subject of single degree of freedom vibration is reviewed quickly but thoroughly. The topic of relative motion is also included here.

Chapter 3 discusses the kinetics of single particles. First, Newton's second law is expressed in several coordinates, and moments of forces and angular momentum terms are defined. The principle of work and kinetic energy, conservative forces, power expressions, the principles of linear impulse and momentum, and angular impulse and angular momentum are subsequently derived.

In Chapter 4, systems of particles are studied. The center of mass is defined. Equations of motion, moment equations referenced to a fixed point or the center of mass, and the principles of work and kinetic energy and impulse and momentum are rederived for the multiparticle system. A discussion of impact as well as a sample problem on rocket propulsion is included.

Chapter 5 reviews the kinematics of rigid body dynamics. The concept of degrees of freedom, translation, rotation about a fixed point, rotation about a fixed axis, 2 degrees of rotational freedom, angular acceleration, 3 degrees of rotational freedom, general motion, general plane motion, mechanisms, the method of instantaneous centers, time derivatives of vectors referenced to rotating frames, vector transformations, orthogonality conditions, infinitesimal rotation matrices, and motion relative to the rotating earth are all discussed. Many sample problems are presented here. It is my conviction that most students have not mastered this material from their elementary courses. Without attaining these analytical skills, further work in three-dimensional motion and with Lagrange's equations (based on velocity expressions) will be encumbered.

Chapter 6 concerns itself with rigid body plane motion. The general equations developed previously for a system of particles are now restricted and simplified to the rigid body. The mass moment and product of inertia matrices about a fixed point and the center of mass are defined. Principal axes are also introduced. The general equations of motion for a rigid body are then limited to plane motion applications. Several important sample problems review this topic. Similarity transformations for the inertia matrix, the inertia ellipsoid, the parallel axis theorem, and the eigenvalue and eigenvector problem are discussed. Moment equations about an arbitrary point, the work energy principle, and the principle of impulse and momentum for a rigid body are all studied.

In Part Two, *Analytical Mechanics*, Chapter 7 concentrates on general rigid body dynamics in three dimensions. The general moment equations for body-fixed axes that are not principal axes are discussed first. Using principal axes, Euler's moment equations are then derived, followed by a derivation of the modified Euler equations for bodies of revolution. Eulerian angles are introduced. The geometric solution to the general force-free motion of a rigid body is considered. Exact solutions for force-free motion of axially symmetric shapes are analyzed. The discussion of the motion of a symmetrical top closes the chapter's theoretical development. The last section in the chapter reviews and describes gyroscopic instruments.

Chapter 8 introduces energy methods. Discussions of independent generalized coordinates, constraint equations, virtual work, and generalized forces lead to the development of Lagrange's equation. Cyclic coordinates, conservation theorems, and Lagrange multipliers are also studied. The routhian is then introduced and used to solve the classic two-body problem in detail. Hamilton's principle, followed by a discussion of the hamiltonian and Hamilton's equations, closes this essential chapter.

Chapter 9 reviews the theory of small oscillations. Normal modes of vibration, the stiffness and flexibility matrix, the characteristic equation derived from the eigenvalue problem, mode shapes, orthogonality conditions, and the generalized mass and generalized stiffness matrices are studied. Normal coordinates are introduced. Forced vibrations with damping present are treated using state variables and the system matrix. Numerical methods are discussed in great detail here. The chapter ends with a consideration of small oscillations about steady motion.

The last chapter, Chapter 10, introduces the topics of computer simulation.

First the theory and history of the analog computer are reviewed. This is followed by an elementary discussion of digital and hybrid computer simulation. Finally, an extensive and thorough summary of the features of the continuous system simulation language (CSSL), version 4, is presented. The reader who has physical access to this language can study this chapter earlier and apply its powerful numerical routines wherever useful throughout the book.

There are many people who assisted with this book. I would like to express my appreciation to all of them. I would like to thank my advanced dynamic classes at California State University, Long Beach, for their constructive criticisms. Special thanks go to students Ms. Patricia Baker, Ms. Deborah Selnick, and Mr. Che-Kuang Lin for proofreading the original manuscript and to Mr. Robert Luher who solved the numerical homework problems.

The original manuscript was typed at Chalmers University of Technology, Gothenburg, Sweden. I am indebted to Mrs. Gunilla Ekman and Mrs. Irmgard Molin for their help. I am still impressed by how they read my scribbled handwriting, in what for them is a foreign language, and produced near-perfect copy. I would also like to offer thanks to the faculty, staff, and students at Chalmers University of Technology. I am extremely grateful to Professor Bengt Holmberg for inviting me to visit at the Division of Mechanics which he chairs, and for the generous permission he gave to me to freely use in my book problems from his *Kompendium i Mekanik MK*, Chalmers Tekniska Högskola, 1982, and from past final exams in mechanics, written and administered by the faculty and graduate assistants of the division. His contribution greatly enhanced the breadth and quality of the homework exercises given at the end of each chapter.

The final draft was done at California State University, Long Beach. Here, thanks are due to Professor Hillar Unt, chairman of the mechanical engineering department, for his support of the project, and to Mrs. Helen Tyler and Ms. Laju Tejwani who performed a superb job in typing the manuscript.

Finally, I would like to thank those people who took the time to review and offer suggestions for the manuscripts development. Professor Leroy S. Fletcher, Series Editor, was especially encouraging and helpful.

Bruce J. Torby

ADVANCED DYNAMICS FOR ENGINEERS

PART ONE: REVIEW OF NEWTONIAN MECHANICS

INTRODUCTORY REMARKS 1

The engineer is very interested in dynamics because it is the branch of mechanics that deals with bodies in motion. Its two main divisions are the study of motion itself (kinematics) and the study of how external forces acting upon a body influence its motion (kinetics). Because shock and vibration considerations are so important in engineering design, a knowledge of dynamics is essential for the practicing mechanical engineer. In fact, dynamic studies are required for most engineering practice in all branches of the profession. They can be used, for example, by the structural designer of buildings who is concerned with earthquake loading or by the electrical control engineer who needs to understand the equations of motion governing a mechanical system so that overall electromechanical system response can be improved. The aerospace engineer, for example, investigates the effects of aerodynamic flutter on structural loading or designs for shock tolerance in the landing legs of a lunar capsule. It is the intent of this book on dynamics to first review elementary classical dynamic concepts and then introduce the more advanced topics that enhance understanding and facilitate for the engineer the analysis of more complex systems. This chapter reexamines briefly the fundamental laws of dynamics.

1.1 NEWTON'S LAWS

The foundation which the study of dynamics is built upon is the work of Sir Isaac Newton (1642–1727). His *Principia Mathematica Philosophiae Naturalis* (London, 1687) describes the laws of motion for a particle. This work was a breakthrough in many ways for early science. Though the field of statics was highly developed at the time, dating back to the works of Archimedes (287?–212 B.C.), it was Newton's description of the relationship between acceleration and force on a body that set the basis for the study of accelerating objects. The following is a modern day paraphrasing of Newton's laws for a *particle*. In the subject of dynamics, a particle is defined to be any rigid object that does not rotate. Its size, therefore, can be neglected.

1. If no forces act on a particle, the particle will remain at rest or continue motion in a *straight line* with *constant speed.*
2. The resultant force acting on a particle is proportional to the acceleration of the particle in magnitude and is parallel to it in direction $\mathbf{F} \sim \mathbf{a}$.
3. For each action (force) exerted upon one body by another, there is an equal, opposite, and collinear force reaction upon the other body.

4 REVIEW OF NEWTONIAN MECHANICS

The second law can be written as

$$\mathbf{F} = m\mathbf{a} \tag{1.1}$$

where m, the proportionality constant, is defined to be the mass of the particle. A fundamental premise in *classical*, or newtonian, mechanics is that the mass in equation (1.1) is constant. Static equilibrium is defined by

$$\mathbf{F} = 0 \tag{1.2}$$

Inherent in the statement of the second law is the introduction of the derivative. Among his many accomplishments, Newton is credited with the independent development of the differential calculus. The time derivative of the velocity vector, the acceleration, is defined to be

$$\mathbf{a} = \lim_{\Delta t \to 0} \frac{\Delta \mathbf{v}}{\Delta t} \tag{1.3}$$

This concept of a limiting process, or an instantaneous average, is used to describe an "instantaneous state of change."

1.2 THE LAW OF GRAVITATION

If one considers the total body of knowledge available to him at that time, the attainments of Newton are truly inspiring. The promulgation of the law of universal gravitation is yet another example. Based on the experimental studies of planetary motion, it states that the force of gravitational attraction acting between two particles is proportional to their masses, M and m, and inversely proportional to the square of the separation distance r. The force of attraction acts along the line joining the particles. The proportionality constant is G, the universal constant of gravity. G is approximately equal to 6.673×10^{-11} m^3/kg·s^2.

$$\mathbf{F} = -\frac{GMm}{r^3} \mathbf{r} \tag{1.4}$$

1.3 UNITS

Historically, there has been much confusion concerning units. The international scientific and engineering community have recently decided upon the following set of units, called SI units, for the primary dimensions of mass M, length L, and time T present in Newton's second law, equation (1.1).

F: newton (N = 1 kg·m/s^2), ML/T^2
m: kilogram (kg), M
l = length: meter (m), L
a = acceleration: meters per second squared (m/s^2), L/T^2

This is an "absolute" set of units based on mass, which is an invariant, i.e., the mass quantity is independent of gravitational attraction. In this system, g, the acceleration due to gravity, can be taken as 9.81 m/s² when close to the earth's surface. Hence, the weight of a freely falling body near the earth's surface, or the magnitude of the gravitational force acting upon it, is

$$W = mg \tag{1.5}$$

The SI units are similar to the units of the universally known metric system. In United States–British units,

F: pound (lb stands for pound-force, not pound-mass or poundal)
m: slug (1 lb · s²/ft)
l: feet (ft)
a: feet per second squared (ft/s²)

TABLE 1.1 Conversion from American Units to Common SI Units

	American Units	SI Units
Acceleration	ft/s²	0.3048 m/s²
Angle	rad	rad
Angular velocity	rad/s	rad/s
Angular acceleration	rad/s²	rad/s²
Area	ft²	0.0929 m²
Density	slug/ft³	515.2 kg/m³
Energy	ft · lb	1.356 J (joules)
Force	lb	4.448 N (newtons)
Frequency	1/s, Hz	1/s, Hz (hertz)
Impulse	lb · s	4.448 N · s
Length	ft	0.3048 m
Mass	slug	14.59 kg
	lb$_{mass}$	0.4536 kg
	ton	907.2 kg
Moment of a force	lb · ft	1.356 N · m
Moment of inertia	in⁴ (area)	0.4162 (10²) cm⁴
	lb · ft · s² (mass)	1.356 kg · m²
Momentum	slug · ft/s	4.448 kg · m/s
Power	ft · lb/s	1.356 W (watts)
	hp	745.7 W
Pressure, stress	lb/in²	6.895 kPa (kilopascal)
Velocity	ft/s	0.3048 m/s
	mi/h (mph)	0.4470 m/s
Volume	ft³	0.02832 m³
	gal	3.785 liters
Work	ft · lb	1.356 J

6 REVIEW OF NEWTONIAN MECHANICS

TABLE 1.2   Multiplication Factors and SI Unit Prefixes

SI Prefix	Symbol	Factor
atto	a	10^{-18}
centi	c	10^{-2}
deci	d	10^{-1}
deka	da	10^{1}
femto	f	10^{-15}
giga	G	10^{9}
hecto	h	10^{2}
kilo	k	10^{3}
mega	M	10^{6}
micro	μ	10^{-6}
milli	m	10^{-3}
nano	n	10^{-9}
pico	p	10^{-12}
tera	T	10^{12}

This is a "relative" set of units dependent upon the local force of gravitational attraction. Since the pound is defined by the *weight* of a platinum *standard pound*, it could vary according to the placement of the standard on the earth's surface. By agreement, the placement of the standard has been established to be at 45° latitude and at sea level.

In these units, the number of slugs can be found by using equation (1.5) with g taken to be 32.2 ft/s²

$$\text{Number of slugs} = m = \frac{W}{32.2} \tag{1.6}$$

Some useful conversions between the two sets of units are

1 ft = 0.3048 m
1 lb = 4.448 N
1 slug = 24.59 kg

Sample Problem 1.1
A derived expression for the acceleration of a particle is given by the following relation:

$$a = \frac{m\alpha}{\beta} \ln\left(1 + \frac{v}{\alpha}\right)$$

Find the basic dimensions (M, L, T) for the constants α and β, and express these dimensions in SI units.

Analysis: The basic dimensions of a, m, and v are

$$a = \text{acceleration}, L/T^2$$
$$m = \text{mass}, M$$
$$v = \text{speed}, L/T$$

For $\ln x$, x must be dimensionless. Hence α has the same dimension as v

$$\alpha = \frac{L}{T} \quad \text{m/s}$$

Acceleration a has the dimension L/T^2, then

$$\beta = \frac{m\alpha}{a} \ln\left(1 + \frac{v}{\alpha}\right)$$

$$\beta = \frac{ML/T}{L/T^2} = MT \quad \text{kg} \cdot \text{s} \qquad \blacksquare$$

1.4 INERTIAL REFERENCE

A cursory glance at the first law leads to the impression that it is a restatement of the concepts in the second law. The second law states that if the sum of the forces acting on a particle is zero, then the acceleration must be zero. Since the acceleration is the derivative of a vector quantity, having both magnitude and direction, the changes in magnitude (speed) and path must be zero. On closer inspection, and in review of Newton's original work, it becomes apparent that this is not what is meant at all. What Newton was describing in his first law was an inertial or absolute frame of reference for measurement. An inertial reference coordinate system would neither accelerate nor rotate. If the sum of the forces acting on a particle is zero and the observer has a frame of reference that accelerates and rotates, the observer will see a relative path *that is curved*. Newton realized this and specifically assumed that an absolute frame of reference existed in the universe, even though he was unable to identify it. His first law does not specify a location of a unique origin of an absolute coordinate system. In fact, it is not required. Only the frame is characterized. An absolute reference frame in our physical world can only be approximated. For most engineering applications, the surface of an assumed nonrotating earth is used as an absolute coordinate system reference. In other applications where the earth's rotation becomes significant, a coordinate system placed at the center of an assumed stationary earth, pointing to the celestial north pole and the vernal equinox, is used. Another commonly employed reference frame assumes that the center of mass of the solar system is fixed. The coordinate reference axes are at the center-of-mass pointing to two "fixed" stars.

It is impressive that Newton had the ability to visualize an absolute path without ever experiencing inertial observations. In his studies of planetary motion, he himself always accelerated, always rotated. For him to understand

8 REVIEW OF NEWTONIAN MECHANICS

the need of having an inertial reference and then to also see what motion would look like in this supposed system, viewed from a point off the earth's surface, is indeed an accomplishment for the seventeenth century.

Sample Problem 1.2
Study the cart and package shown in the figure. The cart's mass is m_C and the package has mass m_P. When a constant force F is applied to the cart, it is observed that the package slides along the cart's length s in $\sqrt{s/2}$ seconds starting from rest. What is the acceleration of the package? Assume that friction acts between the cart and package but not between the cart and floor.

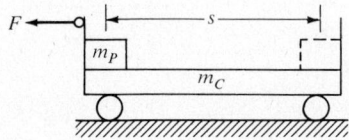

Analysis: Since two masses are involved, two sets of equations of motion are necessary.

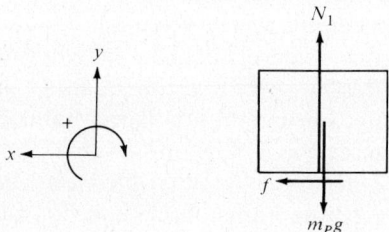

From the free-body diagram shown of the package,

$$\Sigma F_x = m_P(a_P)_x \qquad f = m_P(a_P)_x \qquad \text{(a)}$$
$$\Sigma F_y = m_P(a_P)_y \qquad N_1 - m_P g = 0 \qquad N_1 = m_P g$$

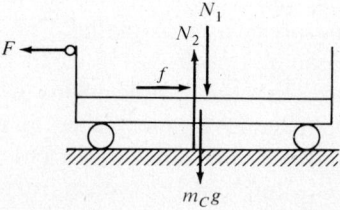

For the cart,

$$\Sigma F_x = m_C(a_C)_x \qquad F - f = m_C(a_C)_x \qquad \text{(b)}$$
$$\Sigma F_y = m_C(a_C)_y \qquad N_2 - N_1 - m_C g = 0 \qquad N_2 = (m_P + m_C)g$$

Combining equations (a) and (b) yields

$$F = m_P(a_P)_x + m_C(a_C)_x \qquad \text{(c)}$$

where both accelerations are *absolute*.

The motion described of the package, however, is relative to the cart.

$$a_{P/C} = a_P - a_C$$

With F and f constant, a_P and a_C are also constant. Then

$$a_{P/C} = \ddot{x}_{P/C} = \text{constant}$$
$$\dot{x}_{P/C} = a_{P/C} t + C_1$$
$$x_{P/C} = \frac{a_{P/C} t^2}{2} + C_1 t + C_2$$

The initial conditions have the package starting from rest; hence $C_1 = C_2 = 0$. At $t = \sqrt{s}/2$,

$$s = x_{P/C} = \frac{a_{P/C} t^2}{2}$$

or

$$a_{P/C} = \frac{2s}{s/4} = a_P - a_C = 8$$

Then

$$a_P = \frac{F + 8 m_C}{m_P + m_C}$$

∎

PARTICLE DYNAMICS: KINEMATICS

2.1 DEFINITIONS

Newton's second law, equation (1.1), the starting point from which all is to follow, introduces the concept of acceleration. It is, therefore, appropriate to begin the study of dynamics with the subject of kinematics, the geometry of motion. If **r** is defined to be the position vector of a particle measured from an absolute fixed reference point (see Figure 2.1), then the *absolute instantaneous velocity* is defined as

$$\mathbf{v} \equiv \lim_{\Delta t \to 0} \frac{\Delta \mathbf{r}}{\Delta t} = \frac{d\mathbf{r}}{dt} = \dot{\mathbf{r}} \tag{2.1}$$

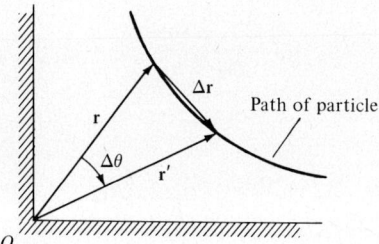

Figure 2.1

It is an "instantaneous average." The magnitude of the velocity v is called the *speed*. In the limit, as can be seen in Figure 2.1, the chord $\Delta \mathbf{r}$ becomes tangent to the path so that

$$\mathbf{v} = v \mathbf{i}_t \tag{2.2}$$

where $\mathbf{i}_t$ is a unit vector in the tangential direction.

The *absolute instantaneous acceleration* is defined by

$$\mathbf{a} \equiv \lim_{\Delta t \to 0} \frac{\Delta \mathbf{v}}{\Delta t} = \frac{d\mathbf{v}}{dt} = \dot{\mathbf{v}} \tag{2.3}$$

Also

$$\dot{\mathbf{v}} = \ddot{\mathbf{r}} = \frac{d^2 \mathbf{r}}{dt^2}$$

Since **v** is a vector, it can change both in magnitude and/or direction (Figure 2.2).

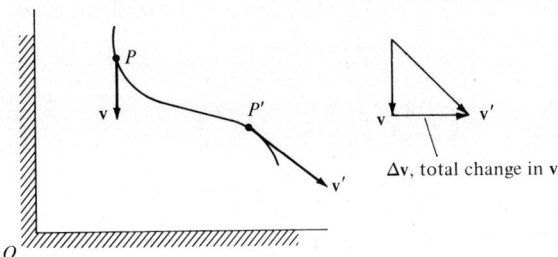

Figure 2.2

2.2 CARTESIAN COORDINATES

In the cartesian coordinate system (Figure 2.3), the rectangular components of the vectors **r**, **v**, and **a** are expressed by

$$\mathbf{r} = x\mathbf{i} + y\mathbf{j} + z\mathbf{k} \tag{2.4}$$

and
$$\mathbf{v} = \dot{x}\mathbf{i} + \dot{y}\mathbf{j} + \dot{z}\mathbf{k} \tag{2.5}$$

$$\mathbf{a} = \ddot{x}\mathbf{i} + \ddot{y}\mathbf{j} + \ddot{z}\mathbf{k} \tag{2.6}$$

where $\dot{x} = v_x$, $\dot{y} = v_y$, $\ddot{x} = \dot{v}_x = a_x$, $\ddot{y} = \dot{v}_y = a_y$, etc.

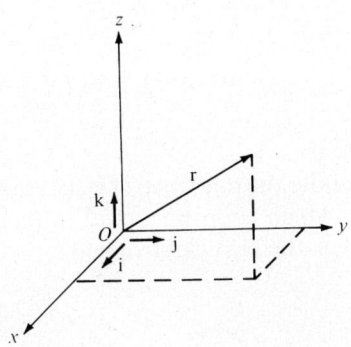

Figure 2.3

The magnitude of **a** is

$$|\mathbf{a}| = (a_x^2 + a_y^2 + a_z^2)^{1/2} \tag{2.7}$$

For integration purposes, the independent variable may be changed to distance.

$$a_x = \dot{v}_x = \frac{dv_x}{dt}$$

$$a_x = \frac{dv_x}{dt}\frac{dx}{dx} = v_x \frac{dv_x}{dx} \tag{2.8}$$

12   REVIEW OF NEWTONIAN MECHANICS

The same change may be made for all the scalar components of acceleration. For straight line motion

$$a = v \frac{dv}{ds} \qquad (2.9)$$

Sample Problem 2.1
In a series of test-tank experiments it is found that a model boat, with a given initial speed v_0, will stop after a distance $S = 5 \ln(1 + v_0/4)$. Determine the initial resistive force of the water.

Analysis:

$$F = m\ddot{x} = mv \frac{dv}{dx} = -f(v)$$

$$S = -\int_{v_0}^{0} \frac{mv\, dv}{f(v)}$$

$$S = \int_{0}^{v_0} \frac{mv\, dv}{f(v)} = 5 \ln\left(1 + \frac{v_0}{4}\right) \qquad (a)$$

Taking the derivatives of both sides of equation (a) with respect to v_0 yields

$$\frac{mv_0}{f(v_0)} = \frac{5}{4(1 + v_0/4)}$$

$$f(v_0) = \frac{4}{5} m \left(v_0 + \frac{v_0^2}{4}\right) \qquad \blacksquare$$

Sample Problem 2.2
A particle falls from rest under the influence of a constant gravitational acceleration g and air resistance, which is proportional to its speed, so that $a = g - cv$. After t seconds, how far has the particle dropped?

Analysis:

$$a = g - cv$$

$$a = \frac{dv}{dt} = \ddot{x} = g - c\dot{x}$$

$$\ddot{x} + c\dot{x} = g$$

$$D(D + c)x = g$$

where D is the differential operator. For the complementary solution

$$D = 0 \qquad D = -c$$

$$x_c = A + Be^{-ct}$$

A trial solution for x_p of the form $a_1 t$ yields

$$x_p = \frac{gt}{c}$$

$$x = A + Be^{-ct} + \frac{gt}{c}$$

At $t=0$, $x=0$ and $\dot{x}=0$. It follows then that

$$0 = A + B$$

and

$$0 = -cB + \frac{g}{c}$$

or

$$B = \frac{g}{c^2} \qquad A = -\frac{g}{c^2}$$

Thus

$$x = \frac{g}{c^2}(e^{-ct} - 1) + \frac{gt}{c}$$

∎

Some special motions will now be discussed.

A. **Uniform rectilinear motion, $a=0$**

$$v = \text{constant} = v_0 = \frac{ds}{dt} \tag{2.10}$$

$$s = v_0 t + s_0 \tag{2.11}$$

B. **Uniformly accelerated rectilinear motion, $a=$ constant**

$$a = \frac{dv}{dt} = v\frac{dv}{ds} = \text{constant}$$

$$v = \int_0^t a\, dt = at + v_0 \tag{2.12}$$

$$\frac{ds}{dt} = v$$

$$s = \frac{at^2}{2} + v_0 t + s_0 \tag{2.13}$$

$$v^2 = v_0^2 + 2a(s - s_0) \tag{2.14}$$

C. **Freely falling body, $a = g = 9.81$ m/s^2**

$$v_y = -9.81t + v_{y,0} \tag{2.15}$$

$$y = -\frac{9.81 t^2}{2} + v_{y,0} t + y_0 \tag{2.16}$$

D. Motion of a projectile in a uniform gravity field

$$\mathbf{r} = x\mathbf{i} + y\mathbf{j} \tag{2.17}$$

$$\mathbf{v} = v_x\mathbf{i} + v_y\mathbf{j} \tag{2.18}$$

$$\mathbf{a} = a_x\mathbf{i} + a_y\mathbf{j} \tag{2.19}$$

In cartesian coordinates (Figure 2.4), the acceleration becomes

$$a_x = 0 \tag{2.20}$$

$$a_y = -g \tag{2.21}$$

Therefore

$$v_x = v_{x,0} = v_0 \cos \gamma \tag{2.22}$$

$$x = v_{x,0} t + x_0 \tag{2.23}$$

$$v_y = -gt + v_{y,0} = -gt + v_0 \sin \gamma \tag{2.24}$$

$$y = -\frac{gt^2}{2} + v_{y,0} t + y_0 \tag{2.25}$$

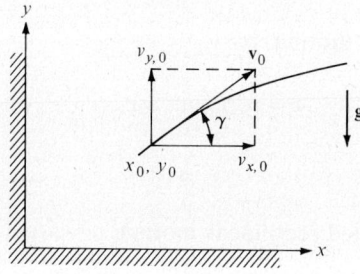

Figure 2.4

Sample Problem 2.3

A cam follower is to perform the motion indicated in the figure. The cam rotates at 600 rpm. Plot the follower displacement and acceleration versus degrees of cam rotation curves for the first 150° of rotation.

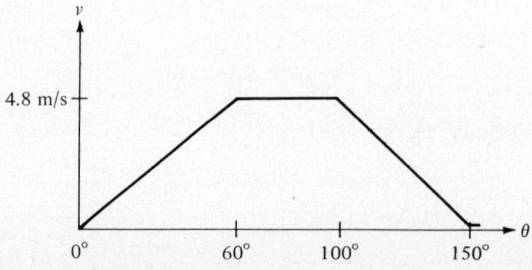

Analysis: The cam rotates at 600 rpm $= 3600°/s$. The time for the constant accelerated motion through the first $60°$ is $\frac{60}{3600} = \frac{1}{60}$ s.

$$v = at + v_0$$
$$4.8 = a(\tfrac{1}{60}) + 0$$
$$a = 288 \text{ m/s}^2$$
$$s = \tfrac{288}{2}(\tfrac{1}{60})^2 + 0 + 0 = 0.04 \text{ m}$$

For constant accelerated motion, the second-degree displacement curve has an initial slope $ds/dt = v_0 = 0$.

From 60 to $100°$, the velocity is constant and equal to 4.8 m/s. The time required to rotate $40°$ is

$$t = \tfrac{40}{3600} = \tfrac{1}{90} \text{ s}$$
$$s = v_0 t + s_0$$
$$= 4.8(\tfrac{1}{90}) + 0.04 = 0.0933 \text{ m}$$

For constant velocity motion, the displacement curve is a straight line, with a constant slope equal to the velocity.

$$\frac{ds}{dt} = v = 4.8 \text{ m/s}$$

For the last $50°$ of rotation,

$$t = \tfrac{50}{3600} = \tfrac{1}{72} \text{ s}$$
$$v = at + v_0$$
$$0 = a(\tfrac{1}{72}) + 4.8$$
$$a = -72(4.8) = -345.6 \text{ m/s}^2$$
$$s = \frac{at^2}{2} + v_0 t + s_0 = 0.1266 \text{ m}$$

Again the motion has a second-degree displacement curve with an initial slope now equal to 4.8 m/s. The final slope is zero. The rate of change of slope, a, is now negative. The solution is shown below.

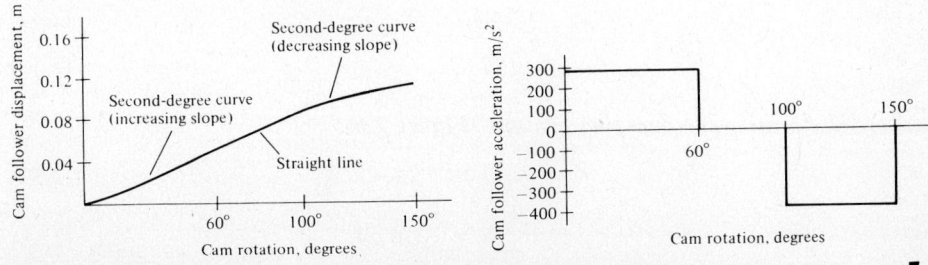

E. Simple harmonic motion: One degree of freedom

Free vibration, no damping present (Figure 2.5)

With free vibration, there is no driving force. Employing Newton's second law,

$$F_x = -kx = ma_x = m\ddot{x}$$

or

$$\ddot{x} + \frac{k}{m}x = 0 \quad (2.26)$$

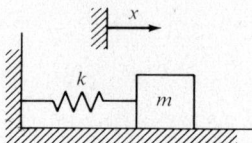

Figure 2.5

Equation (2.26) describes simple harmonic motion. The restoring spring force is linearly proportional to x, that is, $F_x = kx$, k has units of N/m. Dividing the solution into a particular and a complementary part,

$$x = x_c + x_p$$

With no driving force, $x_p = 0$ and $x_c = x$, or

$$\ddot{x}_c + \frac{k}{m}x_c = 0$$

$$x = A \cos \omega_n t + B \sin \omega_n t \quad (2.27)$$

Here

$$\omega_n = \sqrt{\frac{k}{m}} = \text{circular frequency, rad/s} \quad (2.28)$$

$$f = \frac{\omega_n}{2\pi} = \text{natural frequency, cycles/s or hertz (Hz)} \quad (2.29)$$

$$T = \frac{1}{f} = \text{period, s} \quad (2.30)$$

Equation (2.27) can be rewritten as

$$x = x_{max} \sin(\omega_n t + \phi) \quad (2.31)$$

Then

$$v_{max} = \omega_n x_{max} \quad (2.32)$$

and

$$a_{max} = -\omega_n^2 x_{max} \quad (2.33)$$

Forced vibration, no damping present (Figure 2.6).

$$F_x = F \sin \omega t - kx = m\ddot{x} \quad (2.34)$$

$$\ddot{x} + \frac{k}{m}x = \frac{F}{m} \sin \omega t \quad (2.35)$$

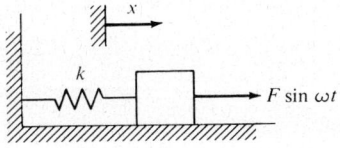

Figure 2.6

Then with
$$x_c = A \cos \omega_n t + B \sin \omega_n t$$
$$\omega_n = \sqrt{\frac{k}{m}}$$

and
$$x_p = \frac{(F/k) \sin \omega t}{1 - (\omega/\omega_n)^2} \tag{2.36}$$

The total solution becomes

$$x = A \cos \omega_n t + B \sin \omega_n t + \frac{(F/k) \sin \omega t}{1 - (\omega/\omega_n)^2} \tag{2.37}$$

The two constants A and B are evaluated from the initial conditions using equation (2.37), but they play a minor role. The complementary solution, also called the transient solution, is practically damped out by frictional effects. The particular solution x_p, or steady-state solution, is the key descriptor of the motion. If the force is constant, the static solution is given by

$$x_s = \frac{F}{k} \tag{2.38}$$

With the transient solution neglected, the *magnification factor* (M.F.) is defined as the ratio of the maximum steady-state response to the static response of the

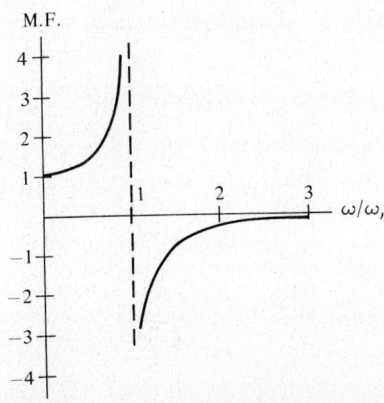

Figure 2.7

system. Then x_p maximum divided by x_s is

$$\text{M.F.} = \frac{1}{1-(\omega/\omega_n)^2} \quad (2.39)$$

Note the consequences of driving a system at its natural frequency (see Figure 2.7).

Free vibration with viscous damping present (Figure 2.8)

$$F_x = -kx - c\dot{x} = m\ddot{x} = ma_x$$

Rewritten, this becomes

$$\ddot{x} + \frac{c}{m}\dot{x} + \frac{k}{m}x = 0 \quad (2.40)$$

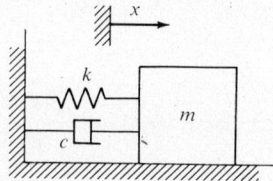

Figure 2.8

Again $x_p = 0$ for free motion, and x_c is obtained by substituting a trial solution of the form $x_c = Ae^{st}$. This leads to

$$\left(s^2 + \frac{c}{m}s + \frac{k}{m}\right)Ae^{st} = 0 \quad (2.41)$$

Since $Ae^{st} = 0$ is a trivial solution,

$$s^2 + \frac{c}{m}s + \frac{k}{m} = 0 \quad (2.42)$$

Equation (2.42) is called the *characteristic equation*. Its roots are

$$s_{1,2} = -\frac{c}{2m} \pm \sqrt{\left(\frac{c}{2m}\right)^2 - \frac{k}{m}} \quad (2.43)$$

The solution for equation (2.40) falls into three categories depending on the value of c. When $(c/2m)^2 = k/m$, the system is critically damped.

$$c_c = 2m\omega_n, \frac{\text{lb} \cdot \text{s}}{\text{ft}} \quad (2.44)$$

Case A: Critically Damped, $c = c_c = 2m\omega_n$
From equation (2.43),

$$s_{1,2} = -\frac{c}{2m}$$

For repeated roots, the solution for equation (2.40) takes on the form

$$x = (A + Bt)e^{-ct/2m} \qquad (2.45)$$

The solution is nonvibratory. A system regains its equilibrium position in the shortest possible time when critical damping occurs.

Case B: Heavily Damped, $c > c_c$

For the heavily damped case, the solution to equation (2.40) becomes

$$x = Ae^{s_1 t} + Be^{s_2 t} \qquad (2.46)$$

where s_1 and s_2 are both negative real numbers. This represents a decaying nonvibratory solution.

Case C: Damped Vibration, $c < c_c$

With the aid of equation (2.43), the solution to equation (2.40) becomes

$$x = e^{-ct/2m}(A \sin \omega_d t + B \cos \omega_d t) \qquad (2.47)$$

where $\quad \omega_d = \sqrt{\dfrac{k}{m} - \left(\dfrac{c}{2m}\right)^2} \quad$ Damped natural frequency, rad/s $\qquad (2.48)$

$$\omega_d = \omega_n \sqrt{1 - \left(\dfrac{c}{c_c}\right)^2} \qquad (2.49)$$

The total motion would have the general form shown in Figure 2.9.

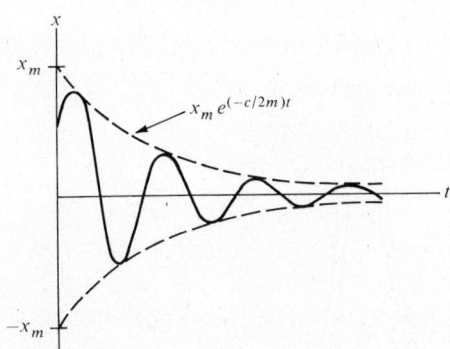

Figure 2.9

Forced damped vibrations

Now
$$\ddot{x} + \dfrac{c}{m}\dot{x} + \dfrac{k}{m}x = \dfrac{F}{m}\sin \omega t \qquad (2.50)$$

Ignoring the transient solution, the steady-state solution can be obtained with a trial solution of

$$x = A \sin(\omega t - \phi) \qquad (2.51)$$

Substituting equation (2.51) into equation (2.50) yields

$$\left(\frac{k}{m}-\omega^2\right)A(\sin\omega t\cos\phi-\cos\omega t\sin\phi)+\frac{c}{m}A\omega(\cos\omega t\cos\phi+\sin\omega t\sin\phi)$$
$$=\frac{F}{m}\sin\omega t \quad (2.52)$$

After equating like cofactors on both sides of the equation, one obtains

$$\left(\frac{k}{m}-\omega^2\right)(-A\cos\omega t\sin\phi)+\frac{c}{m}\omega A\cos\omega t\cos\phi=0$$

or

$$\phi=\tan^{-1}\frac{(c/m)\omega}{(k/m-\omega^2)} \quad (2.53)$$

where

$$\sin\phi=\frac{(c/m)\omega}{\sqrt{(k/m-\omega^2)^2+[(c/m)\omega]^2}}$$

$$\cos\phi=\frac{(k/m-\omega^2)}{\sqrt{(k/m-\omega^2)^2+[(c/m)\omega]^2}}$$

Solving for A from equation (2.52),

$$A=\frac{F/m}{\sqrt{(k/m-\omega^2)^2+[(c/m)\omega]^2}} \quad (2.54)$$

The magnification factor is

$$\text{M.F.}=\frac{1}{\sqrt{[1-(\omega/\omega_n)^2]^2+[2(c/c_c)(\omega/\omega_n)]^2}} \quad (2.55)$$

This relationship is shown in Figure 2.10.

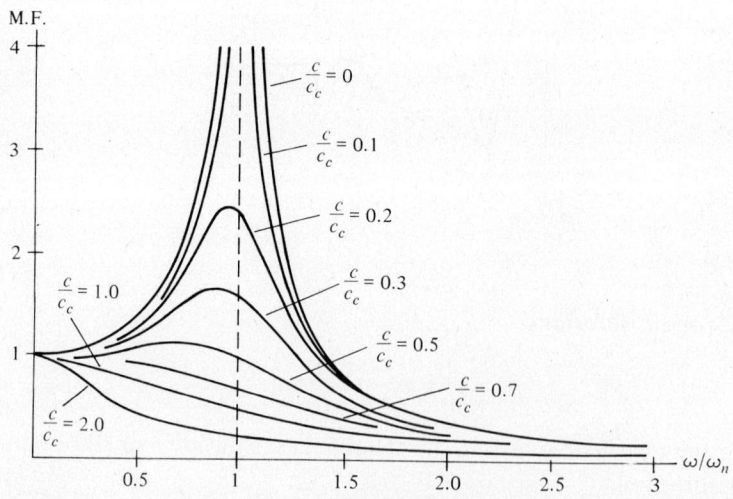

Figure 2.10

Sample Problem 2.4

Shown in the figure is a simple pendulum of length $l = 10$ cm. The rod is assumed massless, but the pendulum bob has a mass of 0.5 kg. If the pendulum is released from rest at a position 3° from the vertical, find the period of the motion and the maximum velocity occurring.

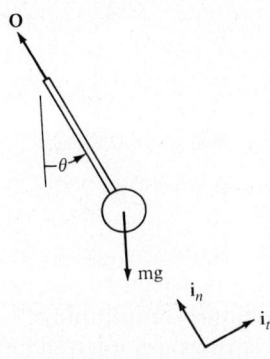

Analysis: Since the particle's motion is circular, normal and tangential coordinates will be used for the motion's description. Referring ahead to Section 2.3, and employing the free-body diagram shown,

$$F_t = ma_t = m\dot{v} = ml\ddot{\theta} = -mg \sin\theta$$

Then
$$ml\ddot{\theta} + mg \sin\theta = 0$$

or
$$\ddot{\theta} + \frac{g}{l}\sin\theta = 0$$

For small θ, $\sin\theta \sim \theta$; so

$$\ddot{\theta} + \frac{g}{l}\theta = 0 \qquad (a)$$

The above equation describes simple harmonic motion.

$$\omega_n = \sqrt{\frac{g}{l}} = \sqrt{\frac{9.81}{0.1}} = 9.904 \text{ rad/s}$$

22 REVIEW OF NEWTONIAN MECHANICS

$$f = \frac{\omega_n}{2\pi} = 1.576 \text{ cycles/s}$$

$$T = \frac{1}{f} = \frac{1}{1.576} = 0.634 \text{ s}$$

The solution to equation (a) is

$$\theta = \theta_{max} \sin(\omega_n t + \phi)$$

$$\dot{\theta} = \theta_{max} \omega_n \cos(\omega_n t + \phi)$$

At $t=0$, $\dot{\theta}=0$ and $\theta=3°$, or 0.052 rad. Then

$$\dot{\theta} = 0 = \theta_{max} \omega_n \cos(\omega_n t + \phi)$$

or

$$\phi = \frac{\pi}{2}$$

Also

$$\theta_0 = 0.052 = \theta_{max} \sin\left[\omega_n(0) + \frac{\pi}{2}\right]$$

So

$$\theta_{max} = 0.052$$

Finally,

$$\theta = 0.052 \sin\left(\omega_n t + \frac{\pi}{2}\right)$$

$$\dot{\theta} = 0.052\omega_n \cos\left(\omega_n t + \frac{\pi}{2}\right)$$

$$\dot{\theta}_{max} = 0.052\omega_n$$

and

$$v_{max} = l\dot{\theta}_{max} = 0.1(0.052)(9.904) = 0.0518 \text{ m/s} \quad \blacksquare$$

Sample Problem 2.5

The spring-mass system shown is driven by the excitation force $F = F_0 \sin \omega t$. Find an expression for the maximum amplitude of x. The distance x is measured from the position where the springs are unstretched. Neglect frictional effects.

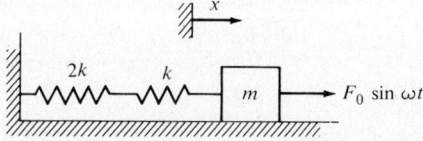

Analysis: The springs carry equal forces; therefore

$$\delta_{total} = \frac{F}{2k} + \frac{F}{k} = \frac{3F}{2k}$$

The equivalent spring constant becomes

$$k_{eq.} = \tfrac{2}{3}k$$

Then from the equation of motion

$$m\ddot{x} + \tfrac{2}{3}kx = F_0 \sin \omega t$$

$$x_m = \frac{3F_0/2k}{1-(\omega/\omega_n)^2} \quad \text{[see equation (2.36)]}$$

with
$$\omega_n = \sqrt{\frac{2k}{3m}}$$
∎

Sample Problem 2.6

An electric motor weighing 37 kg rotates at an operating speed of 1800 rpm. It is supported by two springs having equal spring constants of 280 kN/m. Due to manufacturing tolerances, the rotor imbalance is known to be 2.25 N · m. If the motor is constrained to move vertically, find the maximum amplitude of the motion and the magnification factors for the case where damping is zero and the case where $c/c_c = 0.2$. Find the phase angles.

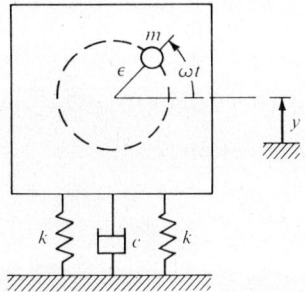

Analysis: The rotor imbalance can be thought of as a small additional mass set at a distance ε from the shaft's axis of rotation. Summing forces in the y direction to obtain the equation of motion for the motor, one obtains

$$(M-m)\ddot{y} = -2ky - c\dot{y} + T \sin \omega t - (M-m)g \tag{a}$$

and for the unbalanced mass

$$m(\ddot{y} - \varepsilon\omega^2 \sin \omega t) = -T \sin \omega t - mg \tag{b}$$

The displacement of the small unbalanced mass is

$$y_{\text{total}} = y + \varepsilon \sin \omega t$$

and its acceleration is given by

$$\ddot{y}_{\text{total}} = \ddot{y} - \varepsilon\omega^2 \sin \omega t$$

After substituting equation (b) into equation (a), equation (c) results:

$$M\ddot{y} + c\dot{y} + 2ky = \varepsilon m\omega^2 \sin \omega t - Mg \tag{c}$$

The product εmg is called the rotor imbalance. In this problem it equals 2.25 N · m; εm equals 2.25/9.81.

REVIEW OF NEWTONIAN MECHANICS

Equation (c) describes forced damped vibrations with $F = \varepsilon m \omega^2$.

$$\ddot{y} + \frac{c}{M}\dot{y} + \frac{2ky}{M} = \frac{F}{M}\sin \omega t - g$$

$$\omega = \frac{2\pi}{60}(1800) = 188.5 \text{ rad/s}$$

$$\ddot{y} + \frac{c}{M}\dot{y} + \frac{2ky}{M} = \frac{2.25 \text{ N} \cdot \text{m}}{37 \text{ kg}(9.81 \text{ N/kg})} 188.5^2 \sin 188.5t - 9.81$$

$$\ddot{y} + \frac{c\dot{y}}{M} + \frac{2ky}{M} = 220.3 \sin 188.5t - 9.81 \quad \text{m/s}^2$$

Ignoring the transient solution, the amplitude of the vibration is given by equation (2.54), and the magnification factor by equation (2.55).

$$A = \frac{F/M}{\sqrt{(k/M - \omega^2)^2 + [(c/m)\omega]^2}}$$

with no damping

$$A = \left|\frac{220.3}{\sqrt{(k/M - \omega^2)^2}}\right| = \frac{220.3}{2(280)(10^3)/37 - 188.5^2} = 10.8 \text{ mm}$$

The phase angle is given by equation (2.53):

$$\phi = \tan^{-1}\frac{(c/M)\omega}{k/M - \omega^2}$$

With no damping, $c = 0$, or $\phi = 0°$ or $180°$. Since $\omega^2 > k/M$, $\phi = 180°$.

The magnification factor for the nondamped case is

$$\text{M.F.} = \frac{1}{1 - (\omega/\omega_n)^2}$$

$$\omega_n = \sqrt{\frac{k}{M}} = \sqrt{\frac{2(280)(10^3)}{37}} = 123 \text{ rad/s}$$

$$\text{M.F.} = \frac{1}{1 - (188.5/123)^2} = 0.74 \quad \text{(phase angle included)}$$

For the damped case with $c/c_c = 0.2$,

$$c = 0.2c_c \qquad c_c = 2M\omega_n$$

$$c = 0.2(2)(37)(123) = 1820 \text{ N} \cdot \text{s/m}$$

$$A = \frac{220.3}{\sqrt{(k/M - \omega^2)^2 + [(c/M)\omega]^2}}$$

$$= \frac{220.3}{\{[2(280)(10^3)/37 - 188.5^2]^2 + [0.2(2)(123)(188.5)]^2\}^{1/2}} = 9.8 \text{ mm}$$

$$\phi = \frac{\tan^{-1}(c/M)\omega}{k/M - \omega^2} = -24°, 156°$$

$$= 156° \quad \text{(since the denominator is negative)}$$

$$\text{M.F.} = \frac{1}{\{[1-(188.5/123)^2]^2 + [2(0.2)(188.5/123)]^2\}^{1/2}} = 0.675 \quad \blacksquare$$

Maximum deflection can now be found by adding the static deflection to the amplitude found. For a homework problem, it would be useful to solve this example once more with $\omega = 1300$ rpm. This value lies closer to the system's natural frequency.

2.3 NORMAL AND TANGENTIAL COORDINATES

Another description of motion can be obtained by using normal and tangential coordinates (Figure 2.11). Recalling equation (2.2), the acceleration is given by

$$\mathbf{a} = \dot{v}\mathbf{i}_t + v\dot{\mathbf{i}}_t \tag{2.56}$$

Here $\dot{\mathbf{i}}_t$ represents the derivative of the unit tangential vector. It is easily shown that

$$\dot{\mathbf{i}}_t = \dot{\theta}\mathbf{i}_n$$

The vector $\mathbf{i}_n$ is defined to be the unit vector pointing in the direction of change of the unit tangential vector, i.e., toward the center of curvature, Figure 2.12

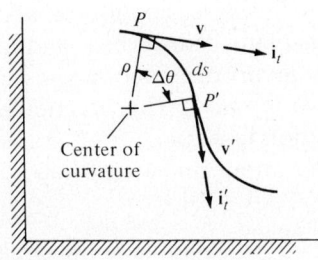

Figure 2.11

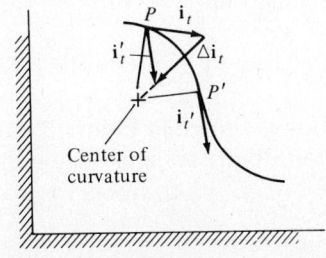

Figure 2.12

describes this more clearly. Equation (2.56) then becomes

$$\mathbf{a} = \dot{v}\mathbf{i}_t + v\dot{\theta}\mathbf{i}_n \tag{2.57}$$

With ρ defined as the *radius of curvature*,

$$ds = \rho \, d\theta$$

so

$$v = \rho\dot{\theta} \tag{2.58}$$

Alternative forms of equation (2.57) are

$$\mathbf{a} = \dot{v}\mathbf{i}_t + \frac{v^2}{\rho}\mathbf{i}_n \tag{2.59}$$

$$\mathbf{a} = \dot{v}\mathbf{i}_t + \rho\dot{\theta}^2\mathbf{i}_n \tag{2.60}$$

$$\mathbf{a} = a_t\mathbf{i}_t + a_n\mathbf{i}_n \tag{2.61}$$

The tangential acceleration a_t of equation (2.61) describes a change in the speed of the particle; a_n is the normal or centripetal acceleration that affects a change in the *direction* of the velocity. a_n is always present except for the case of straight line motion. Summarizing

$$a_t = \dot{v} \tag{2.62}$$

$$a_n = v\dot{\theta} = \rho\dot{\theta}^2 = \frac{v^2}{\rho} \tag{2.63}$$

With just two components, these coordinates can also describe three-dimensional motion. This is because the motion is always described and contained in the instantaneous plane defined by $\mathbf{i}_t$ and $\mathbf{i}_n$. The plane twists, but by definition it must always contain the velocity and acceleration of the particle. The plane is called the *osculating plane*. The direction perpendicular to this plane is called the *binormal* direction $\mathbf{i}_B \equiv \mathbf{i}_t \times \mathbf{i}_n$.

The most frequent and important use of these coordinates is for *circular motion*. Here

$$\rho = R \quad \text{Radius of the circle} \tag{2.64}$$

$$v = R\dot{\theta} \tag{2.65}$$

$$a_t = \dot{v} = R\ddot{\theta} \tag{2.66}$$

$$a_n = \frac{v^2}{R} = R\dot{\theta}^2 = v\dot{\theta} \tag{2.67}$$

A case of circular motion is shown in Figure 2.13a. Figure 2.13b shows the vector **a** described in two coordinate systems. The vector itself does not change due to the coordinate rotation, but its components do.

$$a = \sqrt{a_n^2 + a_t^2} = \sqrt{a_x^2 + a_y^2} \tag{2.68}$$

PARTICLE DYNAMICS: KINEMATICS 27

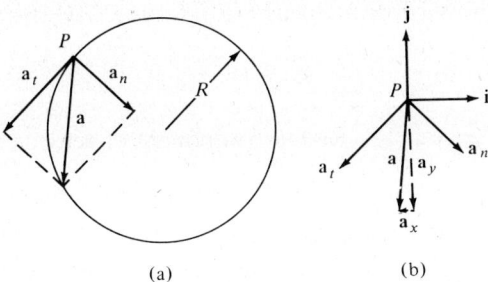

(a)   (b)

Figure 2.13

2.4 RADIAL AND TRANSVERSE COMPONENTS (POLAR COORDINATES)

A set of coordinates that is especially useful when a particle's position is described by polar coordinates is the *radial* and *transverse* pair. In this coordinate system, the position vector **r** is expressed by

$$\mathbf{r} = r\mathbf{i}_r \tag{2.69}$$

Note that unlike the normal direction, the $\mathbf{i}_r$ direction points *away* from a fixed origin as shown in Figure 2.14. The derivative $\dot{\mathbf{r}}$ becomes

$$\mathbf{v} = \dot{\mathbf{r}} = \dot{r}\mathbf{i}_r + r\dot{\mathbf{i}}_r \tag{2.70}$$

One can show that the derivative of the unit vector $\mathbf{i}_r$ is given by

$$\dot{\mathbf{i}}_r = \dot{\theta}\mathbf{i}_\theta \tag{2.71}$$

and

$$\dot{\mathbf{i}}_\theta = -\dot{\theta}\mathbf{i}_r \tag{2.72}$$

In Figure 2.14, $\mathbf{i}_r$ is in the radial direction, and $\mathbf{i}_\theta$ is in the transverse direction. The transverse direction is defined to be positive in the direction of change of $\mathbf{i}_r$, or in the direction of increasing θ. In general, $\mathbf{i}_\theta$ is *not* tangent to the path. It is perpendicular at all times to $\mathbf{i}_r$. Figure 2.14 also shows the normal and tangential directions for the curve so that comparisons can be made. The direction perpendicular to the plane of motion is the **k** direction, $\mathbf{k} \equiv \mathbf{i}_r \times \mathbf{i}_\theta$. If

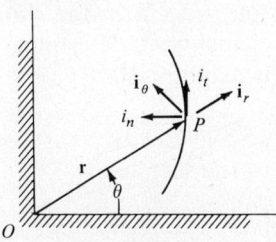

Figure 2.14

equation (2.71) is now substituted into equation (2.70), the following result is obtained.

$$\mathbf{v} = \dot{r}\mathbf{i}_r + r\dot{\theta}\mathbf{i}_\theta \tag{2.73}$$

Defining

$\quad v_r = \dot{r} \quad$ Radial component of velocity $\hfill (2.74)$

$\quad v_\theta = r\dot{\theta} \quad$ Transverse component of velocity $\hfill (2.75)$

then equation (2.73) can be rewritten as

$$\mathbf{v} = v_r \mathbf{i}_r + v_\theta \mathbf{i}_\theta \tag{2.76}$$

The acceleration **a** is given by

$$\mathbf{a} = \ddot{r}\mathbf{i}_r + \dot{r}\dot{\mathbf{i}}_r + \dot{r}\dot{\theta}\mathbf{i}_\theta + r\ddot{\theta}\mathbf{i}_\theta + r\dot{\theta}\dot{\mathbf{i}}_\theta \tag{2.77}$$

With the substitution of equations (2.71) and (2.72), this becomes

$$\mathbf{a} = (\ddot{r} - r\dot{\theta}^2)\mathbf{i}_r + (r\ddot{\theta} + 2\dot{r}\dot{\theta})\mathbf{i}_\theta \tag{2.78}$$

The acceleration components can be defined as

$\quad a_r = \ddot{r} - r\dot{\theta}^2 \quad$ Radial component of acceleration $\hfill (2.79)$

$\quad a_\theta = r\ddot{\theta} + 2\dot{r}\dot{\theta} \quad$ Transverse component of acceleration $\hfill (2.80)$

so that equation (2.78) may be rewritten as

$$\mathbf{a} = a_r \mathbf{i}_r + a_\theta \mathbf{i}_\theta \tag{2.81}$$

The reader is cautioned to note that the radial component of acceleration does not represent the change in the radial component of speed. The derivative of v_r, or $\ddot{r}$, is just one part of the two-part expression for a_r. The fact that the coordinate system, or **r**, rotates at a rate $\dot{\theta}$ means that the transverse component of velocity must also rotate at a rate $\dot{\theta}$. The rate of change of the transverse component of velocity caused by rotation accounts for the extra negative term.

A final expression similar to equation (2.68) is

$$a = \sqrt{a_r^2 + a_\theta^2} = \sqrt{a_x^2 + a_y^2} \tag{2.82}$$

Sample Problem 2.7

Pin B at the end of rod OB describes circular motion with radius R. At the same time, it also slides along rod CE in the slot provided. If rod OB has a constant angular velocity ω clockwise, determine the angular velocity of rod CE for the instant the pin is at point D.

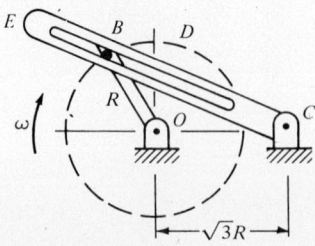

PARTICLE DYNAMICS: KINEMATICS   29

Analysis: At point D, pin B has a velocity

$$v_B = R\dot{\phi} = R\omega$$
$$\mathbf{v}_B = R\omega \mathbf{i}_t$$

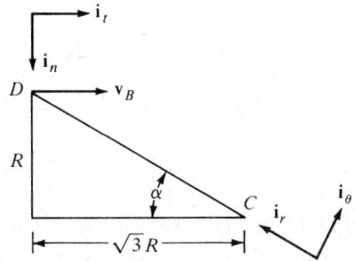

From the known geometry, $\alpha = 30°$ for this instant:

$$(v_B)_r = -v_B \cos\alpha = -\frac{R\omega\sqrt{3}}{2} = \dot{r}$$

$$(v_B)_\theta = v_B \sin\alpha = \frac{R\omega}{2} = r\dot{\theta} = 2R\dot{\theta}$$

or

$$\dot{\theta} = \frac{R\omega/2}{2R}$$

$$\dot{\theta} = \frac{\omega}{4} \quad \text{(clockwise)} \qquad \blacksquare$$

2.5 CYLINDRICAL COORDINATES

For three-dimensional motion, the radial and transverse coordinate set can be simply expanded to include a third direction. This is shown in Figure 2.15.

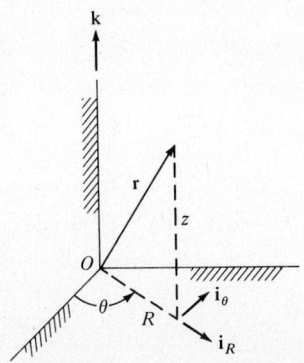

Figure 2.15

In this set

$$\mathbf{r} = R\mathbf{i}_R + z\mathbf{k} \tag{2.83}$$

$$\mathbf{v} = \dot{R}\mathbf{i}_R + R\dot{\theta}\mathbf{i}_\theta + \dot{z}\mathbf{k} \tag{2.84}$$

$$\mathbf{a} = (\ddot{R} - R\dot{\theta}^2)\mathbf{i}_R + (R\ddot{\theta} + 2\dot{R}\dot{\theta})\mathbf{i}_\theta + \ddot{z}\mathbf{k} \tag{2.85}$$

The results are so similar to equations (2.73) and (2.78) that only a few brief comments will be made here. First, the coordinate axes are such that they rotate only about the vertical at a rate $\dot{\theta}$. In other words, $\mathbf{i}_R$ and $\mathbf{i}_\theta$ always remain in the horizontal plane. Second, $\mathbf{i}_R$ does not equal to $\mathbf{i}_r$, which by definition lies along the position vector emanating from the origin. Instead, $\mathbf{i}_R$ is a unit vector along the direction of the horizontal component of the position vector. The magnitude of the horizontal component of $\mathbf{r}$ is R.

2.6 SPHERICAL COORDINATES

The last coordinate set to be described is spherical coordinates. The coordinates are effective in depicting three-dimensional motion in terms of distance from the origin and two defined angles. Figure 2.16 illustrates these coordinate axes. Here

$$\mathbf{i}_\theta \equiv \mathbf{i}_r \times \mathbf{i}_\phi$$

with
$$\mathbf{r} = r\mathbf{i}_r \tag{2.86}$$

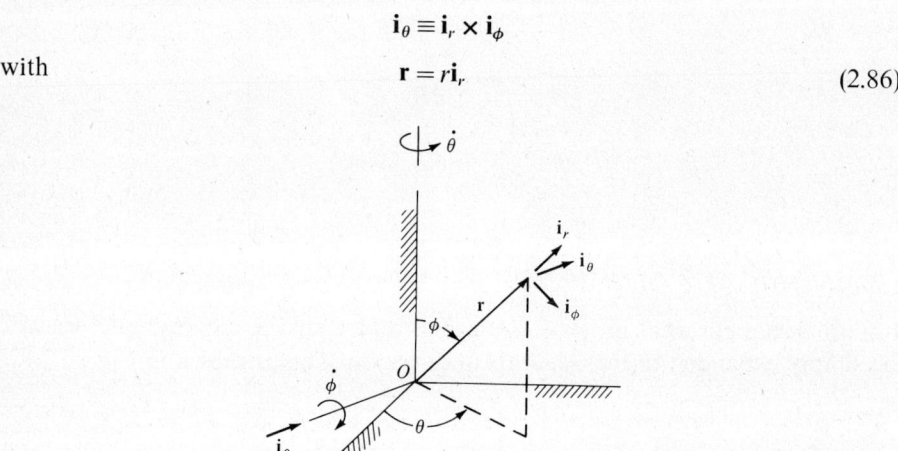

Figure 2.16

The velocity expression can be shown to be

$$\mathbf{v} = \dot{r}\mathbf{i}_r + r\dot{\phi}\mathbf{i}_\phi + r\dot{\theta}\sin\phi\,\mathbf{i}_\theta \tag{2.87}$$

and the acceleration is

$$\mathbf{a} = (\ddot{r} - r\dot{\phi}^2 - r\dot{\theta}^2\sin^2\phi)\mathbf{i}_r + (r\ddot{\phi} + 2\dot{r}\dot{\phi} - r\dot{\theta}^2\sin\phi\cos\phi)\mathbf{i}_\phi$$
$$+ (r\ddot{\theta}\sin\phi + 2\dot{r}\dot{\theta}\sin\phi + 2r\dot{\phi}\dot{\theta}\cos\phi)\mathbf{i}_\theta \tag{2.88}$$

The coordinate system rotates about the vertical direction and about a revolving horizontal axis. This horizontal axis, perpendicular to the plane formed by the vertical direction and **r**, rotates at the rate $\dot{\theta}$. As shown in Figure 2.16, the turning rate of the coordinate axes themselves about the horizontal axis is $\dot{\phi}$. $\boldsymbol{\Phi}$ rotates at a rate $\dot{\theta}$ and acts along the $\mathbf{i}_\theta$ direction. It does not act along $\mathbf{i}_\phi$.

$$\dot{\boldsymbol{\Phi}}_\phi = \dot{\phi}\mathbf{i}_\theta \tag{2.89}$$

The rate of turning of the coordinate system, or **r**, about the vertical axis is $\dot{\theta}$. Expressed solely in these coordinates, $\dot{\theta}$ becomes

$$\dot{\theta} = \dot{\theta}(\cos\phi\mathbf{i}_r - \sin\phi\mathbf{i}_\phi) \tag{2.90}$$

2.7 RELATIVE MOTION

Shown in Figure 2.17 are the absolute position vectors to a particle at P and to an observer at B. Also shown is the relative position vector from B to P.

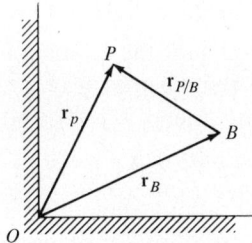

Figure 2.17

It is seen that

$$\mathbf{r}_{P/B} = \mathbf{r}_P - \mathbf{r}_B \tag{2.91}$$

$$\mathbf{r}_P = \mathbf{r}_B + \mathbf{r}_{P/B} \tag{2.92}$$

The derivative of equation (2.92) yields an expression for the velocity $\mathbf{v}_P$

$$\dot{\mathbf{r}}_P = \mathbf{v}_P = \mathbf{v}_B + \mathbf{v}_{P/B} \tag{2.93}$$

where the derivative of the relative position vector is defined as

$$\mathbf{v}_{P/B} \equiv \dot{\mathbf{r}}_{P/B} \tag{2.94}$$

Also

$$\mathbf{a}_P = \dot{\mathbf{v}}_P = \mathbf{a}_B + \mathbf{a}_{P/B} \tag{2.95}$$

Figure 2.18 depicts in more detail the relative position vector and two coordinate sets at B and O. The coordinate set at O is fixed. Treating plane motion only, and assuming parallel coordinate axes for this instant, in the primed coordinate set

$$\mathbf{r}_{P/B} = x'\mathbf{i}' + y'\mathbf{j}' \tag{2.96}$$

32 REVIEW OF NEWTONIAN MECHANICS

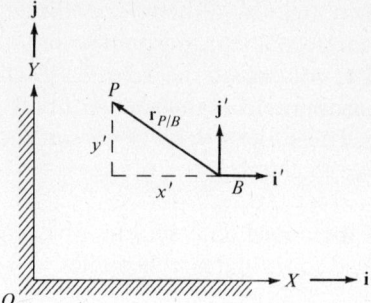

Figure 2.18

With the $\mathbf{i}'$, $\mathbf{j}'$ parallel to the nonrotating $\mathbf{i}$, $\mathbf{j}$, equation (2.96) also can be written as

$$\mathbf{r}_{P/B} = (X_P - X_B)\mathbf{i} + (Y_P - Y_B)\mathbf{j} = x'\mathbf{i} + y'\mathbf{j}$$

Taking the derivative of equation (2.96) yields

$$\mathbf{v}_{P/B} = \dot{x}'\mathbf{i}' + \dot{y}'\mathbf{j}' + x'\dot{\mathbf{i}}' + y'\dot{\mathbf{j}}' \qquad (2.97)$$

If $\mathbf{i}'$, $\mathbf{j}'$ not only are parallel to $\mathbf{i}$, $\mathbf{j}$ for the moment when the instantaneous velocities are calculated but also remain that way, i.e., the reference frame for $\mathbf{i}'$, $\mathbf{j}'$ translates but does not rotate, then $\dot{\mathbf{i}}' = \dot{\mathbf{j}}' = 0$, and equation (2.97) becomes

$$\mathbf{v}_{P/B} = \dot{x}'\mathbf{i}' + \dot{y}'\mathbf{j}' = \dot{x}'\mathbf{i} + \dot{y}'\mathbf{j} \qquad (2.98)$$

One can conclude from the above that different frames of references, *independent of their linear velocities*, will provide the same measurement of the velocity of P relative to B if they do not rotate with respect to each other.

In equations (2.94) and (2.97), $\mathbf{v}_{P/B}$ is defined as the velocity of P relative to the reference frame at B as seen by any nonrotating observer. In equation (2.97), the first two terms to the right of the equal sign describe the velocity of P relative to *a rotating reference frame at B*. The last two terms on the right represent the rate of change of the relative position vector brought about by the rotation of the frame.

More will be said in Chapter 5 about this problem of motion in a frame

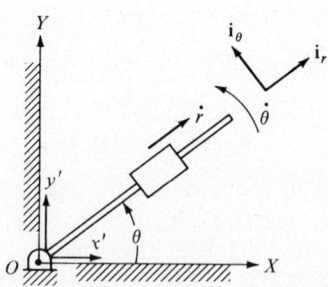

Figure 2.19

that itself rotates. For now a simple illustrated example may suffice. Figure 2.19 shows a slider moving outward along a rotating rod at a rate $\dot{r}$. The rod rotates at an angular speed $\dot{\theta}$. The stationary observer in nonrotating reference frame OXY sees the spiral motion of the particle. Another observer in frame $Ox'y'$, which *rotates* with the rod at the rate $\dot{\theta}$, will only see straight line motion. Using polar coordinates in this frame for convenience, the velocity he or she measures is $\dot{r}\mathbf{i}_r$. For a nonrotating observer $\mathbf{v} = \dot{r}\mathbf{i}_r + r\dot{\theta}\mathbf{i}_\theta$. The second term on the right is caused by the relative rotation of the frames.

2.8 NUMERICAL SOLUTIONS TO ORDINARY DIFFERENTIAL EQUATIONS WITH INITIAL CONDITIONS

This section will briefly investigate several numerical methods, all of them *self-starting*, that can be utilized to numerically solve ordinary differential equations when initial conditions are given.

A. Euler's Method

The ordinary differential equation

$$y' = f(y, t) \tag{2.99}$$

with the given initial condition that $y = y_0$ at $t = 0$, has a solution valid about $t = 0$ expressed by the Taylor series expansion

$$y_{i+1} = y_i + y'_i \Delta t + y''_i \frac{\Delta t^2}{2} + \cdots \tag{2.100}$$

or more specifically,

$$y = y(0) + y'(0) \Delta t + y''(0) \frac{\Delta t^2}{2} + \cdots \tag{2.101}$$

Approximating the solution by keeping only linear terms, equation (2.101) becomes

$$y = y(0) + f(y(0), 0)\Delta t \tag{2.102}$$

where y is the numerical value of the solution at a time Δt later. If this new value of time and y are now used instead of the initial values, the solution can be advanced still another time step Δt. In general, the numerical solution can be expressed by

$$y_{i+1} = y_i + y'_i \Delta t \tag{2.103}$$

Equation (2.103) is known as Euler's forward integration method. It is a very simple method to program for a computer. It has the advantage of being a fast algorithm, and it does not require evaluation of auxiliary terms before starting; i.e., it is a self-starting method. In effect, the approximation is being made that the solution has constant slope during a step size Δt. The method allows for

34 REVIEW OF NEWTONIAN MECHANICS

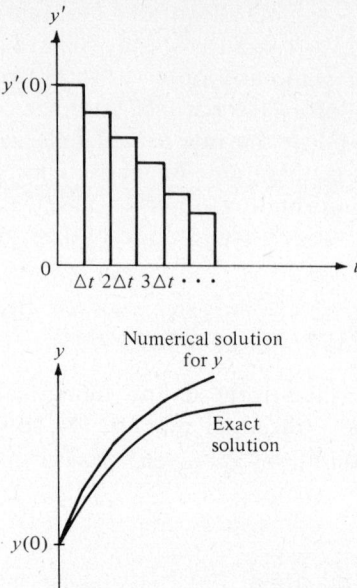

Figure 2.20

an abrupt change of the slope after each step. Because of this inherent feature, Euler's method is a preferred integration routine whenever there are sudden discontinuities for the variables present in a problem (see Figure 2.20).

Euler's method has several undesirable features. Since only the linear terms in the Taylor series expression are kept, restricting the solution curve to piecewise linear segments as shown in Figure 2.20, the subsequent shortening of the series solution leads to *truncation error*. The estimated truncation error for Euler's method is of the order of $(\Delta t)^2$

$$(E)_i \simeq y_i'' \frac{\Delta t^2}{2} \tag{2.104}$$

Obviously, for increased accuracy, the step size in a program employing Euler's method should be kept small. Unfortunately, if the step size is chosen too small and if the solution is to cover a long time interval, many applications of Euler's formula are required. Each introduction of Euler's formula produces *round-off error*. Over a long computer run, round-off error will cause just as much difficulty as truncation error. Picking a proper step size then is very critical. It may require several repetitive runs with increasing and/or decreasing step size until two successive solutions agree. In practice, a step size that allows at least 10 samples per period of the highest system frequency component (of interest) is used. Of course, prior knowledge of the system's frequency components may not always be available or easy to approximate. Further, certain systems, called *stiff systems*, have widely varying time constants. In treating these problems, it is impossible to find one single step size that is suitable for the entire run. Special algorithms are needed. Still, because of its simplicity, Euler's method is often preferred.

For a second-order equation of the form

$$y'' = f(y', y, t) \tag{2.105}$$

numerical solutions can be obtained by first transforming the equation to a set of two first-order equations

$$y' = v \tag{2.106}$$

$$v' = f(v, y, t) \tag{2.107}$$

Then
$$y_{i+1} = y_i + v_i \Delta t \tag{2.108}$$

and
$$v_{i+1} = v_i + f(v_i, y_i, t_i) \Delta t \tag{2.109}$$

[Hamilton's equations, (8.96) and (8.97), provide first-order equations of motion for a system directly. Hence, they are more suitable for digital computer calculation than the equations of motion provided by Newton's laws or those obtained by Lagrange's method, equation (8.41).]

Equations (2.108) and (2.109) can now numerically be solved simultaneously on the computer. Note that if at $t=0$ both $v(0)$ and $y(0)$ equal zero, the development of the solution for y becomes delayed one time-step. If $\dot{v}=0$ as well at $t=0$, the solution takes two time-steps to begin. These starting errors can be smoothed if additional Taylor series terms are carried during the first step calculations. These errors are usually neglected when small step sizes are employed.

Sample Problem 2.8
Write a digital computer program that employs Euler's method to numerically solve for the x motion of the given spring-mass system.

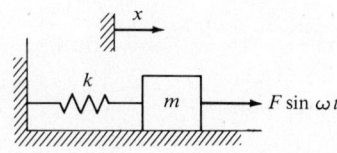

$$m = 37 \text{ kg} \qquad k = 560 \text{ kN/m}$$

$$F = 8150 \text{ N} \qquad \omega = 188.5 \text{ rad/s}$$

At $t=0$, $x=0$ and $v=0$. Let $t_{\text{final}} = 0.06$ s and $\Delta t = 0.003$ s.

Analysis: The equation of motion is

$$m\ddot{x} + kx = F \sin \omega t$$

36 REVIEW OF NEWTONIAN MECHANICS

or
$$\ddot{x} + \frac{k}{m} x = \frac{F}{m} \sin \omega t$$

$$\ddot{x} + 15\,135 x = 220.3 \sin 188.5 t$$

$$x_{i+1} = x_i + v_i \Delta t$$

$$v_{i+1} = v_i + (220.3 \sin 188.5 t_i - 15\,135 x_i) \Delta t$$

The solution cannot start exactly at $t=0 (i=1)$ since all terms are zero. One can ignore this problem for small Δt or try to smooth out the solution curve at $t=0$. Taking the latter approach, and changing the superscript for clarity, one can write

$$x_2 = x_1 + v_1 \Delta t + \frac{a_1 \Delta t^2}{2} + \frac{\dot{a}_1 \Delta t^3}{6} + \cdots$$

The given initial conditions are that $x_1 = v_1 = 0$ at $t=0$.

The first significant term of the series is $\dot{a}_1 \Delta t^3/6$. If one assumes a linearly varying acceleration between time increments, then

$$\frac{\dot{a}_1 \Delta t^3}{6} \sim \frac{\ddot{x}_2 \Delta t^3}{\Delta t\; 6} = \frac{\dot{v}_2 \Delta t^3}{\Delta t\; 6}$$

$$\dot{v}_2 = \ddot{x}_2 = 220.3 \sin 188.5 t_2 - 15\,135 x_2$$

$$x_2 \sim \frac{\ddot{x}_2 \Delta t^2}{6} = (220.3 \sin 188.5 t_2 - 15\,135 x_2) \frac{\Delta t^2}{6}$$

$$x_2 \sim \frac{220.3 (\Delta t)^2 \sin 188.5(\Delta t)}{6 + 15\,135 (\Delta t)^2}$$

With $\Delta t = 0.003$ s,

$$x_2 \sim 3 \times 10^{-3} \text{ mm}$$

$$\dot{v}_2 = \ddot{x}_2 \sim 2.13 \text{ m/s}$$

$$v_2 = \frac{\dot{a}_1 \Delta t^2}{2} = \frac{a_2 \Delta t}{2} = \frac{\dot{v}_2 \Delta t}{2} = 3.2 \text{ mm/s}$$

The method can now proceed with

$$v_3 = v_2 + \dot{v}_2 \Delta t$$

$$x_3 = x_2 + v_2 \Delta t$$

See Figures 2.21 and 2.22 for the solutions. ∎

B. The Modified Euler Method (Euler-Trapezoidal Method)

The modified Euler method is another self-starting algorithm that is very easy to implement on a computer. This method improves on the basic Euler approach by retaining the next higher order term in the Taylor series expansion for

```
00110 PROGRAM EULER(INPUT,OUTPUT)
00120 XDDF(T,X)=A*SIN(B*T)-C*X
00130 DATA A/220.3/,B/188.5/,C/15135./,DT/.0001/,TFIN/.06/
00140 DATA DP/.001/
00150 TP=0.
00160 X=0.
00170 XD=0.
00180 XDD=0.
00190 T=0.
00200 PRINT 7,
00210 7 FORMAT(6X,"TIME",6X,"X",16X,"XD",15X,"XDD"//)
00220 PRINT 8,T,X,XD,XDD
00230 8 FORMAT(6X,F5.4,6X,E11.4,6X,E11.4,6X,E11.4)
00240 T=T+DT
00250 XN=(A*DT*DT*SIN(B*DT))/(6.+C*DT*DT)
00260 XDDN=XDDF(T,XN)
00270 XDN=XDDN*DT/2.
00280 9 TP=TP+DP
00290 PRINT 8,T,XN,XDN,XDDN
00300 IF(T.GT.(TFIN-0.00001)) GO TO 10
00310 14 X=XN
00320 XD=XDN
00330 XDD=XDDN
00340 XN=X+XD*DT
00350 XDN=XD+XDD*DT
00360 T=T+DT
00370 XDDN=XDDF(T,XN)
00380 IF(T.GT.(TP-0.00001))GO TO 9
00390 GO TO 14
00400 10 STOP
00410 END
```

Figure 2.21

TIME	X	XD	XDD
.0000	0.	0.	0.
.0001	.6921E-08	.2076E-03	.4152E+01
.0010	.5168E-05	.1883E-01	.4120E+02
.0020	.4734E-04	.7792E-01	.8038E+02
⋮	⋮	⋮	⋮
.0600	.2362E-01	.4266E+00	-.5972E+03

Figure 2.22

y_{i+1}. The truncation error is therefore reduced so that the estimated error now becomes

$$(E)_i \simeq y_i''' \frac{\Delta t^3}{12} \qquad (2.110)$$

The first three terms of the Taylor series expansion are

$$y_{i+1} = y_i + y_i' \Delta t + y_1'' \frac{\Delta t^2}{2} \qquad (2.111)$$

y_i'' can be approximated using a *forward-difference* formula

$$y_i'' = \frac{y_{i+1}' - y_i'}{\Delta t} \qquad (2.112)$$

Then
$$y_{i+1} = y_i + \frac{\Delta t}{2}(y_{i+1}' + y_i') \qquad (2.113)$$

At this point, one must bear in mind that y_{i+1}' is a future value that is not yet known. An estimate to its value can be made by *predicting* the value of y_{i+1} using the basic Euler method.

$$P(y_{i+1}) = y_i + y_i' \Delta t \qquad (2.103)$$

and with the given defining functional relationship

$$y' = f(y, t) \qquad (2.99)$$

a *predicted* value of y_{i+1}' may also be obtained.

$$P(y_{i+1}') = f(P(y_{i+1}), t_{i+1}) \qquad (2.114)$$

The modified Euler method then *corrects* these values using equation (2.113) rewritten as

$$C(y_{i+1}) = y_i + \frac{\Delta t}{2}[P(y_{i+1}') + y_i'] \qquad (2.115)$$

and with equation (2.99)

$$C(y_{i+1}') = f(C(y_{i+1}), t_{i+1}) \qquad (2.116)$$

These corrected values become the known values for the next calculation. No iteration to refine the answer should be attempted.[1] The method is graphically illustrated in Figure 2.23.

The modified Euler method is one example of the many *predictor-corrector* methods available for the numerical solution of ordinary differential equations. Its multiple advantages of being self-starting, fast, fairly accurate, and easy to

---
[1] Forman S. Acton, *Numerical Methods That Work*, Harper and Row, New York, 1970, pp. 133–134.

PARTICLE DYNAMICS: KINEMATICS   39

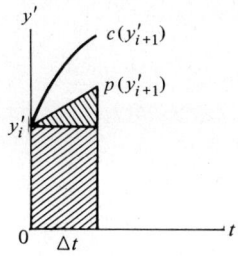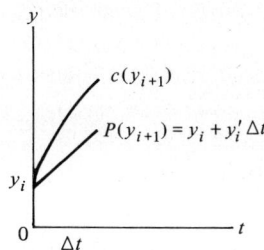

Figure 2.23

program make it a very popular approach. Solution verification and step-size selection are accomplished in the same manner as was done for the Euler method.

Sample Problem 2.9
Repeat Sample Problem 2.8 using the modified Euler method.

Analysis: The second-order equation

$$m\ddot{x} + kx = F \sin \omega t$$

can be rewritten as the two first-order equations

$$\dot{x} = v \qquad \dot{v} = \frac{-kx}{m} + \frac{F}{m} \sin \omega t$$

Then
$$P(x_{i+1}) = x_i + v_i \, \Delta t$$
$$P(v_{i+1}) = v_i + \dot{v}_i \, \Delta t$$
$$P(\dot{v}_{i+1}) = -\frac{k}{m} P(x_{i+1}) + \frac{F}{m} \sin \omega(t_i + \Delta t)$$
$$C(x_{i+1}) = x_{i+1} = x_i + \frac{\Delta t}{2} [v_i + P(v_{i+1})]$$
$$C(v_{i+1}) = v_{i+1} = v_i + \frac{\Delta t}{2} [\dot{v}_i + P(\dot{v}_{i+1})]$$
$$C(\dot{v}_{i+1}) = \dot{v}_{i+1} = -\frac{k}{m} (x_{i+1}) + \frac{F}{m} \sin \omega(t_i + \Delta t)$$

and the method proceeds to the next value. The Fortran program and solution are shown in Figures 2.24 and 2.25, respectively.   ∎

C.   The Runge-Kutta Fourth-Order Method

The Runge-Kutta fourth-order method is used very often in practice to numerically solve ordinary differential equations with initial condition given. It is a popular method because it is self-starting, it is very accurate, and it is easy to program. It is called a *fourth-order method* because it is derived under the

```
00100 PROGRAM MODEUL(INPUT,OUTPUT)
00110 DIMENSION X(3),PX(3)
00120 XDD(T,X)=A*SIN(B*T)-C*X
00130 DATA A/220.3/,B/188.5/,C/15135./,DT/.001/,TFIN/.06/
00140 DO 1 I=1,3
00150 X(I)=0.
00160 PX(I)=0.
00170 1 CONTINUE
00180 T=0.
00190 PRINT 7,
00200 7 FORMAT(6X,'TIME',6X,'X',16X,'XD',15X,'XDD'//)
00210 PRINT 8,T,(X(I),I=1,3)
00220 8 FORMAT(6X,F4.3,6X,E11.4,6X,E11.4,6X,E11.4)
00230 5 DO 2 I=1,2
00240 PX(I)=X(I)+X(I+1)*DT
00250 2 CONTINUE
00260 T=T+DT
00270 PX(3)=XDD(T,PX(1))
00280 DO 3 I=1,2
00290 X(I)=X(I)+(X(I+1)+PX(I+1))*DT/2.
00300 3 CONTINUE
00310 X(3)=XDD(T,X(1))
00320 PRINT 8,T,(X(I),I=1,3)
00330 IF(T.GT.(TFIN-0.00001)) GO TO 4
00340 GO TO 5
00350 4 STOP
00360 END
```

Figure 2.24

TIME	X	XD	XDD
.000	0.	0.	0.
.001	0.	.2064E-01	.4128E+02
.002	.4128E-04	.8167E-01	.8047E+02
.003	.1632E-03	.1800E+00	.1156E+03
.004	.4010E-03	.3106E+00	.1447E+03
.005	.7840E-03	.4667E+00	.1664E+03
⋮	⋮	⋮	⋮
.058	.2364E-01	.1454E+01	-.5777E+03
.059	.2481E-01	.8655E+00	-.5940E+03
.060	.2537E-01	.2695E+00	-.5935E+03

Figure 2.25

specific criteria that it match the terms of a Taylor series expansion for y_{i+1} up to terms including Δt^4. Namely,

$$y_{i+1} = y_i + y_i' \Delta t + y_i'' \frac{\Delta t^2}{2} + y_i''' \frac{\Delta t^3}{6} + y_i^{(4)} \frac{\Delta t^4}{24} + \cdots \qquad (2.117)$$

The general Runge-Kutta fourth-order recurrence formula is given by

$$y_{i+1} = y_i + a_1 k_1 + a_2 k_2 + a_3 k_3 + a_4 k_4 \qquad (2.118)$$

where
$$\begin{aligned}
k_1 &= \Delta t \, f(t_i, y_i) \\
k_2 &= \Delta t \, f(t_i + b_1 \Delta t, y_i + C_{11} k_1) \\
k_3 &= \Delta t \, f(t_i + b_2 \Delta t, y_i + C_{21} k_1 + C_{22} k_2) \\
k_4 &= \Delta t \, f(t_i + b_3 \Delta t, y_i + C_{31} k_1 + C_{32} k_2 + C_{33} k_3)
\end{aligned} \qquad (2.119)$$

If equation (2.119) is substituted into equation (2.118), and if the entire expression is then expanded in a Taylor series about y_i, the resulting series can be matched like-order term to like-order term with the series expression, equation (2.117). Relationships for the 13 constants can then be established. The results are

$$\begin{aligned}
b_1 &= C_{11} \\
b_2 &= C_{21} + C_{22} \\
b_3 &= C_{31} + C_{32} + C_{33} \\
a_1 &+ a_2 + a_3 + a_4 = 1 \\
a_2 b_1 &+ a_3 b_2 + a_4 b_3 = \tfrac{1}{2} \\
a_2 b_1^2 &+ a_3 b_2^2 + a_4 b_3^2 = \tfrac{1}{3} \\
a_2 b_1^3 &+ a_3 b_2^3 + a_4 b_3^3 = \tfrac{1}{4} \\
a_3 b_1 C_{22} &+ a_4 (b_1 C_{32} + b_2 C_{33}) = \tfrac{1}{6} \\
a_3 b_1^2 C_{22} &+ a_4 (b_1^2 C_{32} + b_2^2 C_{33}) = \tfrac{1}{12} \\
a_3 b_1 b_2 C_{22} &+ a_4 b_3 (b_1 C_{32} + b_2 C_{33}) = \tfrac{1}{8} \\
a_4 b_1 C_{22} C_{33} &= \tfrac{1}{24}
\end{aligned} \qquad (2.120)$$

With 11 equations and 13 unknowns, there is latitude in the selection of coefficients. The Runge-Kutta fourth-order recurrence formula in *classic form* has

$$y_{i+1} = y_i + \tfrac{1}{6}(k_1 + 2k_2 + 2k_3 + k_4) \qquad (2.121)$$

$$\begin{aligned}
k_1 &= \Delta t \, f(t_i, y_i) \\
k_2 &= \Delta t \, f\left(t_i + \frac{\Delta t}{2}, y_i + \frac{k_1}{2}\right) \\
k_3 &= \Delta t \, f\left(t_i + \frac{\Delta t}{2}, y_i + \frac{k_2}{2}\right) \\
k_4 &= \Delta t \, f(t_i + \Delta t, y_i + k_3)
\end{aligned} \qquad (2.122)$$

42 REVIEW OF NEWTONIAN MECHANICS

The per-step truncation error is of order Δt^5. The four k values geometrically define the slopes of the curve at the starting point, at the midpoint (two values), and at the right-hand point, multiplied by Δt. A part of equation (2.121) then represents the weighted average of the slopes. This average, multiplied by Δt, accounts for the change in y_i. The change in y_i is then added to y_i to produce a new y_{i+1}. (See Sample Problem 2.10 for treatment of a second-order equation.) See Figure 2.26.

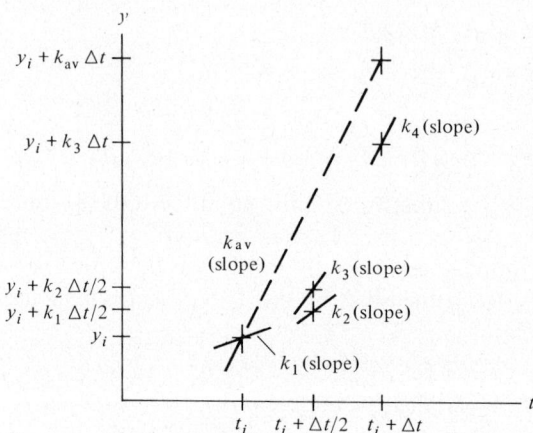

Figure 2.26

Although the method has strong advantages, it does have at least two weaknesses. Because of the complexity of equations (2.121) and (2.122), it is very difficult to estimate per-step truncation error, and thus the method is usually used only with fixed-step-size computation. This is inefficient and time-consuming. It is preferable to be able to run a program at the largest possible step size that maintains a user-defined level of acceptable accuracy. Other more powerful integration routines allow for variable step size and error control. The reader is referred to a textbook on numerical analysis for a further discussion of this topic. The second weakness also concerns running time. Because of the four calls to the function required per step [see equation (2.122)], the Runge-Kutta method is considered to offer a relatively slow solution. For *real-time* applications, e.g., in a control system loop, the Runge-Kutta method may prove unacceptable. When speed, rather than accuracy or cost, is the most critical factor, the user should consider using an *analog computer* for integration purposes (see Chapter 10). Even though it is slow, the Runge-Kutta method still has proved to be a recommended, simple, accurate, and reliable means for numerical analysis of ordinary differential equations.

Sample Problem 2.10
Repeat Sample Problem 2.8 using the Runge-Kutta fourth-order recurrence formula. Use a step size of 0.003 s.

Analysis: The second-order equation

$$m\ddot{x} + kx = F \sin \omega t$$

can be rewritten as the two first-order equations

$$\dot{x} = v \qquad \dot{v} = -\frac{k}{m}x + \frac{F}{m}\sin \omega t$$

then

$$x_{i+1} = x_i + \tfrac{1}{6}(m_1 + 2m_2 + 2m_3 + m_4)$$
$$v_{i+1} = v_i + \tfrac{1}{6}(k_1 + 2k_2 + 2k_3 + k_4) \qquad (a)$$

where

$$m_1 = \Delta t \ g(t_i, x_i, v_i) = \Delta t \ v_i$$

$$m_2 = \Delta t \ g\left(t_i + \frac{\Delta t}{2}, x_i + \frac{m_1}{2}, v_i + \frac{k_1}{2}\right) = \Delta t \left(v_i + \frac{k_1}{2}\right) \qquad (b)$$

$$m_3 = \Delta t \ g\left(t_i + \frac{\Delta t}{2}, x_i + \frac{m_2}{2}, v_i + \frac{k_2}{2}\right) = \Delta t \left(v_i + \frac{k_2}{2}\right)$$

$$m_4 = \Delta t \ g(t_i + \Delta t, x_i + m_3, v_i + k_3) = \Delta t \ (v_i + k_3)$$

After substituting equations (b), the equations for the k's become

$$k_1 = \Delta t \ f(t_i, x_i, v_i)$$

$$k_2 = \Delta t \ f\left(t_i + \frac{\Delta t}{2}, x_i + \frac{\Delta t \ v_i}{2}, v_i + \frac{k_1}{2}\right) \qquad (c)$$

$$k_3 = \Delta t \ f\left(t_i + \frac{\Delta t}{2}, x_i + \frac{\Delta t \ (v_i + k_1/2)}{2}, v_i + \frac{k_2}{2}\right)$$

$$k_4 = \Delta t \ f\left(t_i + \Delta t, x_i + \Delta t \left(v_i + \frac{k_2}{2}\right), v_i + k_3\right)$$

and equations (a) become

$$x_{i+1} = x_i + \Delta t \ v_i + \frac{\Delta t}{6}(k_1 + k_2 + k_3) \qquad (d)$$
$$v_{i+1} = v_i + \tfrac{1}{6}(k_1 + 2k_2 + 2k_3 + k_4)$$

In effect, the explicit dependence on m has been eliminated from equations (a) and (b) because of the relation $\dot{x} = v$. The reader should also note that equation (c) is in general form. In this specific problem, the function describing $\dot{v}$ is not a function of v, that is, $f = f(t, x)$ only. The Fortran program and the numerical solution are shown in Figures 2.27 and 2.28, respectively.

At this point, the reader should compare all three solutions provided. Note that the Runge-Kutta method provides near agreement with the exact solution. The least accurate numerical solution came from Euler's method. A step size of 0.0001 was needed to produce acceptable accuracy. The effects of round-off error are clearly demonstrated in the output. ∎

44 REVIEW OF NEWTONIAN MECHANICS

```
00100 PROGRAM RUNGKUT(INPUT,OUTPUT)
00110 DIMENSION X(3),XN(3),RK(4)
00120 XDD(T,X)=A*SIN(B*T)-C*X
00130 DATA A/220.3/,B/188.5/,C/15135./,DT/.003/,TFIN/.06/
00140 DO 1 I=1,3
00150 X(I)=0.
00160 XN(I)=0.
00170 1 CONTINUE
00180 T=0.
00190 PRINT 7
00200 7 FORMAT(6X,"TIME",6X,"X",16X,"XD",15X,"XDD"//)
00210 PRINT 8,T,(XN(I),I=1,3)
00220 8 FORMAT(6X,F4.3,6X,E11.4,6X,E11.4,6X,E11.4)
00230 5 RK(1)=DT*X(3)
00240 RK(2)=DT*XDD(T+DT/2.,X(1)+DT*X(2)/2.)
00250 RK(3)=DT*XDD(T+DT/2.,X(1)+(X(2)+RK(1)/2.)*DT/2.)
00260 RK(4)=DT*XDD(T+DT,X(1)+(X(2)+RK(2)/2.)*DT)
00270 XN(1)=X(1)+DT*(X(2)+(RK(1)+RK(2)+RK(3))/6.)
00280 XN(2)=X(2)+(RK(1)+2.*(RK(2)+RK(3))+RK(4))/6.
00290 T=T+DT
00300 XN(3)=XDD(T,XN(1))
00310 PRINT 8,T,(XN(I),I=1,3)
00320 IF(T.GT.(TFIN-0.00001)) GO TO 9
00330 DO 4 I=1,3
00340 X(I)=XN(I)
00350 4 CONTINUE
00360 GO TO 5
00370 9 STOP
00380 END
```

Figure 2.27

Sample Problem 2.11
It is known that the spring-mass damper system shown is critically damped. If the system is released from rest with $x = 0.02$ m, determine the time that it takes for x to return to the position $x = 0.002$ m. A trial-and-error or Newton-Raphson method with computer implementation is required.

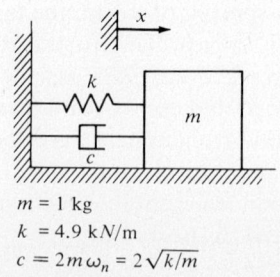

$m = 1$ kg
$k = 4.9$ kN/m
$c = 2m\omega_n = 2\sqrt{k/m}$

TIME	X	XD	XDD	
.000	0.	0.	0.	
.003	.1844E-03	.1799E+00	.1153E+03	− x_{exact} = .1826E−03
.006	.1366E-02	.6391E+00	.1787E+03	
.009	.4089E-02	.1165E+01	.1567E+03	
.012	.8154E-02	.1490E+01	.4632E+02	
.015	.1259E-01	.1384E+01	−.1225E+03	
.018	.1592E-01	.7500E+00	−.2957E+03	
.021	.1663E-01	−.3332E+00	−.4123E+03	
.024	.1371E-01	−.1618E+01	−.4239E+03	
⋮	⋮	⋮	⋮	
.042	−.2563E-01	.1021E+01	.6078E+03	
.045	−.1997E-01	.2689E+01	.4805E+03	
.048	−.1008E-01	.3783E+01	.2336E+03	
.051	.1868E-02	.4033E+01	−.6959E+02	
.054	.1321E-01	.3387E+01	−.3507E+03	
.057	.2146E-01	.2023E+01	−.3382E+03	
.060	.2498E-01	.2978E+00	−.5875E+03	− x_{exact} = .2501E−01

Figure 2.28

Analysis: From equation (2.45)

$$x = (A + Bt)e^{-ct/2m} = (A + Bt)e^{-\omega_n t}$$

At $t = 0$,

$$\dot{x} = 0 = -\omega_n A + B \quad \text{or} \quad B = \omega_n A$$

Then
$$x = A(1 + \omega_n t)e^{-\omega_n t}$$

Also, since $x(0) = 0.02$ m, then $A = 0.02$ m. The equation for x becomes

$$x = 0.02(1 + 70t)e^{-70t} \tag{a}$$

When $x = 0.002$, equation (a) becomes

$$0.1 = (1 + 70t)e^{-70t} \tag{b}$$

Equation (b) can be solved using a trial-and-error approach or by a numerical method such as Newton-Raphson. In keeping with the spirit of this book, which greatly admires the many accomplishments of Sir Isaac Newton, the latter approach will be taken.

One can define

$$F(t) = (1 + 70t)e^{-70t} - 0.1 = 0 \tag{c}$$

Expanding $F(t)$ in a Taylor series about point t_i leads to

$$F(t_i + \Delta t) = F(t_i) + F'(t_i)\Delta t + F''(t_i)\frac{\Delta t^2}{2} + \cdots \tag{d}$$

REVIEW OF NEWTONIAN MECHANICS

Retaining terms only up to first order in the expansion

$$F(t_i + \Delta t) \simeq F(t_i) + F'(t_i)\Delta t \tag{e}$$

We can now solve for the increment of time, Δt, required to make the function $F(t_i + \Delta t)$ identically zero.

$$0 = F(t_i + \Delta t) = F(t_i) + F'(t_i)\Delta t$$

$$\Delta t = -\frac{F(t_i)}{F'(t_i)}$$

$$t_{i+1} = t_i - \frac{F(t_i)}{F'(t_i)} \tag{f}$$

The repeated application of equation (f) until convergence between the successive values t_{i+1} and t_i occurs is known as the *Newton-Raphson* method. The method is very useful for determining the roots of any equation of the form $F(x) = 0$, provided a single root can be isolated. In the vicinity of the root, the slope $F'(x)$ and the change of slope $F''(x)$ must not change sign if convergence is to occur. Also, as seen by equation (f), the function $F(x)$ must be finite in the neighborhood of the root.

```
00100 PROGRAM NEWTON (INPUT,OUTPUT)
00110 F(T)=(1.+A*T)*EXP(-A*T)-B
00120 FD(T)=-A*A*T*EXP(-A*T)
00130 DATA A/70./,B/.1/,T0/.04/,NM/50./,EPSI/.0001/
00140 T=T0
00150 3 TN=T-F(T)/FD(T)
00160 IF(ABS(TN-T).LE.EPSI) GO TO 6
00170 IF(N.GE.NM) GO TO 5
00180 N=N+1
00190 T=TN
00200 GO TO 3
00210 5 PRINT 8
00220 8 FORMAT(5X,"NMAX REACHED WITHOUT CONVERGENCE")
00230 6 PRINT 7,T
00240 7 FORMAT(3X,"T=",E11.4)
00250 STOP
00260 END



        T=  .5556E-01
```

Figure 2.29

For this problem then,

$$F(t) = (1 + 70t)e^{-70t} - 0.1 = 0$$
$$F'(t) = -70(1 + 70t)e^{-70t} + 70e^{-70t} = -4900te^{-70t}$$
$$t_{i+1} = t_i - \frac{0.1 - (1 + 70t)e^{-70t}}{4900te^{-70t}}$$

In the computer analysis, an initial guess of $t = 0.04$ s was used. Why can't an initial value of $t = 0$ be taken? The solution is given in Figure 2.29. ∎

PROBLEMS

2.1 A falling satellite having an initial speed v_0 experiences a resistive force from the atmosphere that can be approximated as $F_0 = \alpha \dot{x} e^{kx}$. If the attractive force of the earth is neglected, how far into the atmosphere will the satellite travel before it stops?

2.2 When a car traveling with an initial speed v_0 locks its brakes, it experiences a friction force due to air resistance equal to Cv and a constant force from the ground caused by sliding. If the coefficient of sliding friction is known to be μ, find an expression for the time it takes for the automobile to stop.

2.3 What is the maximum value of initial speed a particle can have at $x = 0$ if it is attracted by a force at the origin equal to ae^{-bx} and it is desired to have the particle return to $x = 0$?

2.4 For a projectile in a uniform gravity field having initial speed v_0, find the elevation angle needed at launch so that it may pass through a specified point x, y. What is the equation of the curve, the envelope of all possible trajectories, that can only be reached by one launch angle?

2.5 Two massless springs just fit in their unstretched states between two fixed supports. Spring AB is attached to one support and to a small mass having weight W that lies between the springs. Spring CD is attached to the other wall at D. The two spring constants for springs AB and CD are k and $2k$, respectively. If spring AB is compressed a distance e with its attached mass and then released from rest, determine the subsequent maximum acceleration of the mass. Neglect friction and the weight's size.

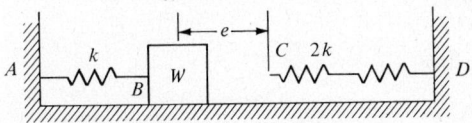

2.6 The simple pendulum shown has mass m and length L. A horizontal spring with constant k is attached to the pendulum at a distance of one-third the length of the massless rod from the pivot point. The spring is unstretched when the pendulum is vertical. Find the period of motion for the pendulum assuming θ to be very small.

48 REVIEW OF NEWTONIAN MECHANICS

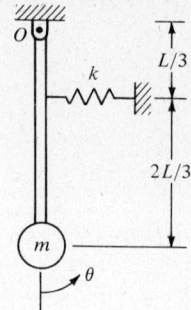

2.7 In problem 2.6, what would be the effect on the period if the pivot point was moved to the middle of the rod?

2.8 The needle in an electrical instrument is suspended at its center of mass. Its motion clockwise from the zero position is resisted by a torsional spring with spring constant k_θ. In addition, torsional damping is built into the instrument; the damping constant is c_θ. If a constant moment M_0 is supplied to the needle's shaft, determine the angle of the new equilibrium position. How long does it take from the instant of application of the moment until the needle turns after having passed the equilibrium position? The mass moment of inertia of the needle is given by c_θ^2/k_θ.

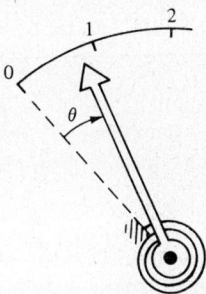

2.9 The weight shown, having mass m, moves on a frictionless horizontal surface. Attached to one side is a spring having spring constant k. A dashpot having damping constant $c = \sqrt{3km}$ opposes motion on the other side. If the mass is suddenly given a speed to the right of $v_0 = a\sqrt{k/m}$, find the speed of the weight the first time the weight passes through the equilibrium position ($x = 0$).

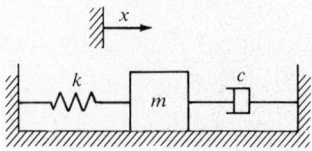

2.10 In problem 2.9, determine the logarithm of the ratio of any two successive amplitudes separated by a time equal to the damped period. This ratio δ, called the *logarithmic decrement*, provides a simple means to determine the damping present in a system.

PARTICLE DYNAMICS: KINEMATICS   49

2.11   A gun barrel of mass m has a recoil spring with a spring constant $4k$. It is desired to regain the equilibrium position of the gun as soon as possible after the recoil. What must the damping be?

2.12   Repeat Sample Problem 2.6 for the case when $\omega = 1300$ rpm.

2.13   For the system shown, driven by the force $F = F_0 \sin \omega t$, find an expression for the maximum amplitude of x. The distance x is measured from the unstretched position of the spring.

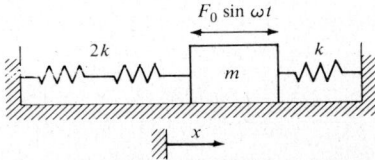

2.14   For the system given, determine the amplitude of the steady-state vibration. What is the magnification factor? Let x be measured from the unstretched position of the two springs.

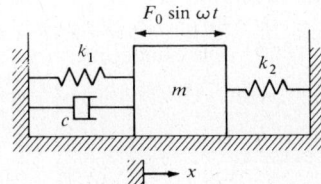

2.15   If in problem 2.14 the force was removed and instead the support on the left was driven harmonically so that $y = y_0 \sin \omega t$, determine the resulting absolute amplitude ratio of $|x/y_0|$.

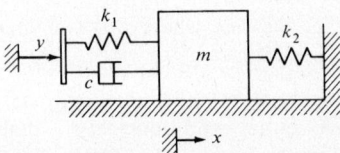

2.16   A simple schematic of a vibration-measuring instrument is shown in the figure. If z is the relative motion of the mass and pen stylus relative to the moving case, determine the absolute value of the amplitude ratio z/y_0 for a given base motion $y = y_0 \sin \omega t$. Such an instrument having $\omega/\omega_n \to \infty$ is called a *seismometer*. Explain why when $\omega/\omega_n \to 0$, the instrument is called an *accelerometer*.

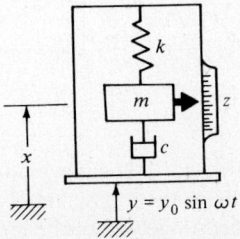

*2.17 If the viscous damper in Sample Problem 2.11 is engaged only after the recoil stroke of x_0, write a computer program using the Newton-Raphson method to determine the time it takes for the barrel to return to a position $x = 0.1 x_0$; let time start from the moment recoil begins.

*2.18 Write a computer program that will yield values of x with increasing time for the given system. Let $F_0 = 50$ N, $k_1 = 1000$ N/m, $k_2 = 2000$ N/m, $c = 20$ N·s/m, $\omega = 100$ rad/s, and $m = 1.5$ kg. Use a communication interval of 0.01 s. Take $t = 0.8$ s as the final time. All initial conditions are assumed to be zero.

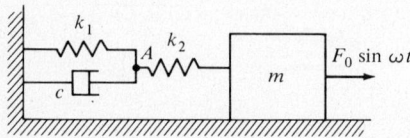

*2.19 Write a computer program that will yield values of x with increasing time for the given system. The excitation applied to the mass, $F(t)$, is shown in the figure to the right. Assume all initial conditions to be zero. The following values are given: $t_1 = 0.25$ s, $F_0 = 10$ N, $k = 800$ N/m, $c = 25$ N·s/m, $m = 2.0$ kg. Use a communication interval of 0.05 s. Take $t = 1.0$ s as the final time.

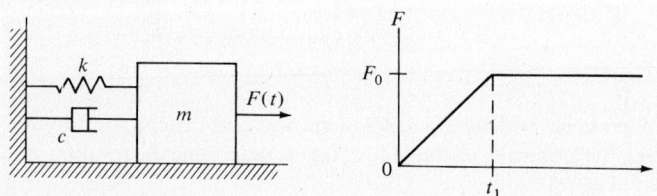

*2.20 Repeat problem 2.19 for the case where $F(t)$ is a single rectangular pulse having magnitude F_0 and duration time t_1.

*2.21 Write a computer program that will numerically solve the equations of motion for the system given. As initial conditions, let $x = 0$ and $\dot{x} = 0.24$ m/s. The springs each have stiffness k, and the free gaps both have dimensions 0.01 m. Let $m = 20$ kg and $k = 1.250$ kN/m. Use a communication interval of 0.1 s, and terminate the program at $t = 2.0$ s.

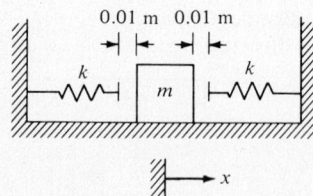

*2.22 Write a computer program that will numerically solve the equation of motion for the system given. Coulomb friction, having constant magnitude F and always opposing motion, is assumed to act here. The initial conditions at $t = 0$ are $x_0 = 0$ and $\dot{x}_0 = 0.25$ m/s. The distance x is measured from the unstretched position of the spring. The mass m is 20 kg, and the spring stiffness k is equal to 2.0 kN/m. Use a communication interval of 0.05 s, and terminate the program at $t = 2.0$ s. The friction coefficient μ is equal to 0.1.

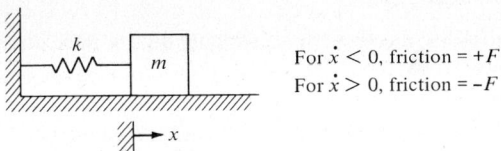

2.23 If the motion of a particle is given in meters by $y = 2t$ and $x = t^2$, find the radius of curvature of the particle's path when $t = 4$ s.

2.24 If a projectile is fired from a cannon with an initial speed v_0 at an elevation angle γ, find what the radius of curvature of the trajectory is immediately after the projectile leaves the barrel. What is the radius of curvature at the zenith?

2.25 In an experiment, a particle is free to slide inside a smooth pipe of length L that rotates at a constant angular speed ω. If the particle is injected somehow into the pipe at its midpoint, with no radial velocity component, determine the relative speed of the particle with respect to the pipe when it leaves.

2.26 If in the previous problem $\theta = \omega t$, find the value of the particle's normal acceleration when it just leaves the tube.

2.27 A particle is constrained to move along a curved wire having the equation $y = -x^2 + \frac{4}{3}$. At the same time, it also slides along the slot in rod OA as the rod turns with constant angular velocity ω clockwise. For the instant when OA makes a 30° angle with the vertical, find the radial and transverse components of the velocity. Find a unit vector in the tangential direction.

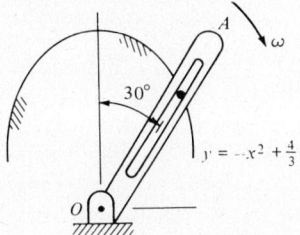

2.28 A satellite is in an elliptic orbit around the earth. When the satellite passes the semiminor axis of its orbit, it has a speed v_0. What is the radius of curvature at that point?

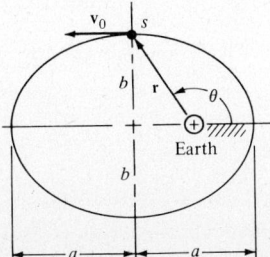

2.29 A particle of weight w slides down a smooth helical wire having constant helix angle α and radius r. If the particle starts from rest, derive an expression for the radial, transverse, and vertical accelerations as a function of the number of complete revolutions made.

2.30 A wheel of radius R rolls without slipping on a surface having a radius of curvature $4R$. Determine the velocity of point P on the rim as a function of the absolute rotation angle ϕ. At $t=0$, $\theta = \phi = 0$.

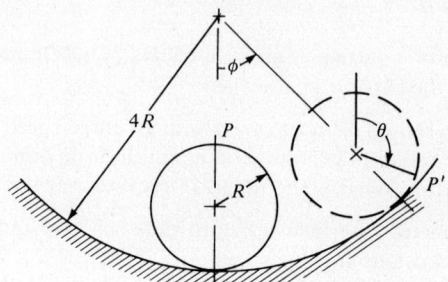

PARTICLE DYNAMICS: KINETICS

3.1 EQUATIONS OF MOTION

With all the basic relationships for acceleration described, it is now possible to express Newton's second law in component form. The second law restated is

$$\mathbf{F} = m\mathbf{a} \qquad (1.1)$$

In terms of rectangular components, Newton's second law can be written as

$$F_x = ma_x = m\ddot{x} \qquad (3.1)$$
$$F_y = ma_y = m\ddot{y} \qquad (3.2)$$
$$F_z = ma_z = m\ddot{z} \qquad (3.3)$$

For the radial and transverse directions, the components are

$$F_r = ma_r = m(\ddot{r} - r\dot{\theta}^2) \qquad (3.4)$$
$$F_\theta = ma_\theta = m(r\ddot{\theta} + 2\dot{r}\dot{\theta}) \qquad (3.5)$$

and for the normal and tangential directions, the components appear as

$$F_n = ma_n = m\frac{v^2}{\rho} = mv\dot{\theta} = m\rho\dot{\theta}^2 \qquad (3.6)$$
$$F_t = ma_t = m\dot{v} \qquad (3.7)$$

Similarly, in a cylindrical coordinate set the component terms are

$$F_r = ma_R = m(\ddot{R} - R\dot{\theta}^2) \qquad (3.8)$$
$$F_\theta = ma_\theta = m(R\ddot{\theta} + 2\dot{R}\dot{\theta}) \qquad (3.9)$$
$$F_z = ma_z = m\ddot{z} \qquad (3.10)$$

and finally, for spherical axes, the resolved components are

$$F_r = ma_r = m(\ddot{r} - r\dot{\phi}^2 - r\dot{\theta}^2 \sin^2\phi) \qquad (3.11)$$
$$F_\theta = ma_\theta = m(r\ddot{\theta} \sin\phi + 2\dot{r}\dot{\theta} \sin\phi + 2r\dot{\phi}\dot{\theta} \cos\phi) \qquad (3.12)$$
$$F_\phi = ma_\phi = m(r\ddot{\phi} + 2\dot{r}\dot{\phi} - r\dot{\theta}^2 \sin\phi \cos\phi) \qquad (3.13)$$

The reader must bear in mind that these are instantaneous values of forces and acceleration. In general, they are not constant, and if displacement and passage of time occur in a problem, integration of the equations of motion must be performed. This problem will be discussed in the next section.

The moment of the resultant force acting upon a particle about a fixed origin O is also of interest. Since instantaneous particle dynamic problems can be readily solved without resorting to moment equations, the necessity for this introduction will only become apparent later when rigid body motion, which includes rotation, is discussed. The moment equals to

$$\mathbf{M}_O = \mathbf{r} \times \mathbf{F} \tag{3.14}$$

$$= \mathbf{r} \times m\mathbf{a} \tag{3.15}$$

$$= \frac{d}{dt}(\mathbf{r} \times m\mathbf{v}) \quad \mathrm{N \cdot m} \tag{3.16}$$

Equation (3.15) can be derived from equation (3.16) if one recalls that $d\mathbf{r}/dt = \mathbf{v}$. Since $\mathbf{r}$ is defined to be the moment arm to the particle from the fixed origin, its derivative is the particle's absolute velocity.

Defining the *angular momentum*, or alternatively the *moment of linear momentum*, of the particle about a fixed point to be (Figure 3.1)

$$\mathbf{H}_O \equiv \mathbf{r} \times m\mathbf{v} \quad \mathrm{kg \cdot m^2/s} \tag{3.17}$$

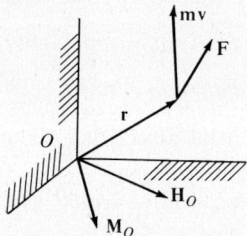

Figure 3.1

where the linear momentum is defined as

$$\mathbf{p} \equiv m\mathbf{v} \quad \mathrm{kg \cdot m/s} \tag{3.18}$$

Equation (3.16) can be rewritten as

$$\mathbf{M}_O = \dot{\mathbf{H}}_O \tag{3.19}$$

Equation (3.19) states that the sum of the moments of forces acting upon a particle about a fixed origin is equal to the time rate of change of angular momentum for the particle about the fixed origin.

3.2 INTEGRALS OF THE MOTION

A. Principle of Work and Kinetic Energy

Since Newton's second law only provides instantaneous description of a system's dynamic behavior, integration of the equations of motion becomes necessary if a change in position and time occurs. Here, in this section, integrals of the equation of motion will be taken with respect to displacement or time. The first integral to be discussed is one where the independent variable is displacement.

The equation of motion is

$$\mathbf{F} = m\mathbf{a} \qquad (1.1)$$

then

$$\mathbf{F} \cdot d\mathbf{r} = m \frac{d\mathbf{v}}{dt} \cdot d\mathbf{r}$$

or

$$\int_1^2 \mathbf{F} \cdot d\mathbf{r} = \int_1^2 m\mathbf{v} \cdot d\mathbf{v} \qquad (3.20)$$

The integral on the left side of equation (3.20) is defined as the *work done by a force* acting through a displacement from position 1 to position 2. Note that it is a path (or line) integral. It represents the sum of a series of scalar products taken along the path of the particle as it travels from the initial to its final position. Infinitesimal work is only performed by the component of force that acts along the direction of the displacement (Figure 3.2). In other words,

$$W_{1 \to 2} \equiv \int_1^2 \mathbf{F} \cdot d\mathbf{r} = \lim_{\substack{n \to \infty \\ \Delta r_i \to 0}} \sum_{i=1}^n \mathbf{F}_i \cdot \Delta \mathbf{r}_i$$

$$= \int_1^2 F \cos \alpha \, dr$$

The work done then is

$$W_{1 \to 2} \equiv \int_1^2 \mathbf{F} \cdot d\mathbf{r} = \int_1^2 m\mathbf{v} \cdot d\mathbf{v} \qquad \text{J (joule)} \qquad (3.21)$$

The right-hand side of equation (3.21) is not only a scalar product but also an exact differential.

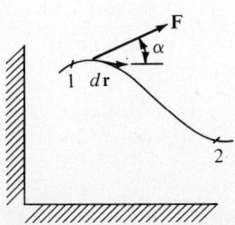

Figure 3.2

$$m\mathbf{v}\cdot d\mathbf{v} = d(\tfrac{1}{2}m\mathbf{v}\cdot\mathbf{v}) = d\left(\frac{mv^2}{2}\right)$$

or
$$\int_1^2 d\left(\frac{mv^2}{2}\right) = T_2 - T_1 \tag{3.22}$$

where the scalar function T is defined as *kinetic energy*,

$$T \equiv \frac{mv^2}{2} \quad \text{J}$$

Then, substituting equation (3.22) into equation (3.21), we obtain

$$W_{1\to 2} = T_2 - T_1 \tag{3.23}$$

Equation (3.23), one integral of the motion, is the mathematical description of the *principle of work and kinetic energy*. It is always valid; i.e., it can be used to analyze conservative or nonconservative systems.

The kinetic energy, equation (3.22), is expressed in terms of absolute velocity. Some useful expressions for kinetic energy in different coordinate systems are

$$T = \frac{m(\dot{x}^2 + \dot{y}^2 + \dot{z}^2)}{2} \tag{3.24}$$

$$T = \frac{m(\dot{r}^2 + r^2\dot{\theta}^2)}{2} \tag{3.25}$$

$$T = \frac{m(\dot{R}^2 + R^2\dot{\theta}^2 + \dot{z}^2)}{2} \tag{3.26}$$

$$T = \frac{m(\dot{r}^2 + r^2\dot{\phi}^2 + r^2\sin^2\phi\,\dot{\theta}^2)}{2} \tag{3.27}$$

The rate of doing work is called *power* and is defined by

$$P = \frac{dW}{dt} = \frac{\mathbf{F}\cdot d\mathbf{r}}{dt} = \mathbf{F}\cdot\mathbf{v} \quad \text{W (watt)} \tag{3.28}$$

Also
$$1 \text{ hp} = 745.7 \text{ W} = 550 \text{ ft}\cdot\text{lb/s}$$

The mechanical efficiency is defined to be

$$\eta = \frac{\text{power out}}{\text{power in}} \tag{3.29}$$

and is always less than 1.

B. Conservative Forces

Great simplification was obtained in equation (3.21) by noting that the right-hand side of the equation was an exact differential. The integration led to a scalar function T. The value of the path integral ultimately was determined by evaluating this scalar function T at two points and then taking their difference. It is remarkable that a vector path integral can be evaluated, *independent of the path*, by knowing the value of a scalar function at two points in space. There are even certain forces, called *conservative* forces, that form exact differentials for the left-hand side of the equation. A conservative force is defined to be one where the work performed by it is zero as it acts along a closed path. The work done by the conservative force while the particle moves from one point in space to another is then independent of the path. This can be expressed mathematically by

$$\int_{1}^{2} \mathbf{F} \cdot d\mathbf{r} \bigg|_{\text{Path } A} = \int_{1}^{2} \mathbf{F} \cdot d\mathbf{r} \bigg|_{\text{Path } B}$$

See Figure 3.3. Then

$$\int_{1}^{2} \mathbf{F} \cdot d\mathbf{r} \bigg|_{\text{Path } A} + \int_{2}^{1} \mathbf{F} \cdot d\mathbf{r} \bigg|_{\text{Path } B} = 0$$

$$\oint \mathbf{F} \cdot d\mathbf{r} = 0 \tag{3.30}$$

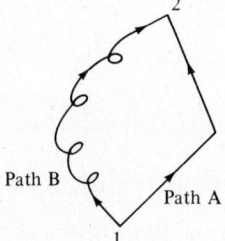

Figure 3.3

The work done by a conservative force during the motion of the particle from one point to another in space can then also be evaluated by finding the difference of two values of a scalar function V, since $\mathbf{F} \cdot d\mathbf{r}$ must be an exact differential.

$$dW = \mathbf{F} \cdot d\mathbf{r} \equiv -dV = -\nabla V \cdot d\mathbf{r} \tag{3.31}$$

$$W_{1 \to 2} = \int_{1}^{2} \mathbf{F} \cdot d\mathbf{r} = -\int_{1}^{2} dV = -(V_2 - V_1) \tag{3.32}$$

Note the minus sign in the definition.

$$V \equiv \text{potential energy, J}$$

If equation (3.32) is combined with equation (3.23), we obtain

$$T_1 + V_1 = T_2 + V_2 = E \equiv \text{mechanical energy, J} \tag{3.33}$$

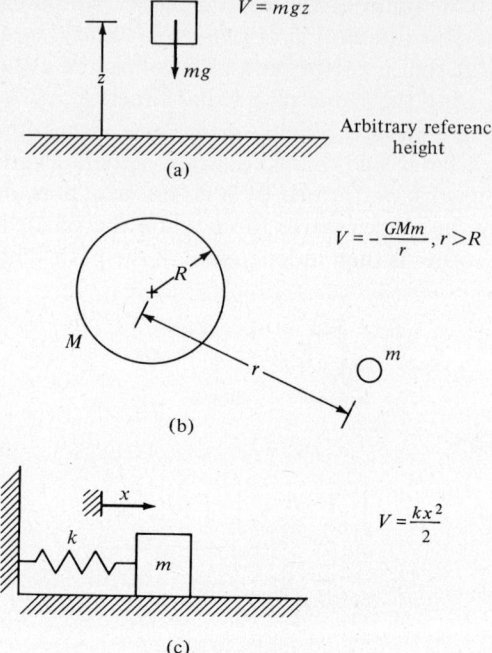

Figure 3.4

Equation (3.33) describes the *principle of conservation of mechanical energy*. This equation, unlike equation (3.23), is valid just for systems where conservative forces act.

Some typical potential functions are shown in Figure 3.4. They are

1. Due to constant gravitational force $F = mg$.

$$V = mgz \tag{3.34}$$

The distance z, positive upward, is measured from an arbitrary reference plane.

2. Due to varying gravitational force $F = -GMm/r^2$.

$$V = -\frac{GMm}{r} \qquad G \text{ is the universal gravity constant} \tag{3.35}$$

The distance r is measured from the center of the attracting body. The equation is valid for a spherical attracting mass but only outside the body, i.e., $r > R$, the radius of the body. V is always negative.

3. Due to a linear restoring spring force $F = -kx$.

$$V = \frac{kx^2}{2} \tag{3.36}$$

The spring constant k is expressed in N/m. The distance x is measured from the undeflected position of the spring. Further, V of a spring is always positive.

The advantage of using potential functions for solving work-energy problems becomes more obvious when considering the work done by a spring with one end fixed and the other end traversing a curved path. The line integral for the work done is complex to evaluate; however, the evaluation of the potential functions and their differences along the path is relatively straightforward.

Equation (3.23), the principle of work and kinetic energy, and equation (3.33), the principle of conservation of mechanical energy, are most useful in problems where a change in position is given or to be found.

Sample Problem 3.1
Shown in the figure is a simple pendulum having mass m and length L. Its pivot point at O is free to move vertically. The attached spring has a stiffness k and an unstretched length L_0. All motion is contained in a plane and is assumed to be friction-free. Write the equations of motion for the particle using normal and tangential components. What is the equation of motion for the z coordinate?

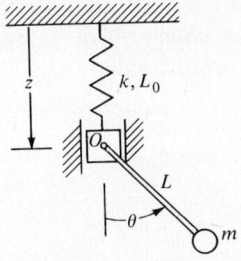

Analysis:

$$\mathbf{a}_m = \mathbf{a}_o + \mathbf{a}_{m/o}$$
$$= \ddot{z}\mathbf{k} + L\ddot{\theta}\mathbf{i}_t + L\dot{\theta}^2\mathbf{i}_n$$
$$= (L\dot{\theta}^2 - \ddot{z}\cos\theta)\mathbf{i}_n + (L\ddot{\theta} - \ddot{z}\sin\theta)\mathbf{i}_t$$

60 REVIEW OF NEWTONIAN MECHANICS

$$\mathbf{F} = m\mathbf{a}_m$$

$$T \cos \theta = k(z - L_0) \tag{a}$$

$$\Sigma F_n = -mg \cos \theta + T = m(L\dot{\theta}^2 - \ddot{z} \cos \theta) \tag{b}$$

$$\Sigma F_t = -mg \sin \theta = m(L\ddot{\theta} - \ddot{z} \sin \theta) \tag{c}$$

Equations (b) and (c) are the equations of motion sought. Rewriting equation (b) with the help of equation (a)

$$-mg \cos^2 \theta + k(z - L_0) = m(L\dot{\theta}^2 \cos \theta - \ddot{z} \cos^2 \theta) \tag{d}$$

Adding the result, equation (d), to equation (c) multiplied by the factor $\sin \theta$ gives

$$m(\ddot{z} - L\dot{\theta}^2 \cos \theta - L\ddot{\theta} \sin \theta) = -k(z - L_0) + mg \tag{e}$$

Equation (e) is the equation of motion rewritten for the z direction. This equation could have been developed independently from a direct force summation on the pendulum bob and massless rod treated together as one entity. ∎

Sample Problem 3.2

(a) Write the equation of motion for the spherical pendulum whose plane of oscillation rotates about the vertical as shown. The mass of the bob is m, and the length of the massless rod is L. (b) Integrate the equation of motion for the ϕ direction. What does the constant of integration physically represent? (c) Repeat part (b) for the θ coordinate.

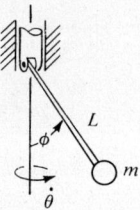

Analysis: The free-body diagram for the pendulum bob is shown below. The coordinate axes used here are not strictly the spherical coordinate set described in Figure 2.16, but they are related to them.

$$\Sigma F_r = ma_r$$

$$-T + mg \cos \phi = m(-L\dot{\phi}^2 - L \sin^2 \phi \, \dot{\theta}^2) \tag{a}$$

$$\Sigma F_\phi = ma_\phi$$

$$-mg \sin \phi = m(L\ddot{\phi} - L \sin \phi \cos \phi \, \dot{\theta}^2) \tag{b}$$

$$\sum F_\theta = ma_\theta$$
$$0 = m(L \sin \phi \, \ddot\theta + 2L\dot\theta\dot\phi \cos \phi) \tag{c}$$

After multiplying equation (c) by $L \sin \phi$, one can integrate the equation as follows:
$$0 = m(L^2 \sin^2 \phi \, \ddot\theta + 2L^2 \dot\theta\dot\phi \sin \phi \cos \phi)$$
$$0 = \frac{d}{dt} m(L^2 \sin^2 \phi) \dot\theta$$

Then $mL^2 \sin^2 \phi \, \dot\theta = h_z$ Angular momentum component along the vertical rotation axis, a constant

Equation (b) can now be rewritten in terms of one dependent variable.
$$mL\ddot\phi - \frac{h_z^2 \cos \phi}{mL^3 \sin^3 \phi} = -mg \sin \phi \tag{d}$$

Multiplying equation (d) by $L\dot\phi$, one has
$$mL^2 \ddot\phi\dot\phi - \frac{h_z^2 \cos \phi \, \dot\phi}{mL^2 \sin^3 \phi} = -mg \sin \phi L\dot\phi \tag{e}$$

Equation (e) can be written as
$$\frac{d}{dt}\left(\frac{mL^2 \dot\phi^2}{2} + \frac{h_z^2}{2mL^2 \sin^2 \phi} - mgL \cos \phi \right) = 0$$

$$\frac{mL^2 \dot\phi^2}{2} + \frac{mL^2 \sin^2 \phi \, \dot\theta^2}{2} - mgL \cos \phi = E \quad \text{Energy of the system, a constant} \quad \blacksquare$$

Sample Problem 3.3
Shown in the figure are two blocks, having identical mass m, connected by an inextensible wire that passes over the smooth massless pulley located at C. Block A is constrained to slide vertically down the smooth guide, whereas body B is forced to travel along the horizontal surface. The coefficient of kinetic friction between block B and the horizontal surface is μ. (a) If body A falls a distance h, prove that the total work done by the tension force, acting on both masses, is zero. (b) If the system starts from rest, find the speed of block A when y is equal to h. Block A starts its motion when it is level with block B. Assume the wire is taut initially.

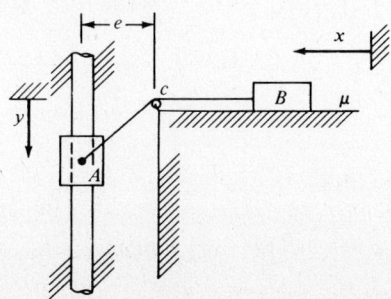

Analysis:

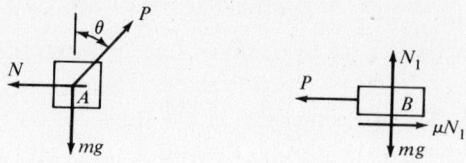

The work done on block A by the tension force for the entire displacement can be expressed by

$$W_{1\to 2} = \int_0^h -P\cos\theta\, dy = -\int_0^h \frac{Py}{(y^2+e^2)^{1/2}}\, dy \qquad (a)$$

Similarly, for body B,

$$W_{1\to 2} = \int_0^s P\, dx \qquad (b)$$

However, with the length of the string constant,

$$dx = d(y^2+e^2)^{1/2} = \frac{y\, dy}{(y^2+e^2)^{1/2}} \qquad (c)$$

Equation (b) then becomes

$$W_{1\to 2} = \int_0^h \frac{Py\, dy}{(y^2+e^2)^{1/2}} \qquad (d)$$

which is identical, except for sign, with equation (a). Hence, the total work done by the tension force acting on both masses is zero. In general, however, the work done by the internal forces of a system, occurring in equal and opposite pairs, is not zero unless the points of application for each force pair have identical displacements. The net work done by the internal forces is called the *internal work*.

The external work done on the system for the motion is

$$W_{1\to 2} = -\int_0^s \mu mg\, dx + \int_0^h mg\, dy$$

$$-\mu mg \int_0^h \frac{y\, dy}{(y^2+e^2)^{1/2}} + \int_0^h mg\, dy = T_2 - T_1 \qquad (e)$$

where

$$T_1 = 0$$

$$T_2 = \frac{mv_A^2}{2} + \frac{mv_B^2}{2} \qquad (f)$$

The integral to equation (e) is

$$-\mu mg\{(h^2+e^2)^{1/2}-e\} + mgh = \frac{m}{2}\{v_A^2+v_B^2\}$$

but for the final position

$$v_B = \frac{y}{(y^2+e^2)^{1/2}} v_A = \frac{h}{(h^2+e^2)^{1/2}} v_A \qquad (g)$$

$$v_B^2 = \frac{h^2}{h^2+e^2} v_A^2$$

Then
$$v_A = \left\{ \frac{2g(h-\mu[(h^2+e^2)^{1/2}-e])(h^2+e^2)}{2h^2+e^2} \right\}^{1/2} \qquad ∎$$

Sample Problem 3.4

The 135-N weight shown rotates about a frictionless vertical circular ring of radius 30 cm. A spring is attached to the mass and to a fixed point offset 7 cm from the center. The length of the unstretched spring is 22 cm, and the spring constant is 8800 N/m. If the mass has a velocity of 2.5 m/s at position A, what is its velocity at its lowest position, B?

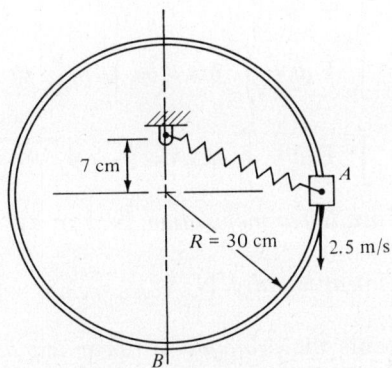

Analysis: Energy is conserved during the particle's motion. From equation (3.33)

$$T_1 + V_1 = T_2 + V_2$$

$$T_1 = \frac{135(2.5^2)}{9.81(2)}$$

$$T_1 = 43.0 \text{ J}$$

The potential energy can be found with the help of equations (3.34) and (3.36). A zero reference level for gravitational potential energy is arbitrarily selected at the center of the circle.

$$V_1 = mgz + \frac{kx^2}{2} = 0 + \frac{8800x^2}{2}$$

$$x = (0.07^2 + 0.3^2)^{1/2} - 0.22 = 0.088 \text{ m}$$

$$V_1 = 34.07 \text{ J}$$

$$V_2 = \frac{8800(0.37 - 0.22)^2}{2} - 0.3(135) = 58.5 \text{ J}$$

$$T_2 = 43.0 + 34.07 - 58.5 = 18.6 \text{ J}$$

$$v_2 = \left[\frac{2(18.6)(9.81)}{135}\right]^{1/2} = 1.64 \text{ m/s}$$

Note that independent of the path, the spring work is the negative difference between the values of the scalar function V at points A and B. ∎

C. Principle of Impulse and Momentum

Linear Impulse and Momentum. The physical problems where time, instead of displacement, is one of the problem variables is best handled by a different integral of motion. When equation (1.1) is integrated with respect to time, the following is obtained

$$\int_1^2 \mathbf{F}\, dt = \int_1^2 m\mathbf{a}\, dt = \int_1^2 m\frac{d\mathbf{v}}{dt} dt$$

or

$$\int_1^2 \mathbf{F}\, dt = \int_1^2 m\, d\mathbf{v} = \mathbf{p}_2 - \mathbf{p}_1 \qquad (3.37)$$

where
$$\mathbf{p} \equiv m\mathbf{v}, \text{ linear momentum, } \mathrm{N\cdot s} \text{ or } \mathrm{kg\cdot m/s}$$

$$\int_1^2 \mathbf{F}\, dt \equiv \text{linear impulse, } \mathrm{N\cdot s}$$

Equation (3.37) represents the *principle of linear impulse and momentum*. Expressed in terms of rectangular components, equation (3.37) becomes

$$\int_1^2 F_x\, dt = mv_{x,2} - mv_{x,1} \qquad (3.38)$$

$$\int_1^2 F_y\, dt = mv_{y,2} - mv_{y,1} \qquad (3.39)$$

$$\int_1^2 F_z\, dt = mv_{z,2} - mv_{z,1} \qquad (3.40)$$

If no forces act in a given direction, linear momentum in that direction is said to be conserved or to be constant.

Sample Problem 3.5

Two masses, m_A and m_B are attached by a weightless wire as shown in the figure. If the system is released from rest, find the velocity of mass B after q s. All pulleys are assumed massless and frictionless.

PARTICLE DYNAMICS: KINETICS 65

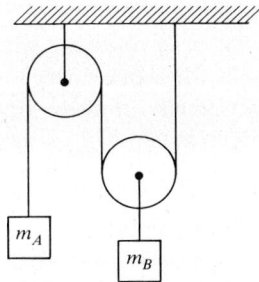

Analysis: Since time is the given problem variable, not displacement, momentum methods will be used. The following impulse and momentum equations are obtained from the free-body diagrams:

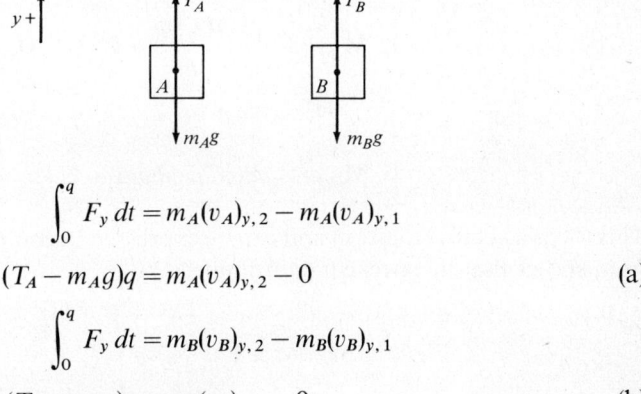

$$\int_0^q F_y\, dt = m_A(v_A)_{y,2} - m_A(v_A)_{y,1}$$

$$(T_A - m_A g)q = m_A(v_A)_{y,2} - 0 \tag{a}$$

$$\int_0^q F_y\, dt = m_B(v_B)_{y,2} - m_B(v_B)_{y,1}$$

$$(T_B - m_B g)q = m_B(v_B)_{y,2} - 0 \tag{b}$$

Because of the kinematic constraint

$$2v_B = -v_A \tag{c}$$

Also
$$2T_A = T_B \tag{d}$$

Equations (a) and (b) then become

$$(T_A - m_A g)q = -2m_A(v_B)_{y,2} = -2m_A v_B \tag{e}$$

$$(2T_A - m_B g)q = m_B v_B \tag{f}$$

Adding equations (e) and (f) together (treating the total system) does not eliminate the linear impulse due to tension even though the *total work* done by the internal tension forces is zero. The sum of equations (e) and (f) is

$$[3T_A - (m_B + m_A)g]q = (m_B - 2m_A)v_B \tag{g}$$

There is a physical explanation for the tension forces that did not cancel in the system equation, equation (g). Referring back to the figure, it is seen that the supporting ceiling exerts a total tension force of $3T_A$ on the system. This total external force does no work since it does not move through a displacement; therefore, it does not appear in the work-energy expression for the system. It does, however, produce a net linear impulse since the only requirement to establish an impulse is that force acts during a passage of time. The forces can be stationary as they are here.

Instead of direct addition of equations (e) and (f), elimination of T_A is called for, leading to

$$(2m_A - m_B)gq = (m_B + 4m_A)v_B$$

or
$$v_B = \frac{(2m_A - m_B)gq}{(4m_A + m_B)}$$ ∎

Angular Impulse and Momentum. The moment equation, equation (3.19), can also be integrated

$$\int_1^2 \mathbf{M}_O\, dt = \int_1^2 \frac{d\mathbf{H}_O\, dt}{dt} = \mathbf{H}_{O,2} - \mathbf{H}_{O,1} \quad (3.41)$$

$$\mathbf{H}_O = \mathbf{r} \times m\mathbf{v} \quad (3.17)$$

$$\int_1^2 \mathbf{M}_O\, dt \equiv \text{angular impulse, N} \cdot \text{m} \cdot \text{s}$$

This too is a vector equation and can be expressed in component form. Recalling from statics that in cartesian coordinates

$$\mathbf{M}_O \equiv \mathbf{r} \times \mathbf{F} = \begin{vmatrix} \mathbf{i} & \mathbf{j} & \mathbf{k} \\ x & y & z \\ F_x & F_y & F_z \end{vmatrix} \quad (3.42)$$

and
$$M_O = rF \sin \beta \quad (3.43)$$

it must follow that

$$\mathbf{H}_O = \mathbf{r} \times m\mathbf{v} = \begin{vmatrix} \mathbf{i} & \mathbf{j} & \mathbf{k} \\ x & y & z \\ mv_x & mv_y & mv_z \end{vmatrix} \quad (3.44)$$

and the magnitude of $\mathbf{H}_O$ is given by

$$H_O = rmv \sin \alpha \quad (3.45)$$

Refer to Figure 3.5.

A more useful coordinate system for discussing angular momentum is the radial and transverse coordinate pair. In this coordinate system (see Figure 3.6),

$$M_O = |\mathbf{r} \times \mathbf{F}| = rF_\theta = rF \sin \beta \quad (3.46)$$

and as shown in Figure 3.7,

$$H_O = rmv_\theta = rmv \sin \alpha \quad (3.47)$$

PARTICLE DYNAMICS: KINETICS 67

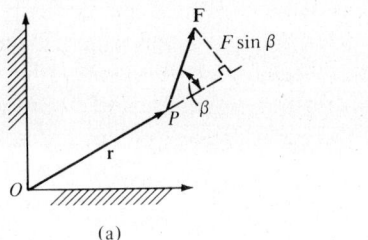

(a)   (b)

Figure 3.5

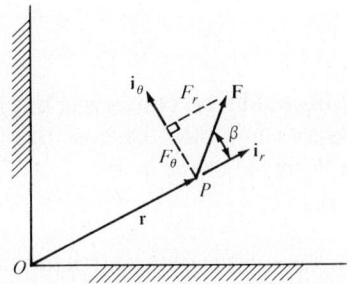

Figure 3.6

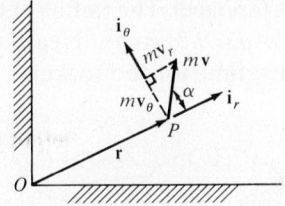

Figure 3.7

Keeping in mind that the definition of v_θ, the transverse component of velocity, is

$$v_\theta = r\dot\theta \qquad (2.75)$$

then equation (3.47) may be rewritten as

$$H_O = mr^2\dot\theta \qquad (3.48)$$

One can see from equation (3.41) that if no moments act about a certain direction, angular momentum in that direction is conserved, or constant. This result is often used when studying *central-force* motion. If the only forces acting on a particle are central, either attractive or repulsive, then angular momentum must be conserved since $\mathbf{M}_O = \mathbf{0}$. Two typical central-force problems are shown in Figure 3.8.

One final comment, the three conservation principles are independent of each other. Energy could be converved in a problem but not momentum. Linear momentum can be conserved while angular momentum is not, or vice versa.

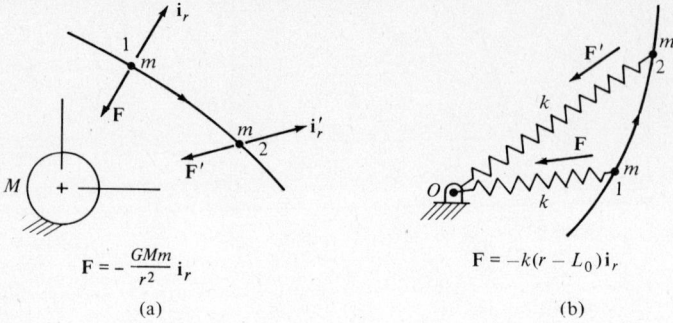

Figure 3.8

Similarly, linear momentum could be conserved but not energy. The simple spring-mass system represents a typical case of the first. Nonelastic impact between two masses is an example of the last.

Sample Problem 3.6
A capsule is launched from the surface of the earth into an elliptic trajectory and attains a maximum altitude of $R/2$ before returning to earth. If the launch velocity v_L equals 7900 m/s, and the launch angle is 60°, find the velocity of the capsule at maximum altitude (apogee). The radius of the earth is 6.371×10^3 km. GM is equal to gR^2, where $g = 9.81$ m/s². Neglect atmospheric effects and assume an instantaneous burn time for the rocket.

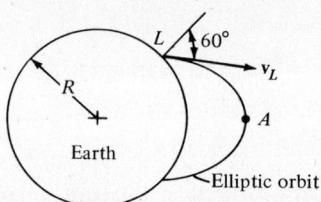

Analysis: This example (see Figure 3.8a describing satellite motion) is a central-force problem.

$$\mathbf{M}_O = \dot{\mathbf{H}}_O = 0$$

$$H_O = rmv_\theta = rmv \sin\alpha = \text{constant}$$

$$H_{O,1} = H_{O,2}$$

$$Rmv_L \sin 60° = \left(R + \frac{R}{2}\right) mv_A \sin 90°$$

$$v_A = \tfrac{2}{3}(7900)(0.866) = 4561 \text{ m/s}$$

The reader may also note that energy as well as momentum is conserved.

PARTICLE DYNAMICS: KINETICS

$$T_1 + V_1 = T_2 + V_2$$

where
$$V = -\frac{GMm}{r}$$

Substituting values with $GM = gR^2$,

$$\frac{mv_L^2}{2} - \frac{mgR^2}{R} = \frac{mv_A^2}{2} - \frac{mgR^2}{\frac{3}{2}R}$$

$$v_L^2 - 2gR = v_A^2 - \frac{4gR}{3}$$

$$v_L^2 - v_A^2 = \frac{2}{3}gR$$

$$7900^2 - 4561^2 = \frac{2}{3}(9.81)(6.371 \times 10^6)$$

which is in agreement with previous results if allowances are made for calculator round-off error. ∎

Sample Problem 3.7

A mass m resting on a horizontal frictionless table is constrained in its motion by an elastic string that has one end fixed. The string has spring constant k and an approximate zero unstretched length. Initially, the mass is set in motion when the string's length is l_0 with a velocity perpendicular to the string of v_0. What is the maximum length that the string attains?

Analysis: This too is a central-force problem. Angular momentum is conserved during the motion from point 1 to point 2.

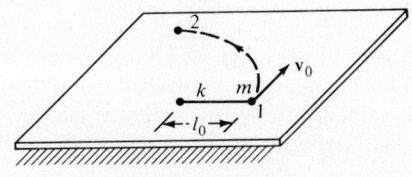

$$H_{O,2} = H_{O,1}$$

$$r_1 m v_1 \sin \alpha_1 = r_2 m v_2 \sin \alpha_2 \tag{a}$$

Here, α_1 is given to be 90°, and α_2 must be 90° at maximum length before the mass travels inward.

Equation (a) rewritten becomes

$$l_0 m v_0 = r_2 m v_2 \tag{b}$$

With one too many unknowns, another equation is needed. Since energy is also conserved

$$T_1 + V_1 = T_2 + V_2$$

or
$$\frac{mv_1^2}{2}+\frac{kx_1^2}{2}=\frac{mv_2^2}{2}+\frac{kx_2^2}{2}$$

$$\frac{mv_0^2}{2}+\frac{kl_0^2}{2}=\frac{mv_2^2}{2}+\frac{kr_2^2}{2}$$

$$r_2^2=\frac{mv_0^2+kl_0^2-mv_2^2}{k} \tag{c}$$

If equation (b) is squared, equation (d) results.

$$l_0^2 v_0^2 = r_2^2 v_2^2 \tag{d}$$

Substituting equation (c) into (d) yields

$$l_0^2 v_0^2 = \frac{(mv_0^2+kl_0^2-mv_2^2)v_2^2}{k}$$

or
$$mv_2^4 - v_2^2(kl_0^2+mv_0^2)+kl_0^2 v_0^2 = 0$$

$$v_2^2 = \left(\frac{(k/m)l_0^2+v_0^2}{2} \pm \frac{v_0^2-(k/m)l_0^2}{2}\right)$$

Since the minus sign leads to minimum velocity but maximum length,

$$v_2^2 = \frac{kl_0^2}{m} \qquad v_2 = l_0\sqrt{\frac{k}{m}}$$

Now from equation (b)

$$r_2 = \frac{l_0 v_0}{l_0}\sqrt{\frac{m}{k}} = v_0\sqrt{\frac{m}{k}} \qquad \blacksquare$$

PROBLEMS

3.1 Two particles both having mass m are attached to a weightless rod of length L. The top particle is constrained to move on a horizontal slot, and the bottom particle only moves vertically. Derive an equation of motion for the system in terms of θ if friction is ignored.

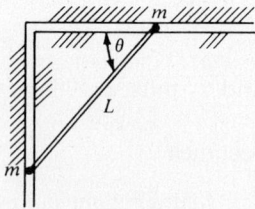

3.2 If the motion in problem 3.1 starts from rest when $\theta = 0°$, integrate the equation of motion to find the angular velocity of the rod when $\theta = 90°$. What physical quantity does the constant of integration represent?

3.3 The pivot point A of a simple pendulum having mass m and length L is free to slide (accelerate) along a horizontal track. Ignoring friction, develop the moment equation for the system about point A and the force equation for the x coordinate. Assume the pivot to be massless.

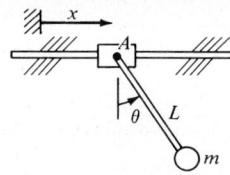

3.4 In problem 3.3, integrate the force equation for the x coordinate. What physical quantity does the constant of integration represent?

*3.5 For problem 3.3 write a computer program that will numerically solve for θ as a function of time given $\ddot{x} = 1.7 \text{ m/s}^2$, $m = 2.0 \text{ kg}$, $L = 0.2 \text{ m}$, $\theta_0 = \dot{\theta}_0 = 0$. Use a communication interval of 0.02 s and a final time of 5 s.

3.6 If in problem 3.3 a spring, having stiffness k and unstretched length L_0, was attached to the wall and to pivot A, develop the new equations of motion. Let x be measured from the wall. Assume the pivot's mass to be 0.1 m.

*3.7 For problem 3.6 write a digital computer program that will numerically solve for the x and θ motion. Let $L_0 = 0.02$ m and $k = 2000$ N/m. Other values are the same as in problem 3.5. Take $x_0 = 0.02$ m, $\dot{x}_0 = 0$, $\dot{\theta}_0 = 0$, and $\theta_0 = 5°$ as initial values. The communication interval should be 0.005 s and the final time 0.6 s.

3.8 The disk shown rotates with constant angular speed ω in a horizontal plane. Placed in a smooth 45° slot in the disk is a small slider having mass m. It is restrained in its motion by a spring connected to it and attached at A. Find the deflection of the spring after steady motion has been obtained with the mass at point B. The spring's stiffness is k.

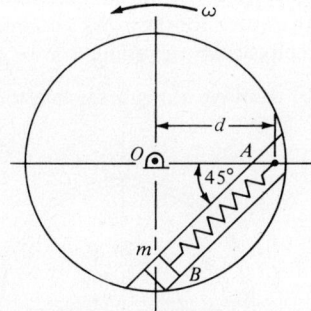

3.9 Develop an expression for the velocity a satellite must have in order to obtain a circular orbit about the earth at an altitude h.

3.10 The mass shown describes oscillatory motion in a vertical plane. Assuming that the spring remains straight, develop the equations of motion for the radial and transverse coordinates. The weight of the particle is w and the spring's stiffness is k.

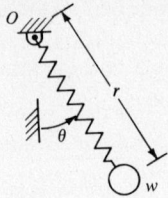

3.11 Repeat problem 3.10 for the case where the mass moves on a smooth horizontal platform. Integrate the equations of motion. What do the constants of the motion physically represent?

3.12 A particle of mass m is constrained to move inside a smooth massless tube. The tube is free to turn in a horizontal plane about a point at A. If the mass is also attached to point A by an internal spring having stiffness k and unstretched length r_0, derive the equations of motion for the system when the tube is driven at constant angular speed ω by the applied torque T.

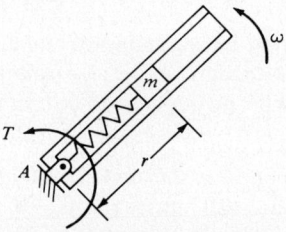

3.13 In problem 3.12, integrate the equations of motion to determine the maximum value of r if at $t=0$, $r=3r_0/4$ and $\dot{r}=0$. What is the limiting value of k?

3.14 If in problem 3.12 $k=3m\omega^2$, show that the same result could be found in problem 3.13 using the principle of work and kinetic energy.

3.15 Repeat problem 3.12 for the case where ω is not constant and the torque T is zero. Integrate the equations of motion for the r and θ coordinates, and describe what the constants of integration physically represent.

3.16 A slider of mass m is free to move along a smooth massless ring of radius R. The

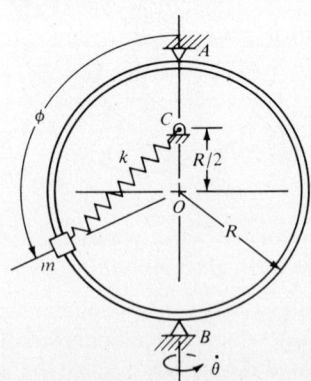

ring itself can rotate about a vertical axis passing through diameter AB. If a spring having stiffness k and unstretched length $3R/4$ connects the slider to a fixed point C that lies on diameter AB at a distance $R/2$ above the ring's center, derive the equations of motion for the slider for the instant when $\phi = 120°$.

3.17 Repeat problem 3.16 for the case where there is no spring attached to the mass. Let ϕ be arbitrary. Integrate the equations of motion for the θ and ϕ coordinates. What do the constants of integration physically represent?

*3.18 Write a computer program, valid for all ϕ, that will numerically solve for the motion of ϕ in problem 3.17 if the ring is turning at constant angular speed ω. Take $m = 2.5$ kg, $R = 0.1$ m, and $\omega = 30.0$ rad/s. Solve for the given initial conditions of $\dot{\phi}_0 = 0$ and $\phi_0 = 60°$. Use a communication interval of 0.0075 s and a final time of 0.5 s.

3.19 Shown in the figure is a conical pendulum having mass m and length L. Its pivot point at O is free to move vertically, and the spring attached to the pivot has a stiffness k and unstretched length L_0. Write the equations of motion for the system. Which momentum is conserved?

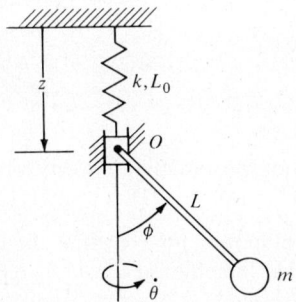

*3.20 Rewrite the equations of motion found in problem 3.19 for the z, ϕ, and θ coordinates. Write a computer program that will numerically solve for z and ϕ, given the initial conditions of $\theta_0 = 0$, $\dot{\theta}_0 = \pi/2$ rad/s, $z_0 = 0.03$ m, $\dot{z}_0 = 0.05$ m/s, $\phi_0 = 0.8$ rad, and $\dot{\phi}_0 = 1.0$ rad/s. Let $m = 2.0$ kg, $L = 0.4$ m, $k = 2.0$ kN/m, and $L_0 = 0.02$ m. Use a communication interval of 0.02 s, and terminate the program at $t = 1.0$ s.

3.21 If $\mathbf{F} = (y - x^2)\mathbf{i} + (x + y^2)\mathbf{j}$, prove that the force is conservative. Develop an expression for the potential energy.

3.22 Repeat problem 3.21 for $\mathbf{F} = (x\mathbf{i} - y\mathbf{j})/(x^2 - y^2)$ if $x^2 > y^2$.

3.23 In problem 3.21, find the work done by the force if a particle travels from point $(0,0)$ to point $(1, 0)$ along the line $y = 0$.

3.24 The particle shown of mass m attached to a string of length $3R$ is released from rest in a horizontal position. It describes circular motion until the string hits an obstruction at A. The distance from point O to point A is $2R$. Determine the angular velocity of the string immediately after impact. What linear impulse was provided by the obstruction?

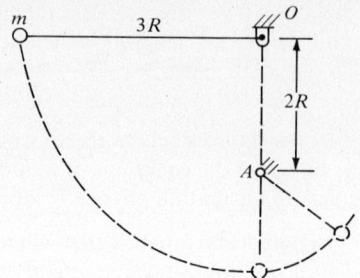

3.25 If in problem 3.24 the obstruction was a peg having radius r, what would be the velocity of the mass after the string has turned through 360°? What would be the angular velocity of the string?

3.26 A particle of mass m is released inside a smooth hemispherical bowl of radius R with a speed equal to v_0. The speed v_0 is tangent to the bowl's surface and parallel with its lip. What is the maximum distance that the particle will drop?

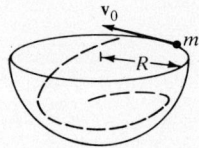

3.27 Derive an expression for the minimum velocity a rocket must have in order to leave the earth's gravitational force field. That speed is called the *escape velocity*.

3.28 Derive the equations of motion for a satellite about the earth. Integrate these equations, and describe what the physical constants represent.

*3.29 Write a computer program that will numerically solve for the position and location of the satellite in problem 3.28. Neglect the earth's rotation. The following data are given: $G = 6.673 \times 10^{-11}$ m³/(kg·s²), $m_e = 5.976 \times 10^{24}$ kg, $r_0 = 6.705 \times 10^3$ kg, $\dot{r}_0 = 0$, $\theta_0 = 0$, and $\dot{\theta}_0 = 1.2457 \times 10^{-3}$ rad/s. Use a communication interval of 250 s, and terminate the program at $t = 7250$ s.

3.30 It is proposed to send a space ship from the Earth's surface to Mars. The Earth and Mars travel around the sun in approximately circular orbits having radii R_1 and R_2, respectively. If the orbit selected is elliptical and tangent to both the Earth's

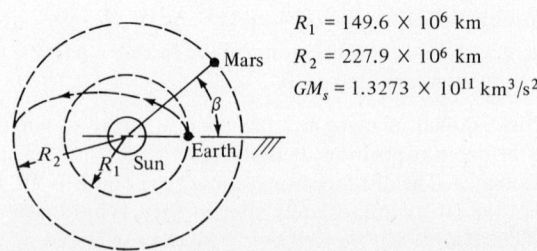

$R_1 = 149.6 \times 10^6$ km
$R_2 = 227.9 \times 10^6$ km
$GM_s = 1.3273 \times 10^{11}$ km³/s²

orbit and to that of Mars, determine the launch velocity necessary. (Such a tangent orbit is called a *Hohmann transfer orbit*.) Neglect all other forces except the sun's gravitational attraction. Both Mars and Earth can be treated as particles.

3.31 If the period of motion of an elliptic orbit is given by $T = 2\pi(a^3/GM)^{1/2}$, where a is the semimajor distance, determine the angle β in problem 3.30 that must exist at launch time.

SYSTEMS OF PARTICLES 4

The great Swiss mathematician and physicist Leonhart Euler (1707–1783) is credited with extending newtonian mechanics from the single particle to the rigid body. To do this requires an averaging of "motions" of all parts of the system. One point having these average values can then be used to identify the entire system. This point is called the *center of mass*.

4.1 THE CENTER OF MASS

A system of particles is shown in Figure 4.1. The center of mass is defined to be a location in the system where all mass could be concentrated to produce an equivalent moment of mass about the origin. Expressed as an equation,

$$\mathbf{R}_{c.m.} \left(\sum_{i=1}^{n} m_i \right) \equiv \sum_{i=1}^{n} m_i \mathbf{r}_i \qquad (4.1)$$

or the location of the center of mass is given by

$$\mathbf{R}_{c.m.} = \frac{\sum_{i=1}^{n} m_i \mathbf{r}_i}{\sum_{i=1}^{n} m_i} \qquad (4.2)$$

The term $\sum_{i=1}^{n} m_i$ is the total mass. Note that in a *uniform* gravitational force field the *center of gravity* and the center of mass are the same. The location of the center of gravity is given by

$$\mathbf{R}_{c.g.} \equiv \frac{\sum_{i=1}^{n} m_i g_i \mathbf{r}_i}{\sum_{i=1}^{n} m_i g_i} \qquad (4.3)$$

and with g_i constant,

$$\mathbf{R}_{c.g.} = \frac{\sum_{i=1}^{n} m_i g \mathbf{r}_i}{\sum_{i=1}^{n} m_i g} = \frac{g \sum_{i=1}^{n} m_i \mathbf{r}_i}{g \sum_{i=1}^{n} m_i} = \mathbf{R}_{c.m.}$$

If the relative position vector from the center of mass to particle i is called $\boldsymbol{\rho}_i$, then

$$\mathbf{r}_i = \mathbf{R}_{c.m.} + \boldsymbol{\rho}_i \qquad (4.4)$$

and

$$\sum_{i=1}^{n} m_i \mathbf{r}_i = \sum_{i=1}^{n} m_i \mathbf{R}_{c.m.} + \sum_{i=1}^{n} m_i \boldsymbol{\rho}_i \qquad (4.5)$$

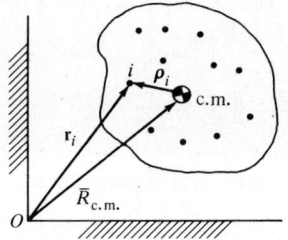

Figure 4.1

Since equation (4.1) must hold,

$$\sum_{i=1}^{n} m_i \boldsymbol{\rho}_i = \mathbf{0} \tag{4.6}$$

Equation (4.6) is an alternative expression for defining the location of the center of mass. It states that the sum of the moments of masses about the center is zero.

By taking the time derivative of equations (4.1) and (4.6), the following are obtained

$$M \mathbf{v}_{c.m.} = \sum_{i=1}^{n} m_i \mathbf{v}_i \tag{4.7}$$

$$M \mathbf{a}_{c.m.} = \sum_{i=1}^{n} m_i \mathbf{a}_i \tag{4.8}$$

$$\sum_{i=1}^{n} m_i \dot{\boldsymbol{\rho}}_i = \mathbf{0} \tag{4.9}$$

$$\sum_{i=1}^{n} m_i \ddot{\boldsymbol{\rho}}_i = \mathbf{0} \tag{4.10}$$

Of these four new expressions, equations (4.7) and (4.8) are the most important, describing system momentum and inertia, respectively. M is the total mass.

In scalar form, equation (4.2) becomes

$$x_{c.m.} = \frac{\sum_{i=1}^{n} m_i x_i}{\sum_{i=1}^{n} m_i} \tag{4.11}$$

$$y_{c.m.} = \frac{\sum_{i=1}^{n} m_i y_i}{\sum_{i=1}^{n} m_i} \tag{4.12}$$

$$z_{c.m.} = \frac{\sum_{i=1}^{n} m_i z_i}{\sum_{i=1}^{n} m_i} \tag{4.13}$$

For a rigid body, a continuum is assumed. Then, in equation (4.2) the summation sign can be replaced by an integral, the mass particle m_i changed to a differential particle dm, and the position vector $\mathbf{r}_i$ replaced by a continuously varying vector $\mathbf{r}$. Equation (4.2) becomes

$$\mathbf{R}_{c.m.} = \frac{\int_{mass} \mathbf{r}\, dm}{\int_{mass} dm} \qquad (4.14)$$

$$\mathbf{R}_{c.m.} = \frac{\int_{vol} \mathbf{r}\rho\, dvol}{\int_{vol} \rho\, dvol} \qquad (4.15)$$

where ρ is the density of the material. For uniform density

$$\mathbf{R}_{c.m.} = \frac{\int_{vol} \mathbf{r}\, dvol}{\int_{vol} dvol} \qquad (4.16)$$

4.2 THE EQUATIONS OF MOTION FOR A SYSTEM OF PARTICLES

It is now possible to derive the equations of motion for a system of particles. Referring to Figure 4.2, it is seen that the forces acting on the ith particle are the external force $\mathbf{F}_i$ and the internal force $\mathbf{f}_{ij}$, $\mathbf{f}_{ik}$, etc. Newton's second law for a particle is

$$\mathbf{F} = m\mathbf{a} \qquad (1.1)$$

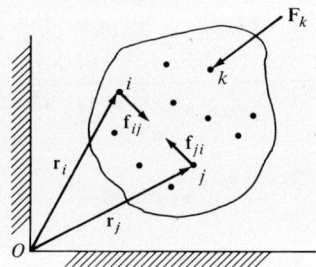

Figure 4.2

Here, this translates to

$$\mathbf{F}_i + \sum_{\substack{j=1 \\ i \ne j}}^{n} \mathbf{f}_{ij} = m_i \mathbf{a}_i \qquad (4.17)$$

The result of summing forces acting on all the particles is

$$\sum_{i=1}^{n} \mathbf{F}_i + \sum_{i=1}^{n} \sum_{\substack{j=1 \\ i \ne j}}^{n} \mathbf{f}_{ij} = \sum_{i=1}^{n} m_i \mathbf{a}_i \qquad (4.18)$$

Using Newton's third law,

$$\mathbf{f}_{ij} = -\mathbf{f}_{ji} \qquad (4.19)$$

The second term in equation (4.18), the sum of the internal forces, cancels. Equation (4.18) then becomes

$$\sum_{i=1}^{n} \mathbf{F}_i = \sum_{i=1}^{n} m_i \mathbf{a}_i \tag{4.20}$$

Recognizing that $\sum_{i=1}^{n} \mathbf{F}_i$ is just the sum of the external forces acting on the system, and $\sum_{i=1}^{n} m_i \mathbf{a}_i$ from equation (4.8) is $M\mathbf{a}_{\text{c.m.}}$, then

$$\mathbf{F}_{\text{ext}} = M \mathbf{a}_{\text{c.m.}} \tag{4.21}$$

This last result pertains to a system of particles or a rigid body. It states that the sum of the external forces acting on a system accelerates a mass M as if all the mass were concentrated at one point. This point, the center of mass, obtains an acceleration equal to $\mathbf{F}_{\text{ext}}/M$.

Recalling equation (3.16), rewriting it for a particle i, and then summing over all the particles, one obtains

$$\sum_{i=1}^{n} \mathbf{M}_{Oi} = \sum_{i=1}^{n} \frac{d}{dt}(\mathbf{r}_i \times m_i \mathbf{v}_i) \tag{4.22}$$

With

$$\sum_{i=1}^{n} \mathbf{M}_{Oi} = \sum_{i=1}^{n} \mathbf{r}_i \times \left(\mathbf{F}_i + \sum_{\substack{j=1 \\ i \neq j}}^{n} \mathbf{f}_{ij} \right) \tag{4.23}$$

the sum of the moments caused by internal forces vanishes

$$\sum_{\substack{i=1 \\ i \neq j}}^{n} \sum_{j=1}^{n} \mathbf{r}_i \times \mathbf{f}_{ij} = 0 \tag{4.24}$$

and the total moment $\mathbf{M}_O$ then is

$$\mathbf{M}_O = \sum_{i=1}^{n} \mathbf{M}_{Oi} = \sum_{i=1}^{n} \mathbf{r}_i \times \mathbf{F}_i \tag{4.25}$$

The contributing moments are due to the externally acting forces only. In terms of the center of mass, one can write [see equation (4.4)]

$$\mathbf{M}_O = \sum_{i=1}^{n} (\boldsymbol{\rho}_i + \mathbf{R}_{\text{c.m.}}) \times \mathbf{F}_i$$

$$\mathbf{M}_O = \mathbf{R}_{\text{c.m.}} \times \mathbf{F}_{\text{ext}} + \sum_{i=1}^{n} \boldsymbol{\rho}_i \times \mathbf{F}_i \tag{4.26}$$

The moment of the external forces about the center of mass clearly is

$$\mathbf{M}_{\text{c.m.}} = \sum_{i=1}^{n} \boldsymbol{\rho}_i \times \mathbf{F}_i \tag{4.27}$$

Hence
$$\mathbf{M}_O = \mathbf{R}_{\text{c.m.}} \times \mathbf{F}_{\text{ext}} + \mathbf{M}_{\text{c.m.}} \tag{4.28}$$

Figure 4.3 shows this moment equivalency for an arbitrary force $\mathbf{F}_i$ acting on a

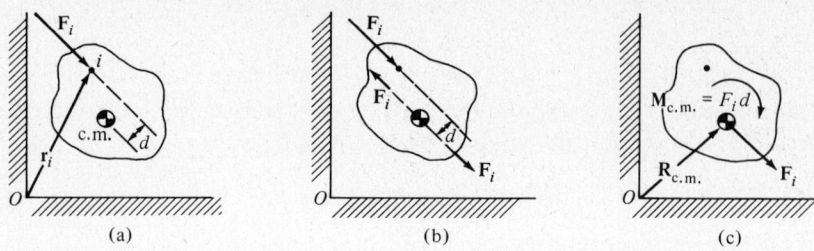

Figure 4.3

system. Referring again to equations (3.17) and (4.22), it is seen that

$$\mathbf{M}_O = \sum_{i=1}^{n} \frac{d}{dt} \mathbf{H}_{Oi} = \frac{d\mathbf{H}_O}{dt} \tag{4.29}$$

where

$$\mathbf{H}_O = \sum_{i=1}^{n} \mathbf{H}_{Oi} = \sum_{i=1}^{n} \mathbf{r}_i \times m_i \mathbf{v}_i \tag{4.30}$$

Equation (4.30) will be discussed further when rigid body motion is studied.

Equation (4.29) is very limited in its application if no fixed point exists in the body. A convenient reference point to use, which moves with the body, is the center of mass. To establish a relationship for the center of mass similar to equation (4.29), one starts with equation (4.22) and substitutes equation (4.4) rewritten in derivative form.

$$\mathbf{v}_i = \dot{\boldsymbol{\rho}}_i + \mathbf{v}_{\text{c.m.}} \tag{4.31}$$

which leads to the result

$$\mathbf{M}_O = \sum_{i=1}^{n} \frac{d}{dt} [(\boldsymbol{\rho}_i + \mathbf{R}_{\text{c.m.}}) \times m_i(\dot{\boldsymbol{\rho}}_i + \mathbf{v}_{\text{c.m.}})] \tag{4.32}$$

By expanding the terms in equation (4.32), and recalling equations (4.6), (4.9), and (4.10), equation (4.32) becomes (with $\mathbf{v}_{\text{c.m.}} \times M\mathbf{v}_{\text{c.m.}} = 0$)

$$\mathbf{M}_O = \mathbf{R}_{\text{c.m.}} \times M\mathbf{a}_{\text{c.m.}} + \sum_{i=1}^{n} \frac{d}{dt}(\boldsymbol{\rho}_i \times m_i \dot{\boldsymbol{\rho}}_i) \tag{4.33}$$

The angular momentum, or moments of linear momentum, for a system of particles about its center of mass is now defined to be

$$\mathbf{H}_{\text{c.m.}} \equiv \sum_{i=1}^{n} \boldsymbol{\rho}_i \times m_i \dot{\boldsymbol{\rho}}_i \tag{4.34}$$

or, equally valid,

$$\mathbf{H}_{\text{c.m.}} = \sum_{i=1}^{n} \boldsymbol{\rho}_i \times m_i \mathbf{v}_i \tag{4.35}$$

The derivation of equation (4.35) from equation (4.34) is left as an exercise

SYSTEMS OF PARTICLES 81

for the reader. Note that the moment arms for the moments of linear momentum are from the center of mass; the velocity terms, however, can be relative or absolute. Equation (4.33) rewritten becomes

$$\mathbf{M}_O = \mathbf{R}_{c.m.} \times M\mathbf{a}_{c.m.} + \frac{d}{dt}\mathbf{H}_{c.m.} \qquad (4.36)$$

By equating equations (4.28) and (4.36), the sought-after result is obtained:

$$\mathbf{M}_{c.m.} = \frac{d\mathbf{H}_{c.m.}}{dt} \qquad (4.37)$$

4.3 INTEGRALS OF THE EQUATIONS OF MOTION

A. Principle of Work and Kinetic Energy

In a development parallel with that for a particle, the integrals of the equations of motion will be derived for a system of particles. Starting with equation (4.17), the differential work done on a particle i due to a displacement $d\mathbf{r}_i$ is given by

$$\mathbf{F}_i \cdot d\mathbf{r}_i + \sum_{\substack{j=1 \\ i \neq j}}^{n} \mathbf{f}_{ij} \cdot d\mathbf{r}_i = m_i \mathbf{a}_i \cdot d\mathbf{r}_i \qquad (4.38)$$

Replacing $d\mathbf{r}_i$ on the left side with $d\boldsymbol{\rho}_i + d\mathbf{R}_{c.m.}$, equation (4.38) becomes

$$\mathbf{F}_i \cdot d\boldsymbol{\rho}_i + \mathbf{F}_i \cdot d\mathbf{R}_{c.m.} + \sum_{\substack{j=1 \\ i \neq j}}^{n} \mathbf{f}_{ij} \cdot d\boldsymbol{\rho}_i + \sum_{\substack{j=1 \\ i \neq j}}^{n} \mathbf{f}_{ij} \cdot d\mathbf{R}_{c.m.} = m_i \mathbf{a}_i \cdot d\mathbf{r}_i \qquad (4.39)$$

If equation (4.39) is summed for all the particles and integrated along a path from 1 to 2, then

$$\int_1^2 \sum_{i=1}^{n} \mathbf{F}_i \cdot d\boldsymbol{\rho}_i + \int_1^2 \sum_{i=1}^{n} \mathbf{F}_i \cdot d\mathbf{R}_{c.m.} + \int_1^2 \sum_{\substack{i=1 \\ i \neq j}}^{n} \sum_{j=1}^{n} \mathbf{f}_{ij} \cdot d\boldsymbol{\rho}_i = \int_1^2 \sum_{i=1}^{n} m_i \frac{d\mathbf{v}_i}{dt} \cdot d\mathbf{r}_i \qquad (4.40)$$

Note that the last term on the left-hand side of equation (4.39) canceled when summed on the index i. The work done in going from point 1 to point 2 then is

$$W_{1 \to 2} = \int_1^2 \sum_{i=1}^{n} \mathbf{F}_i \cdot d\boldsymbol{\rho}_i + \int_1^2 \mathbf{F}_{ext} \cdot d\mathbf{R}_{c.m.} + \int_1^2 \sum_{\substack{i=1 \\ i \neq j}}^{n} \sum_{j=1}^{n} \mathbf{f}_{ij} \cdot d\boldsymbol{\rho}_i \qquad (4.41)$$

The last term on the right in equation (4.41) is due to stretching, or internal relative displacements; the first term is caused by stretching and rotation, and the middle term is produced by system translation. For a rigid body, it can easily be shown that the last term vanishes.

The integral on the right-hand side of equation (4.40) is equal to

$$\int_1^2 \sum_{i=1}^{n} m_i \frac{d\mathbf{v}_i}{dt} \cdot d\mathbf{r}_i = \sum_{i=1}^{n} m_i \frac{\mathbf{v}_i \cdot \mathbf{v}_i}{2} \bigg|_1^2 = T_2 - T_1 \qquad (4.42)$$

where
$$T \equiv \sum_{i=1}^{n} \frac{m_i \mathbf{v}_i \cdot \mathbf{v}_i}{2} = \sum_{i=1}^{n} \frac{m_i v_i^2}{2} \qquad (4.43)$$

In other words, the work-energy principle is still valid for a system of particles, or
$$W_{1 \to 2} = T_2 - T_1 \qquad (4.44)$$

For a conservative system,
$$T_1 + V_1 = T_2 + V_2 = E \qquad (4.45)$$

Sometimes, it is of interest to separate the work and energy expressions into translational and rotational parts. By using the expression $\mathbf{v}_i = \dot{\boldsymbol{\rho}}_i + \mathbf{v}_{\text{c.m.}}$, equation (4.43) becomes

$$T = \sum_{i=1}^{n} m_i \left(\frac{\dot{\boldsymbol{\rho}}_i + \mathbf{v}_{\text{c.m.}}}{2} \right) \cdot (\dot{\boldsymbol{\rho}}_i + \mathbf{v}_{\text{c.m.}})$$

or
$$T = \sum_{i=1}^{n} \frac{m_i \dot{\boldsymbol{\rho}}_i \cdot \dot{\boldsymbol{\rho}}_i}{2} + \sum_{i=1}^{n} \frac{m_i v_{\text{c.m.}}^2}{2} + \mathbf{v}_{\text{c.m.}} \cdot \sum_{i=1}^{n} m_i \dot{\boldsymbol{\rho}}_i \qquad (4.46)$$

The last term reduces to zero with the aid of equation (4.9). T then is

$$T = M \frac{v_{\text{c.m.}}^2}{2} + \sum_{i=1}^{n} \frac{m_i \dot{\rho}_i^2}{2} \qquad (4.47)$$

The first term in equation (4.47) represents the translational part of the kinetic energy expression; the last term represents stretching and/or rotation of the system. For a rigid body, only rotational and translational kinetic energy is permissible. In general though,

$$T = T_{\text{transl}} + T_{\text{rot+str}} \qquad (4.48)$$

The work done under a translation of the center of mass can be expressed by

$$\int_1^2 \mathbf{F}_{\text{ext}} \cdot d\mathbf{R}_{\text{c.m.}} = \int_1^2 M \mathbf{a}_{\text{c.m.}} \cdot d\mathbf{R}_{\text{c.m.}} \qquad (4.49)$$

or
$$W_{1 \to 2, \text{transl}} = \int_1^2 d\left(\frac{M \mathbf{v}_{\text{c.m.}} \cdot \mathbf{v}_{\text{c.m.}}}{2} \right) = \left. \frac{M v_{\text{c.m.}}^2}{2} \right|_1^2$$

It is apparent that
$$W_{1 \to 2, \text{transl}} = (T_2 - T_1)_{\text{transl}} \qquad (4.50)$$

and that
$$W_{1 \to 2, \text{rot+str}} = (T_2 - T_1)_{\text{rot+str}} \qquad (4.51)$$

The results obtained show that the work and energy terms can be separated and treated individually. The rotational work affects only the rotational kinetic energy; the translational work done by all external forces acting through a displacement of the center of mass affects only the translational kinetic energy of the system.

SYSTEMS OF PARTICLES

Sample Problem 4.1

Block B weighing 40 kg rests on a nearly frictionless floor. Block A weighs 30 kg and lies on top of block B. When a force F of 225 N is applied to block A, it moves to the right attaining a speed of 1.5 m/s in 0.6 m. While block B is prevented from moving by the retaining wall, a constant friction force acts between A and B during the motion. Find the restraining force of the wall.

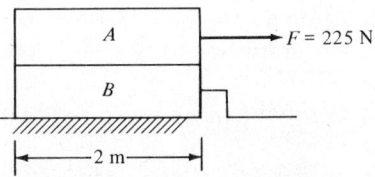

Analysis: There are several ways of doing this problem. The most familiar approach is to draw free-body diagrams for each mass and analyze the forces and accelerations present. It is perhaps more interesting to explore energy methods for the system instead. The easiest solution uses equation (4.50).

$$W_{1\to 2,\,\text{trans}} = \int_1^2 \mathbf{F}_{\text{ext}} \cdot d\mathbf{R}_{\text{c.m.}} = \int_0^{x_{\text{c.m.}}} (-F_{\text{wall}} + 225)\, dx_{\text{c.m.}}$$

$$x_{\text{c.m.}} = \frac{\sum_{i=1}^2 m_i x_i}{\sum_{i=1}^2 m_i} = \frac{30(0.6) + 40(0)}{30 + 40} = 0.26 \text{ m}$$

$$W_{1\to 2,\,\text{trans}} = (-F_{\text{wall}} + 225)(0.26)$$

Friction is internal to the system and plays no role in this equation.

$$W_{1\to 2,\,\text{transl}} = (T_2 - T_1)_{\text{transl}}$$

$$(-F_{\text{wall}} + 225)(0.26) = \frac{30 + 40}{2}\, v_{\text{c.m.}}^2 - 0$$

Also

$$v_{\text{c.m.}} = \frac{\sum_{i=1}^2 m_i v_i}{\sum_{i=1}^2 m_i} = \frac{30(1.5) + 40(0)}{30 + 40} = 0.64 \text{ m/s}$$

$$F_{\text{wall}} = \frac{1}{0.26}[225(0.26) - \tfrac{70}{2} 0.64^2] = 170 \text{ N} \quad\blacksquare$$

B. Principle of Impulse and Momentum

Linear Impulse and Momentum. Starting once more with equation (4.21), but this time integrating with respect to time, one obtains

$$\int_1^2 \mathbf{F}_{\text{ext}}\, dt = \int_1^2 M\mathbf{a}_{\text{c.m.}}\, dt$$

84 REVIEW OF NEWTONIAN MECHANICS

or
$$\int_1^2 \mathbf{F}_{ext} \, dt = \mathbf{P}_2 - \mathbf{P}_1 \tag{4.52}$$

where
$$\mathbf{P} = M\mathbf{v}_{c.m.} = \sum_{i=1}^{n} m_i \mathbf{v}_i = \text{system linear momentum} \tag{4.53}$$

$$\int_1^2 \mathbf{F}_{ext} \, dt = \text{linear impulse acting on system}$$

If no external forces act on a system of particles, total linear momentum is conserved. The conservation of linear momentum does not imply that energy is conserved as well.

In terms of cartesian components equation (4.52) becomes

$$\int_1^2 F_x \, dt = M(v_{c.m.})_{x,2} - M(v_{c.m.})_{x,1}$$

$$\sum_1^2 F_y \, dt = M(v_{c.m.})_{y,2} - M(v_{c.m.})_{y,1} \tag{4.54}$$

$$\int_1^2 F_z \, dt = M(v_{c.m.})_{z,2} - M(v_{c.m.})_{z,1}$$

Sample Problem 4.2
The simple pendulum shown, having mass m and length L, is pivoted at O to a cart that is free to move on a frictionless horizontal surface. The cart's mass is $2m$. The system is initially at rest. If particle A also having mass m, and speed v_0 directly impacts with the pendulum bob and then sticks to it, find for the instant after the impact (a) the location of the center of mass, (b) the velocity of the center of mass, (c) the system's linear momentum, (d) the system's angular momentum about the center of mass, and (e) the new energy of the system.

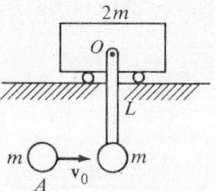

Analysis: Since no external forces act in the x direction, linear momentum is conserved during the impact for the x coordinate.

$$\left(\sum_{i=1}^{n} m_i v_i \right)_{x,1} = \left(\sum_{i=1}^{n} m_i v_i \right)_{x,2}$$

$$2m(0) + m(0) + mv_0 = 2m(0) + (m+m)v$$

$$v = \frac{v_0}{2} \quad \text{Speed of combined mass}$$

SYSTEMS OF PARTICLES 85

$$\mathbf{R}_{c.m.} = \frac{\sum_{i=1}^{n} m_i \mathbf{r}_i}{M} = \frac{2m(0) + 2m(L\mathbf{j})}{4m} = \frac{L}{2}\mathbf{j} \quad (a)$$

$$\mathbf{v}_{c.m.} = \frac{\sum_{i=1}^{n} m_i \mathbf{v}_i}{M} = \frac{2m(v_0/2)\mathbf{i} + 2m(0)}{4m} = \frac{v_0}{4}\mathbf{i} \quad (b)$$

$$M\mathbf{v}_{c.m.} = mv_0\mathbf{i} = \text{initial system momentum} \quad (c)$$

$$\mathbf{H}_{c.m.} = \sum_{i=1}^{n} \boldsymbol{\rho}_i \times m_i \mathbf{v}_i$$

$$= \frac{L}{2}\mathbf{j} \times 2m(0) - \frac{L}{2}\mathbf{j} \times 2m\left(\frac{v_0}{2}\mathbf{i}\right) = \frac{Lmv_0}{2}\mathbf{k} \quad (d)$$

Alternatively,

$$\mathbf{H}_{c.m.} = \frac{L}{2}\mathbf{j} \times 2m\left(-\frac{v_0}{4}\mathbf{i}\right) - \frac{L}{2}\mathbf{j} \times 2m\left(\frac{v_0}{2}\mathbf{i} - \frac{v_0}{4}\mathbf{i}\right)$$

where

$$\dot{\boldsymbol{\rho}}_1 = -\frac{v_0}{4}\mathbf{i}$$

$$\dot{\boldsymbol{\rho}}_2 = \frac{v_0}{4}\mathbf{i}$$

Finally,

$$T = \sum_{i=1}^{n} \frac{m_i v_i^2}{2} = \frac{2m(v_0/2)^2}{2} = \frac{mv_0^2}{4} \quad \blacksquare \quad (e)$$

Sample Problem 4.3

Two masses A and B are separated by a spring compressed a distance δ. The masses travel together on a horizontal frictionless floor with a speed v_0 as shown. If the spring is suddenly released so that the masses fly apart, find the final velocity of mass A. The mass of particle B is twice the mass of particle A; the spring has stiffness k.

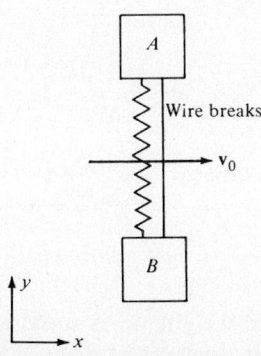

Analysis: Since no external force acts on the system, the velocity of the system's center of mass remains constant. Then

$$W_{1\to 2, \text{rot}+\text{str}} = (T_2 - T_1)_{\text{rot}+\text{str}}$$

$$W_{1\to 2} = -(V_2 - V_1) = \frac{k\delta^2}{2} = \left(\sum_{i=1}^{n} m_i \frac{\dot{\boldsymbol{\rho}}_i \cdot \dot{\boldsymbol{\rho}}_i}{2}\right)_2 - 0$$

$$\frac{k\delta^2}{2} = \frac{m}{2}(\dot{\rho}_A^2 + \rho_A^2 \dot{\theta}^2) + \frac{2m}{2}(\dot{\rho}_B^2 + \rho_B^2 \dot{\theta}^2) \qquad (a)$$

It is obvious that $\dot{\theta}$ must be equal to zero since with no external moments,

$$(\mathbf{H}_{\text{c.m.}})_2 = (\mathbf{H}_{\text{c.m.}})_1 \qquad (b)$$

System momentum is conserved as well; so

$$(m_A + m_B)v_0\mathbf{i} = m_A \mathbf{v}_A + m_B \mathbf{v}_B \qquad (c)$$

For a system whose center of mass travels with constant velocity, it is convenient to transform equation (c) to a translating reference frame attached to the center of mass. For this problem, the transformed equation becomes

$$0 = m_A \dot{\boldsymbol{\rho}}_A + m_B \dot{\boldsymbol{\rho}}_B \qquad (d)$$

or

$$0 = m_A \dot{\rho}_A \mathbf{j} + m_B \dot{\rho}_B \mathbf{j}$$

$$m\dot{\rho}_A = -2m\dot{\rho}_B$$

$$\dot{\rho}_B = \frac{-\dot{\rho}_A}{2}$$

Equation (a) then becomes

$$\frac{k\delta^2}{2} = \frac{m}{2}\dot{\rho}_A^2 + \frac{2m}{2}\left(\frac{-\dot{\rho}_A}{2}\right)^2$$

$$\dot{\rho}_A^2 = \frac{2k\delta^2}{3m}$$

$$\dot{\rho}_A = \delta\sqrt{\frac{2k}{3m}}$$

$$\mathbf{v}_A = v_0\mathbf{i} + \delta\sqrt{\frac{2k}{3m}}\,\mathbf{j}$$

$$\mathbf{v}_B = v_0\mathbf{i} - \delta\sqrt{\frac{k}{6m}}\,\mathbf{j} \qquad \blacksquare$$

Sample Problem 4.4

The railroad coal car of tare weight w_0 is moving at constant speed v while being loaded with coal at the constant rate $\dot{m} = b$ kilogram per second. Determine the force F necessary to sustain the constant speed. Neglect friction.

SYSTEMS OF PARTICLES

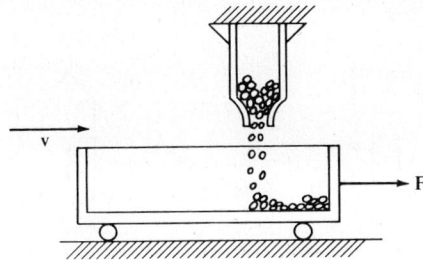

Analysis: This problem involves a *system gaining mass*. As long as there is a *mass balance* at the initial and final times of analysis, equation (4.52) can be applied. After rewriting and combining equations (4.52) and (4.53) with all acting forces assumed constant during a Δt passage of time, the result appears as

$$\mathbf{F}_{ext}\, \Delta t = \left(\sum_{i=1}^{n} m_i \mathbf{v}_i\right)_2 - \left(\sum_{i=1}^{n} m_i \mathbf{v}_i\right)_1$$

where $(\sum_{i=1}^{n} m_i)_2$ must equal $(\sum_{i=1}^{n} m_i)_1$. For this problem one has for the x direction,

$$F\, \Delta t = [(m + \Delta m)v]_2 - [mv + \Delta m(0)]_1$$

where
$$m = \frac{w_0}{g} + \dot{m} t = \frac{w_0}{g} + bt$$

$$\Delta m = \dot{m}\, \Delta t = b\, \Delta t$$

Then
$$F = bv$$

Note that the mass balance agrees for the Δt time period considered.

$$(m + \Delta m)_2 = (m + \Delta m)_1$$

One final comment is necessary. The Δm coal particles falling (entering the system) have no initial speed in the x direction. After *impact* with the resting coal in the train, they attain an instantaneous speed v. Energy is not conserved for the system. ∎

Sample Problem 4.5

Find the maximum altitude reached for a rocket fired from a height h with an initial speed v_0, an initial mass m_0, and a burnout mass m_f. Assume a uniform gravitational force field and that the rocket is fired vertically upward. The burning rate of the fuel is b, and its exit velocity relative to the rocket is v_e. Both quantities can be taken as constant. Neglect air friction, but allow for a final expanded exit gage pressure of the gases p_e and an exit area A_e for the rocket nozzle.

Analysis: The rocket is shown in two figures, for times t and $t + \Delta t$, with external forces depicted.

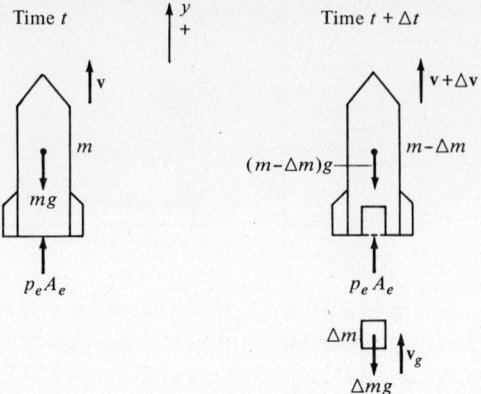

In the positive y direction

$$F_{\text{ext}} \Delta t = [(m - \Delta m)(v + \Delta v) + \Delta m\, v_g] - (mv)$$

With $\Delta m\, \Delta v \simeq 0$,

$$F_{\text{ext}} \Delta t = - \Delta m\, v + \Delta m\, v_g + m\, \Delta v$$

However, $v_g - v = v_e$, and where $v > v_g$, then

$$F_{\text{ext}} \Delta t = - \Delta m\, v_e + m\, \Delta v$$

or

$$-mg + p_e A_e = -\frac{\Delta m}{\Delta t} v_e + m \frac{\Delta v}{\Delta t}$$

$$p_e A_e + b v_e - mg = m \frac{dv}{dt} \qquad (a)$$

The above expression has v_e pointing downward. The combined term $p_e A_e + b v_e$ is called the *static thrust* F_s of the rocket. The *specific impulse* is defined to be

$$I_{\text{sp}} = \frac{F_s}{bg}$$

Equation (a) can now be integrated by changing the independent variable to m.

$$F_s - mg = m \frac{dv}{dm} \frac{dm}{dt} = -mb \frac{dv}{dm}$$

where

$$m = m_0 - bt \quad \text{and} \quad \frac{dm}{dt} = -b$$

During a short burning time, the gravitational force can be neglected. Then

$$dv = -\frac{I_{\text{sp}} g\, dm}{m}$$

and

$$v_f - v_0 = I_{\text{sp}} g \ln \frac{m_0}{m_f} \qquad (b)$$

After burnout, energy is conserved; so

$$T_1 + V_1 = T_2 + V_2$$

$$\frac{m_f v_f^2}{2} + m_f g h = 0 + m_f g h_{\max}$$

$$h_{\max} = h + \frac{[v_0 + I_{sp} g \ln(m_0/m_f)]^2}{2g} \qquad\blacksquare$$

Impact. An important application of equation (4.52) is for the case of impact between two or more objects. An *impact force* is defined to be a very large force acting over a very small time interval. In the limit as Δt approaches zero, it becomes an idealized mathematical quantity that allows for instantaneous changes of mass velocity (infinite acceleration). A baseball bat hitting a ball is an example of impact. During the infinitesimal time of impact, all nonimpact-type forces contribute negligibly to the impulse integral and can therefore be ignored. For example, the weight of the baseball hitting the bat during the instant of impact can be ignored.

When two objects collide, the line of action of the impact forces between the masses may be collinear with a line joining the center of masses of the two objects; this occurrence is called *central impact*. When the opposite is true, it is called *oblique impact*. See Figure 4.4.

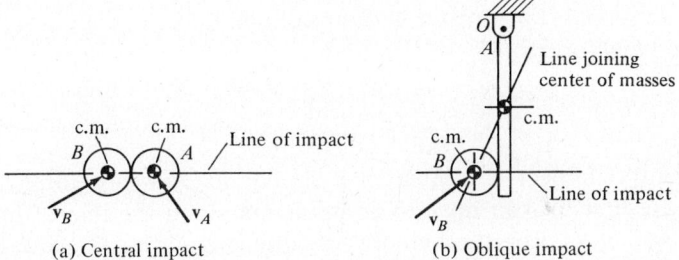

(a) Central impact (b) Oblique impact

Figure 4.4

The discussion of oblique impact will be postponed to Chapter 7 when rigid body rotation will be discussed.

Central impact can be further divided into *direct central* impact and *oblique central* impact. Direct central impact has the velocities of both masses directed along the line of impact before the collision occurs. (Of course, one mass may be at rest before impact in this discussion as well.) Oblique central impact has the velocities of one or both masses directed at an angle to the line of impact for the collision. This is shown in Figure 4.5.

For the cases of direct impact, if no external forces act, linear momentum will be conserved. From Figure 4.6,

$$m_A v_{A,1} + m_B v_{B,1} = m_A v_{A,2} + m_B v_{B,2} \qquad (4.55)$$

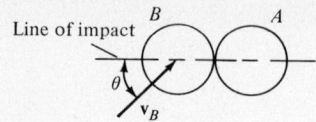

Figure 4.5

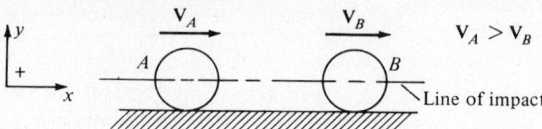

Figure 4.6

where the line of impact and the x axis are taken as collinear. The coefficient of restitution ε is defined to be

$$\varepsilon \equiv -\frac{v_{B,2}-v_{A,2}}{v_{B,1}-v_{A,1}} = -\frac{v_{B/A_2}}{v_{B/A_1}} \qquad (4.56)$$

ε is an experimentally determined constant. Its value depends upon the physical properties of the specific pair of impacting bodies. Equation (4.56) only holds along the line of impact. The value of ε lies between 0 and 1. When $\varepsilon = 0$, it is called a *plastic* impact; the objects stick to each other after impact. When $\varepsilon = 1$, it is called an *elastic* impact. The latter case is the only example of impact where energy is conserved. For impact with an infinite mass, equation (4.56) can be used in modified form:

$$\varepsilon = -\frac{v_{B,2}}{v_{B,1}}$$

Oblique central impact is treated in a similar manner providing the initial velocities are first broken up into components along the line of impact and perpendicular to it. For the velocity components along the line of impact (the x axis is arbitrarily used here), equations (4.55) and (4.56) hold true. If no tangential external forces act on a body, or if the forces that do act are small, its velocity components perpendicular to the line of impact remain constant during the impact. That is to say

$$(v_A)_{y,2} = (v_A)_{y,1}$$
$$(v_B)_{y,2} = (v_B)_{y,1} \qquad (4.57)$$

Angular Impulse and Momentum. The last item to be discussed concerning a system of particles is the principle of angular impulse and momentum. Recall equations (4.29) and (4.37):

$$\mathbf{M}_O = \frac{d\mathbf{H}_O}{dt} \qquad (4.29)$$

SYSTEMS OF PARTICLES 91

$$\mathbf{M}_{c.m.} = \frac{d\mathbf{H}_{c.m.}}{dt} \tag{4.37}$$

The integrals to these equations with respect to time become

$$\int_1^2 \mathbf{M}_O \, dt = \mathbf{H}_{O,2} - \mathbf{H}_{O,1} \tag{4.58}$$

$$\int_1^2 \mathbf{M}_{c.m.} \, dt = \mathbf{H}_{c.m.,2} - \mathbf{H}_{c.m.,1} \tag{4.59}$$

Equations (4.58) and (4.59) express the principle of angular impulse and momentum for the system. With cartesian components one may write

$$\int_1^2 (M_O)_x \, dt = (H_O)_{x,2} - (H_O)_{x,1}$$

$$\int_1^2 (M_O)_y \, dt = (H_O)_{y,2} - (H_O)_{y,1} \tag{4.60}$$

$$\int_1^2 (M_O)_z \, dt = (H_O)_{z,2} - (H_O)_{z,1}$$

Note that angular momentum may be conserved about one axis (if no moment acts about that axis) independent of the other two.

Sample Problem 4.6
A 1-kg ball A released from rest rolls down a frictionless curve until it impacts with a 2-kg block B. The block subsequently slides along a rough surface whose coefficient of sliding friction is 0.5. It comes to rest in 1.6 m. What is the energy lost during the impact?

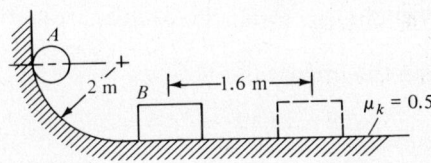

Analysis: This problem emphasizes the point that energy is usually lost during an impact, whereas linear momentum is conserved. The motion must be divided into three parts. The motion of part 1, from the start of motion of A to its impact with B, has energy conserved. Part 2 is the impact itself. The motion of part 3 has work performed by the friction force as it stops the block.

Part 1:
$$T_0 + V_0 = T_1 + V_1$$
$$0 + 1(9.81)(2) = \tfrac{1}{2}(v_{A,1})^2 + 0$$
$$v_{A,1} = \sqrt{2(2)(9.81)} = 6.26 \text{ m/s}$$

Part 2:
$$m_A v_{A,1} + m_B v_{B,1} = m_A v_{A,2} + m_B v_{B,2}$$
$$1(6.26) + 0 = 1 v_{A,2} + 2 v_{B,2} \quad \text{(a)}$$

$$\varepsilon = -\frac{v_{B/A_2}}{v_{B/A_1}}$$

$$\varepsilon = -\frac{v_{B,2} - v_{A,2}}{0 - 6.26} \quad \text{(b)}$$

$$\varepsilon(6.26) = v_{B,2} - v_{A,2}$$

Part 3:
$$W_{1 \to 2} = T_2 - T_1$$
$$-F(1.6) = 0 - \frac{2 v_{B,2}^2}{2}$$
$$-2(9.81)(0.5)(1.6) = -\frac{2 v_{B,2}^2}{2}$$
$$v_{B,2} = 3.96 \text{ m/s}$$

Substituting the value of $v_{B,2}$ directly back into equation (a) yields
$$v_{A,2} = 1(6.26) - 2(3.96) = -1.66 \text{ m/s}$$
[Equation (b) was not used ultimately.]
The initial energy before impact of the system was
$$E_i = T_i = \frac{1(6.26^2)}{2} = 19.59 \text{ J}$$
The final energy immediately after impact is given by
$$E_f = T_f = \tfrac{1}{2} m_A v_{A,2}^2 + \tfrac{1}{2} m_B v_{B,2}^2$$
$$= \tfrac{1}{2}(1)(1.66^2) + \tfrac{1}{2}(2)(3.96^2) = 17.06 \text{ J}$$
The energy lost during the impact is
$$\Delta E = 19.59 - 17.06 = 2.53 \text{ J} \quad \blacksquare$$

Sample Problem 4.7
Mass A moves to the right on a horizontal frictionless surface with a speed v_0 as shown. It impacts with mass B which is at rest. Find the final velocity of mass A after the impact if the coefficient of restitution is ε. The radius of mass A is 1 cm, and that of mass B is 1.5 cm. The mass of B is twice that of mass A.

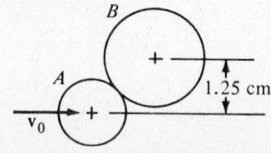

Analysis: This is an oblique central impact problem. For convenience the x direction is selected to be along the line of impact. This line makes an angle of 30° with the original direction of the velocity of A. Equations (4.55) and (4.56) for the x direction and equation (4.57) are applicable here.

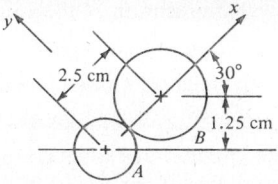

With conservation of system linear momentum,

$$m_A v_{A,1} + m_B v_{B,1} = m_A v_{A,2} + m_B v_{B,2}$$

$$1 v_0 \cos 30° + 0 = 1(v_A)_{x,2} + 2(v_B)_{x,2} \qquad (a)$$

The equation defining the coefficient of restitution states that

$$\varepsilon = -\frac{v_{B/A_2}}{v_{B/A_1}} = -\frac{(v_B)_{x,2} - (v_A)_{x,2}}{0 - v_0 \cos 30°} \qquad (b)$$

For frictionless impact, the y components of velocity remain unchanged:

$$(v_A)_{y,2} = (v_A)_{y,1}$$

$$(v_B)_{y,2} = (v_B)_{y,1}$$

$$(v_A)_{y,1} = -v_0 \sin 30° = (v_A)_{y,2} = -\frac{v_0}{2}$$

$$(v_B)_{y,1} = 0 = (v_B)_{y,2}$$

Combining equations (a) and (b) yields

$$(v_B)_{x,2} = \tfrac{1}{6}(1+\varepsilon)v_0\sqrt{3}$$

or

$$(v_A)_{x,2} = \tfrac{1}{6}(1-2\varepsilon)v_0\sqrt{3}$$

$$v_{A,2} = \left(\frac{1-\varepsilon+\varepsilon^2}{3}\right)^{1/2} v_0$$

$$\psi = \tan^{-1}\frac{\sqrt{3}}{1-2\varepsilon}$$

PROBLEMS

4.1 Three particles of equal mass m are located at $(3, 0, 0)$, $(0, 3, 0)$, and $(0, 0, 3)$. The particle on the x axis has a velocity $2\mathbf{i} + 3\mathbf{j}$, the particle on the y axis has a velocity $4\mathbf{i}$, and the velocity for the particle on the z axis is $3\mathbf{j} + 3\mathbf{k}$. Find (a) the location of the center of mass, (b) the linear momentum of the system, and (c) the velocity of the center of mass.

4.2 In problem 4.1, determine the angular momentum about the origin and about the center of mass. What is the kinetic energy of the system?

4.3 What is the gradient of the gravitational force per unit mass at a height above the earth's surface equal to the radius of the earth? Where is the location of the center of mass relative to the center of gravity?

4.4 Particles A and B rest upon a frictionless horizontal table. They are connected by a massless rod of length L. Suddenly, particle C, moving with a speed v_0 on the table, impacts directly with particle A. The coefficient of restitution for the impact is zero. If all the masses are equal to m, find the instant immediately after impact (a) the location of the combined center of mass, (b) the system's linear momentum, (c) the system's angular momentum referenced to the center of mass, and (d) the system's rotational kinetic energy. Show that $\sum_{i=1}^{n} m_i \rho_i = 0$.

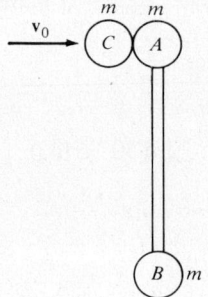

4.5 If the total gravitational potential per unit mass at a given point in space created by n particles is given by the relation $\phi = G \sum_{i=1}^{n} m_i/r_i$ derive the expression for the gravitational potential produced by a solid sphere having radius a.

4.6 In problem 4.5, what is the gravitational field strength, the force per unit mass, created by the sphere?

4.7 The truck shown of mass m_2 has a machine of mass m_1 resting upon it. The machine is fastened to the truck by an elastic member having stiffness k. Neglecting friction, (a) solve for the ensuing motion of mass m_1 if the truck is suddenly given a speed v_0 at $t = 0$, and (b) what is the maximum displacement of the spring?

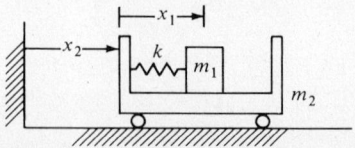

4.8 Repeat problem 4.7(b) for the case when friction acts.

4.9 A simple pendulum of mass m and length L is attached to an oscillating cart of mass m_1. The cart is forced to move so that $x = A \cos \omega t$. Write the general expression for the kinetic and potential energies of the system. The distance x is measured from the unstretched position of the spring.

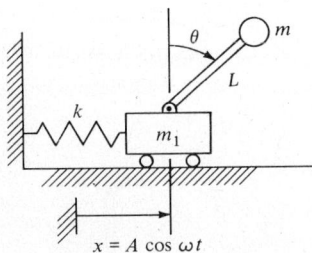

4.10 Derive expressions for the kinetic and potential energies of the double pendulum shown.

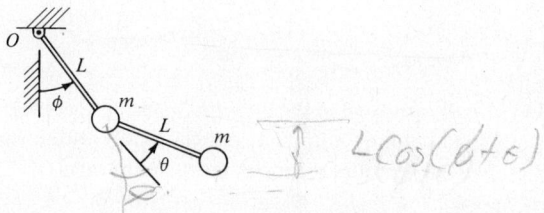

4.11 The simple pendulum having mass m and length L is free to swing in a vertical plane. It is pivoted at A to another mass m_1 which is constrained to have only vertical motion. The spring connecting mass m_1 to the overhead support has stiffness k. If the distance y is measured from the unstretched position of the spring, develop expressions for the kinetic and potential energies of the system.

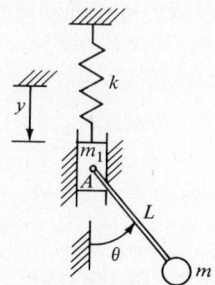

4.12 The two equal masses shown are connected by a weightless rod having length L. Mass A is constrained to move horizontally, while mass B can only move vertically. If the system is initially at rest in a vertical position, calculate the necessary impulse $\hat{F}$ applied at the rod's center that will allow the bar to reach a horizontal position.

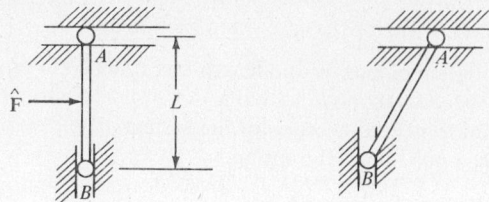

4.13 Shown below is the double pendulum of problem 4.10 at rest in a vertical position. Find the angular velocities of both rods immediately after the impulse $\hat{F}$ strikes the bottom mass.

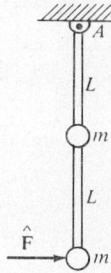

4.14 A ball is thrown in the air with an initial speed v_0 at an angle γ above the horizontal. When landing, it impacts nonelastically with a smooth floor until it stops. Find the total horizontal distance that the ball travels.

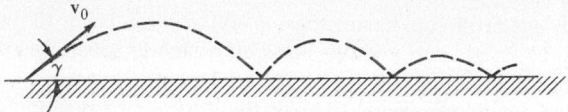

4.15 A particle of mass m traveling with a speed v_0 impacts with a larger mass m_1. The larger mass is connected to a spring having stiffness k. Find the largest deflection of the spring subsequent to the impact as a function of ε.

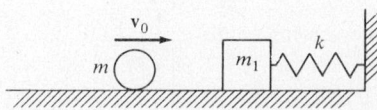

4.16 Two billiard balls A and B, having radii r, lie on a horizontal surface a distance $10r$ apart. It is desired to hit A with a cue stick in such a manner that A travels and impacts with B sending B to the left at an angle of 30°. If friction is neglected, determine the angle ψ required for particle A's path in order to impart the proper motion. If its initial speed is v_0, find particle A's velocity after the impact. The coefficient of restitution is ε.

4.17 Three identical particles lying on a horizontal surface will collide in the manner shown. If the initial velocity of particle A is v_0, and if the collisions are elastic, determine the velocities of the particles after the impact. Neglect all friction. Particles B and C are at rest initially.

4.18 The pendulum shown is made up of two equal masses m firmly attached to a weightless rigid rod of length L. If a third particle having mass $m/2$ impacts directly with the bottom mass, find the angular velocity of the rod immediately after the impact. Initially the pendulum is at rest, and the impacting particle has a velocity given by $v_0 = v_0(\mathbf{i} - \mathbf{j})/\sqrt{2}$. The coefficient of restitution for the impact is ε.

4.19 A barbell consisting of two identical masses separated by a weightless rod of length L is dropped without rotation from a height h. It strikes a rigid ledge as shown. If the coefficient of restitution for the impact is ε, determine the angular velocity of the rod immediately after the impact.

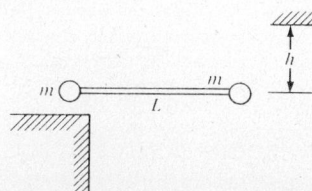

4.20 The end of a chain lying on a table is lifted at a constant speed v_0 by a variable force P. The chain's length is L, and its mass per unit length is γ. Derive an expression for P as a function of time.

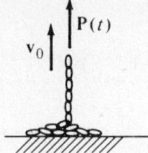

4.21 The same chain from problem 4.20 now lies on a table with a hole in it. As the first link starts to fall through the opening pulling the remaining links along, its speed increases. Find its final speed when the last link of the chain clears the hole. Is energy conserved?

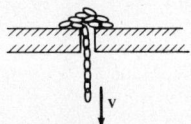

4.22 Repeat problem 4.21, but let the chain have uniform velocity. Ignore friction as the chain slides over the smooth rounded corner.

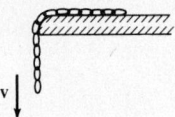

4.23 A stream of water from a turbine nozzle having constant speed v_w strikes a blade that is moving with a constant speed v. The water, with cross-sectional area A, is deflected by the blade through an angle θ. Determine the power developed.

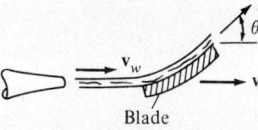

4.24 Develop the equation for the thrust developed by an airplane's jet engine if the plane is traveling with constant speed v_A. The exhaust velocity relative to the plane is given by v_e.

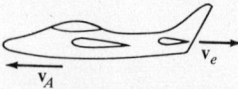

4.25 A two-stage rocket with payload fired from the earth's surface is shown below. Assuming very short burning times and a specific impulse of 200 s for both stages, determine the maximum height the empty last stage and payload achieve. Let the acceleration of gravity vary. The first stage separates immediately after burnout; the second stage then ignites.

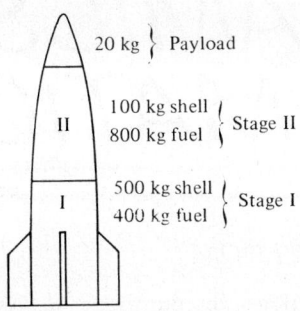

*4.26 For staged rocket vehicles having constant payload ratios $\lambda \equiv M_{i+1}/M_i$ (the index i stands for the stage number) the advantages of staging can be shown by the following expression for burnout velocities.

$$\frac{v_{B\,\text{staged}}}{v_{B\,\text{unstaged}}} = n\,\frac{\ln\left[S(1-G^{-1/n})+G^{-1/n}\right]}{\ln\left[S(1-1/G)+1/G\right]} = K(n)$$

where n = number of stages
  S = structural efficiency factor, ratio of each stage's empty mass to its full mass (In this expression it is assumed equal for each stage.)
  G = mass ratio, ratio of mass of full rocket to payload

Develop enough values for K to plot K versus n. Use the values $S = 0.2$ and $G = 200$. What are your conclusions?

RIGID BODY DYNAMICS: KINEMATICS

5.1 DEGREE OF FREEDOM

For a system of particles, the number of *degrees of freedom* equals the minimum number of independent coordinates needed to completely describe the position of the system. The number of degrees of freedom equals the number of non-independent coordinates used to describe the system minus the number of constraint equations present governing the system's motion. It is a quantity independent of the coordinate system employed; i.e., if a system has 2 degrees of freedom using cartesian coordinates, it will still have 2 degrees of freedom after a coordinate transformation to any other set. For example, in Figure 5.1, a simple pendulum of mass m and length L is shown swinging in a plane. Three cartesian coordinates describe the position of the pendulum bob. The motion has two equations of constraint, namely,

$$z = 0 \quad \text{(plane motion)} \tag{a}$$

and
$$(x^2 + y^2 + z^2)^{1/2} = L \quad \text{(length of the pendulum)} \tag{b}$$

The three coordinates minus the two constraint equations leave 1 degree of freedom for the pendulum. Since this quantity is independent of the coordinate set used, the pendulum's motion could be more easily described using only the angle θ as a coordinate.

The number of degrees of freedom a particle has moving in space is 3. A system of n particles has $3n$ degrees of freedom if the system is free from constraints. Three noncollinear particles have 9 degrees of freedom. If the distances between each of the three particles are fixed, 6 degrees of freedom result. The

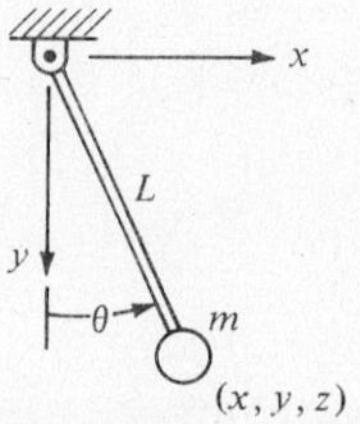

Figure 5.1

position of three noncollinear particles with fixed distances between each other are the minimum needed to define the orientation of a rigid body. A rigid body then has not more than 6 degrees of freedom. For each degree of freedom, there is associated an equation of motion. Three independent equations of motion describe the rigid body's translation; the other three independent equations of motion set forth its rotation.

5.2 TRANSLATION

A rigid body that does not rotate has 3 degrees of translational freedom. Each point in a translating rigid body follows a path parallel to those described by the other points of the body. Lines fixed in the body remain parallel to their original orientations. For a body that translates only, each point in the body has the same velocity and the same acceleration at any instant.

$$\mathbf{v}_A = \mathbf{v}_B = \mathbf{v}_C = \cdots \tag{5.1}$$

$$\mathbf{a}_A = \mathbf{a}_B = \mathbf{a}_C = \cdots \tag{5.2}$$

where $A, B, C, \ldots$ are arbitrary points in the body. A body that translates in a plane has 2 degrees of freedom; a body constrained to move in a straight line has 1 degree of freedom.

5.3 ROTATION ABOUT A FIXED POINT

Euler's well-known theorem in kinematics contributes to our understanding of rigid body motion. It states that if a body has one point fixed, then any motion of the body can be reduced to a simple angular displacement (rotation) about a single axis through the point. For a discussion of this theorem the reader is referred to Section 6.3 (see also Sample Problem 6.5).

Although finite angular displacements have magnitudes and directions as vectors do, they are not vectors since they do not obey the law of vector addition; they are not commutative. On the other hand, infinitesimal angular displacements do act as vectors, and any finite displacement of a rigid body with one point fixed can be built up through a series of infinitesimal rotations. Each infinitesimal rotation acts about an axis called the *instantaneous axis of rotation*. In general this axis itself rotates.

The infinitesimal *linear* displacement of an arbitrary point on a body which rotates about a fixed point can now be found. Referring to Figure 5.2, it is seen that point P describes an infinitesimal circular arc for the instant shown. Its linear displacement is given by

$$|d\mathbf{r}| = dr = r \sin \beta \, d\phi$$

This displacement has the same magnitude and direction as defined by the vector operation

$$d\mathbf{r} = d\boldsymbol{\phi} \times \mathbf{r}$$

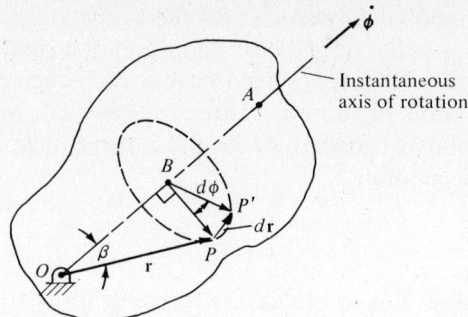

Figure 5.2

Now the instantaneous angular velocity of the body is defined by

$$\omega \equiv \frac{d\phi}{dt} = \dot{\phi} \tag{5.3}$$

or

$$\omega = \dot{\phi} \mathbf{i}_{OA} \tag{5.4}$$

The instantaneous velocity of point P on a rigid body that rotates about a fixed point then becomes

$$\frac{d\mathbf{r}}{dt} = \dot{\mathbf{r}} = \mathbf{v} = \boldsymbol{\omega} \times \mathbf{r} \tag{5.5}$$

The acceleration is given by

$$\mathbf{a} = \frac{d\mathbf{v}}{dt} = \dot{\boldsymbol{\omega}} \times \mathbf{r} + \boldsymbol{\omega} \times \dot{\mathbf{r}} \tag{5.6}$$

or

$$\mathbf{a} = \dot{\boldsymbol{\omega}} \times \mathbf{r} + \boldsymbol{\omega} \times \mathbf{v} \tag{5.7}$$

The instantaneous angular acceleration of a body is defined to be

$$\boldsymbol{\alpha} \equiv \frac{d\boldsymbol{\omega}}{dt} = \dot{\boldsymbol{\omega}} \tag{5.8}$$

Then

$$\mathbf{a} = \boldsymbol{\alpha} \times \mathbf{r} + \boldsymbol{\omega} \times (\boldsymbol{\omega} \times \mathbf{r}) \tag{5.9}$$

A. Rotation about a Fixed Axis

For rotation about a fixed axis, only 1 degree of freedom exists. The velocity and acceleration of a point in the body are given by equations (5.5) and (5.7). Since the axis of rotation is fixed, the point describes circular plane motion. It is convenient then to pick a position vector emanating from the fixed axis but in the plane of motion; i.e., let $\alpha = 90°$ or $\mathbf{r}$ emanate from point B in Figure 5.2. This common example of plane motion leads to the simplification that with

$$\mathbf{v} = \omega \mathbf{i}_{OA} \times \mathbf{r} \tag{5.10}$$

then
$$v = \omega r \tag{5.11}$$

Because the axis of rotation is fixed, the derivative, or change of the angular velocity with time, is caused only by variations in magnitude. That is,
$$\boldsymbol{\alpha} = \dot{\omega} \mathbf{i}_{OA} \tag{5.12}$$

or in scalar form,
$$\alpha = \dot{\omega} = \frac{d^2\phi}{dt^2} \tag{5.13}$$

Note that a change of independent variable is also possible:
$$\alpha = \frac{d\omega}{dt} = \frac{d\omega}{d\phi}\frac{d\phi}{dt} = \omega\frac{d\omega}{d\phi} \tag{5.14}$$

Analogous results to those found earlier for rectilinear motion are easily derived; namely,

1. For uniform rotation, $\alpha = 0$, and
$$\omega = \omega_0 \tag{5.15}$$
$$\phi = \omega_0 t + \phi_0 \tag{5.16}$$

2. For uniformly accelerated rotation, $\alpha = constant$,
$$\omega = \alpha t + \omega_0 \tag{5.17}$$
$$\phi = \frac{\alpha t^2}{2} + \omega_0 t + \phi_0 \tag{5.18}$$
$$\omega^2 = \omega_0^2 + 2\alpha(\phi - \phi_0) \tag{5.19}$$

Returning now to equation (5.9), if equations (5.10) and (5.12) are substituted for the appropriate terms, equation (5.20) results

$$\mathbf{a} = \dot{\omega}\mathbf{i}_{OA} \times \mathbf{r} + \omega\mathbf{i}_{OA} \times (\omega\mathbf{i}_{OA} \times \mathbf{r}) \tag{5.20}$$

or
$$\mathbf{a} = \mathbf{a}_t + \mathbf{a}_n \tag{5.21}$$

where
$$\mathbf{a}_t = \dot{\omega}\mathbf{i}_{OA} \times \mathbf{r} = \boldsymbol{\alpha} \times \mathbf{r} \tag{5.22}$$

then
$$a_t = \alpha r \tag{5.23}$$

Also
$$\mathbf{a}_n = \omega\mathbf{i}_{OA} \times (\omega\mathbf{i}_{OA} \times \mathbf{r}) = \boldsymbol{\omega} \times (\boldsymbol{\omega} \times \mathbf{r}) \tag{5.24}$$

or
$$\mathbf{a}_n = -\omega^2 \mathbf{r} \tag{5.25}$$

and
$$a_n = \omega^2 r \tag{5.26}$$

Since the point is describing circular plane motion of radius r, the results expressed by equations (5.23) and (5.26) could easily have been derived from equations (2.66) and (2.67).

B. Two Degrees of Rotational Freedom

Shown in Figure 5.3 is a body having 2 degrees of rotational freedom. The body is rod *ED* which rotates relative to the wire frame *OFGHIA* with an angular rate $\dot{\theta}$. Its axis of rotation *BC*, however, is not fixed. Instead axis *BC* rotates about the horizontal axis *OA*, with the wire frame, at a rate $\dot{\phi}$. Here also

$$\mathbf{v} = \boldsymbol{\omega} \times \mathbf{r} \qquad (5.5)$$

$$\mathbf{a} = \dot{\boldsymbol{\omega}} \times \mathbf{r} + \boldsymbol{\omega} \times \mathbf{v} \qquad (5.7)$$

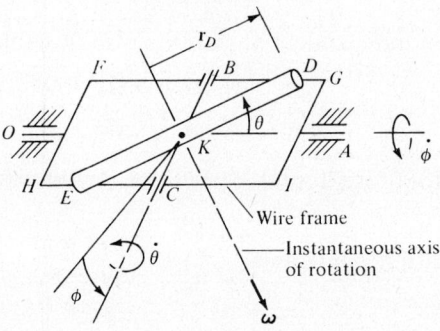

Figure 5.3

but the simplification to plane motion provided by equations (5.11), (5.13), (5.23), and (5.26) is not possible. The angular velocity is the vector sum of two terms:

$$\boldsymbol{\omega} = \dot{\boldsymbol{\phi}} + \dot{\boldsymbol{\theta}} \qquad (5.27)$$

The instantaneous axis of rotation lies in the plane of the wire in the direction of ω.

The angular acceleration of the body is more difficult to calculate. The time rate of change of each of the component vectors of equation (5.27) must be evaluated. One axis *OA* is fixed while the other axis *BC* rotates at an angular rate $\dot{\phi}$. Because axis *BC* rotates, one can describe the angular velocity $\dot{\theta}$ as the relative angular velocity of the rod with respect to the rotating frame. $\dot{\phi}$ is the absolute angular velocity of the frame.

$$\boldsymbol{\omega} = \boldsymbol{\omega}_{\text{rel}} + \boldsymbol{\omega}_{\text{frame}} \qquad (5.28)$$

Considering this further, it is apparent that since $\dot{\phi}$ does not change in direction, a change in $\dot{\phi}$ can only be brought about by magnitude variations. The same is not true for $\dot{\theta}$. It can have directional change as well since it rotates at a rate $\dot{\phi}$. Concentrating on the directional change brought about by the frame's rotation, it is easily seen (refer to Figure 5.4) that

$$d(\dot{\theta}) = d\boldsymbol{\phi} \times \dot{\boldsymbol{\theta}} \qquad \text{(directional change only)} \qquad (5.29)$$

The change in direction of $\dot{\theta}$ due to rotation is calculated then in the same manner

RIGID BODY DYNAMICS: KINEMATICS 105

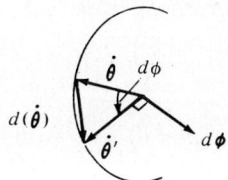

Figure 5.4

as for other vectors that rotate. The directional derivative becomes

$$\frac{d\dot{\theta}}{dt} = \dot{\phi} \times \dot{\theta} \quad \text{(directional change only)} \tag{5.30}$$

Note the order of the cross product.
Returning to equation (5.27),

$$\alpha = \frac{d\omega}{dt} = \underset{\text{Magnitude change}}{\ddot{\phi}\mathbf{i}_{OA}} + \underset{\text{Magnitude change}}{\ddot{\theta}\mathbf{i}_{BC}} + \underset{\text{Directional change of } \dot{\theta}}{\dot{\phi} \times \dot{\theta}} \tag{5.31}$$

or
$$\alpha = \ddot{\phi}\mathbf{i}_{OA} + \ddot{\theta}\mathbf{i}_{BC} + \dot{\phi}\dot{\theta}(\mathbf{i}_{OA} \times \mathbf{i}_{BC}) \tag{5.32}$$

To appreciate this further, let the angular rates be constant for the motion described in Figure 5.3 so that $\ddot{\theta} = \ddot{\phi} = 0$. Then from equation (5.32),

$$\alpha = \dot{\phi}\dot{\theta}(\mathbf{i}_{OA} \times \mathbf{i}_{BC})$$

What results is an angular acceleration of the body along an axis perpendicular to each axis of rotation, in a direction about which the body does not rotate, occurring even when the body rotates at constant angular speed. The angular acceleration would act perpendicular to the instantaneous axis of rotation, i.e., perpendicular to the wire frame, but would not cause a rotation about that axis. Instead it describes the *change* in ω. (See Figure 5.5.)

$$\omega' = \omega + \alpha \, dt \tag{5.33}$$

For the constant angular speed example discussed, the total angular velocity vector ω describes a cone in space as shown in Figure 5.5.

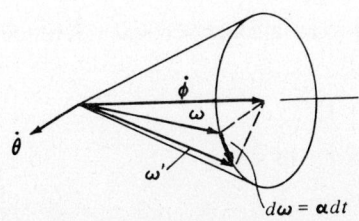

Figure 5.5

106 REVIEW OF NEWTONIAN MECHANICS

To summarize, for 2 degrees of rotational freedom, equations (5.5) and (5.9) still hold, but ω is given by equation (5.27), and α by equation (5.31).

$$\mathbf{v}_D = (\dot{\phi} + \dot{\theta}) \times \mathbf{r}_D \tag{5.34}$$

$$\mathbf{a}_D = (\ddot{\phi}\mathbf{i}_{OA} + \ddot{\theta}\mathbf{i}_{BC} + \dot{\phi} \times \dot{\theta}) \times \mathbf{r}_D + (\dot{\phi} + \dot{\theta}) \times [(\dot{\phi} + \dot{\theta}) \times \mathbf{r}_D] \tag{5.35}$$

If $\mathbf{r}_D$ also changes length, e.g., the body is elastic, additional terms must be added to equations (5.34) and (5.35). Fortunately there exists an easier method to handle these types of problems, and it will be discussed in Section 5.6.

C. Three Degrees of Rotational Freedom

Refer again to Figure 5.3. This time the axis OA will be allowed to rotate also about the vertical. The result is better depicted in Figure 5.6. The motion shown in Figure 5.6 has rod DE rotating relative to the wire frame $OFGHIA$ at an angular rate $\dot{\theta}$ about axis BC. Axis BC rotates about horizontal axis OA at a rate $\dot{\phi}$; $\dot{\phi}$ is the relative angular rate of frame $OFGHIA$ with respect to a vertical outer frame $NPQR$. Outer frame $NPQR$ contains axis OA. It rotates too, providing the third rotational degree of freedom, at an angular rate $\dot{\psi}$.

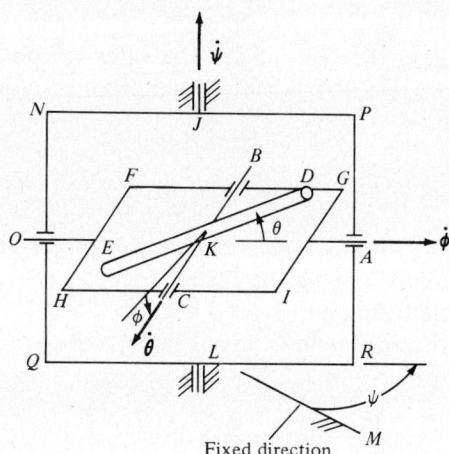

Figure 5.6

The angle ψ is measured from a fixed vertical reference plane $JKLM$. For this case

$$\mathbf{v} = \omega \times \mathbf{r} \tag{5.5}$$

$$\mathbf{a} = \dot{\omega} \times \mathbf{r} + \omega \times \mathbf{v} \tag{5.7}$$

$$\omega = \dot{\psi} + \dot{\theta} + \dot{\phi} \tag{5.36}$$

$$\alpha = \ddot{\psi}\mathbf{i}_{JL} + \ddot{\phi}\mathbf{i}_{OA} + \ddot{\theta}\mathbf{i}_{BC} + (\dot{\psi} + \dot{\phi}) \times \dot{\theta} + \dot{\psi} \times \dot{\phi} \tag{5.37}$$

The reader can now derive a result similar to equation (5.35) if he or she so desires.

Another example will be given to further illustrate the case of 3 degrees of rotational freedom. Please refer to Figure 5.7. Shown in the figure is a top, free to spin about its own symmetry axis at an angular rate $\dot{\psi}$ and nutate (nod) about its line of nodes at an angular rate $\dot{\theta}$. (The line of nodes marks the intersection

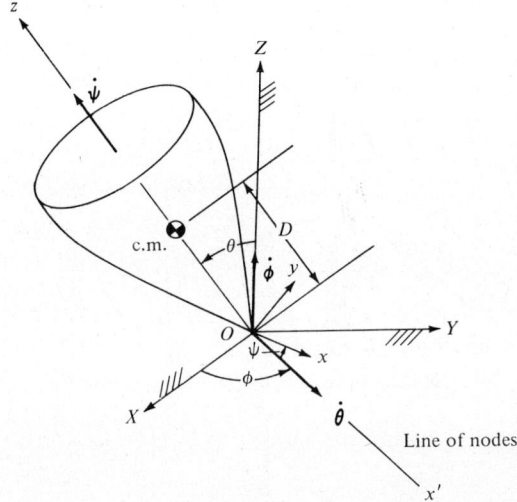

Figure 5.7

between a fixed horizontal plane and the transverse plane of the top perpendicular to the symmetry axis and passing through point O.) Further, its line of nodes, along which $\dot{\theta}$ acts, precesses at an angular rate $\dot{\phi}$ about the fixed vertical direction. Here, with the top assumed stationary at point O,

$$\mathbf{v} = \boldsymbol{\omega} \times \mathbf{r} \tag{5.5}$$

$$\mathbf{a} = \dot{\boldsymbol{\omega}} \times \mathbf{r} + \boldsymbol{\omega} \times \mathbf{v} \tag{5.7}$$

$$\boldsymbol{\omega} = \dot{\boldsymbol{\psi}} + \dot{\boldsymbol{\theta}} + \dot{\boldsymbol{\phi}} \tag{5.38}$$

$$\dot{\boldsymbol{\omega}} = \ddot{\psi}\mathbf{k} + \ddot{\phi}\mathbf{K} + \ddot{\theta}\mathbf{i}' + (\dot{\boldsymbol{\phi}} + \dot{\boldsymbol{\theta}}) \times \dot{\boldsymbol{\psi}} + \dot{\boldsymbol{\phi}} \times \dot{\boldsymbol{\theta}} \tag{5.39}$$

Note the difficulty in describing $\dot{\boldsymbol{\omega}}$ in equation (5.39) using three sets of coordinates: $(\mathbf{I}, \mathbf{J}, \mathbf{K})$, $(\mathbf{i}, \mathbf{j}, \mathbf{k})$, and $(\mathbf{i}', \mathbf{j}', \mathbf{k}')$.

One thing more should be mentioned. For the previous example of the top, the angles shown are in standard form for a 3-degrees-of-rotational-freedom problem and are given the same *Euler angles*. (Some texts may vary slightly from this set.) Section 7.4 describes them in detail.

Sample Problem 5.1

The wheel shown spins about its own axle OC at an angular velocity and acceleration ω_2 and $\dot{\omega}_2$, respectively. The axle itself turns with an angular speed

ω_1 and acceleration $\dot{\omega}_1$ about the vertical direction. (a) Find the total angular velocity and angular acceleration of the wheel. (b) Find the angular velocity and angular acceleration of the wheel relative to a frame rotating with the axle. Express your answers in terms of components in both of the coordinate systems shown.

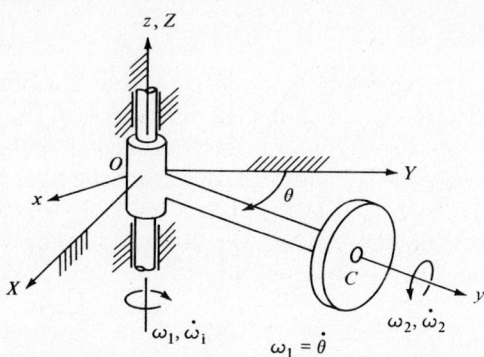

Analysis:

$$\omega_{total} = \omega_1 + \omega_2 = \dot{\theta} + \omega_2 = \omega_t$$
$$= -\omega_1 \mathbf{k} + \omega_2 \mathbf{j} = -\omega_1 \mathbf{K} + \omega_2(\sin\theta \mathbf{I} + \cos\theta \mathbf{J})$$

$$\dot{\omega}_{total} = \underbrace{-\dot{\omega}_1 \mathbf{k}}_{\text{Mag change}} + \underbrace{\dot{\omega}_2 \mathbf{j}}_{\text{Mag change}} - \underbrace{\omega_1 \mathbf{k} \times \omega_2 \mathbf{j}}_{\text{Directional change}}$$

$$= -\dot{\omega}_1 \mathbf{k} + \dot{\omega}_2 \mathbf{j} + \omega_1 \omega_2 \mathbf{i}$$
$$= -\dot{\omega}_1 \mathbf{K} + \dot{\omega}_2 \sin\theta \mathbf{I} + \dot{\omega}_2 \cos\theta \mathbf{J} + \omega_1 \omega_2 (\cos\theta \mathbf{I} - \sin\theta \mathbf{J})$$

For the wheel, relative to the rotating reference frame,

$$\omega_{rel} = \omega_2 \mathbf{j} = \omega_2(\sin\theta \mathbf{I} + \cos\theta \mathbf{J})$$
$$\dot{\omega}_{rel} = \dot{\omega}_2 \mathbf{j} = \dot{\omega}_2(\sin\theta \mathbf{I} + \cos\theta \mathbf{J})$$

Note, the rate of change of the *total* angular velocity vector relative to the rotating reference frame attached to the axle is

$$(\dot{\omega}_t)_{rel} = \dot{\omega}_2 \mathbf{j} - \dot{\omega}_1 \mathbf{k}$$

For a rotating reference frame attached to the wheel,

$$(\dot{\omega}_t)_{rel} = \dot{\omega}_2 \mathbf{j} - \dot{\omega}_1 \mathbf{k} - \omega_2 \times \omega_1 \qquad ■$$

5.4 GENERAL MOTION

Chasles' theorem, an extension of Euler's work, states that the most general displacement of a rigid body may be reduced to that of a translation, followed

RIGID BODY DYNAMICS: KINEMATICS

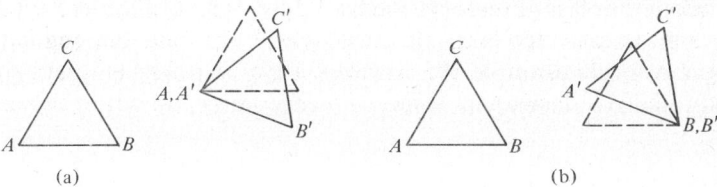

Figure 5.8

by a rotation. This is clearly shown in Figure 5.8a and b. In both figures the triangle starts its general motion at the left and arrives at the same final position and orientation. In Figure 5.8a this motion is broken up into a translation plus a subsequent clockwise rotation about A' to obtain the final position. In Figure 5.8b the motion is broken up into a translation plus a clockwise rotation about B' instead. The effects are equivalent. For the motion shown in Figure 5.8a one can write (see Section 2.7 on relative motion)

$$\mathbf{v}_B = \mathbf{v}_A + \mathbf{v}_{B/A} \tag{5.40}$$

$$\mathbf{a}_B = \mathbf{a}_A + \mathbf{a}_{B/A} \tag{5.41}$$

In equations (5.40) and (5.41) the terms $\mathbf{v}_A$ and $\mathbf{a}_A$ describe the absolute velocity and acceleration, respectively, of a translating reference frame located at A. Because the path that B describes in this frame is a circular arc having its center at A, one can write

$$\mathbf{v}_B = \mathbf{v}_A + \boldsymbol{\omega} \times \mathbf{r}_{B/A} \tag{5.42}$$

$$\mathbf{a}_B = \mathbf{a}_A + \dot{\boldsymbol{\omega}} \times \mathbf{r}_{B/A} + \boldsymbol{\omega} \times (\boldsymbol{\omega} \times \mathbf{r}_{B/A}) \tag{5.43}$$

The results obtained in the last section concerning rotation about a fixed point can be employed, equations (5.5) and (5.9), because the nonrotating observer moving with point A witnesses "fixed-point" rotation in his or her frame of reference. Now, though, instead of an absolute position vector, a relative position vector is used in equations (5.42) and (5.43) to reflect the fact that point A is moving. Similarly, referring to Figure 5.8b, one may write

$$\mathbf{v}_A = \mathbf{v}_B + \mathbf{v}_{A/B} \tag{5.44}$$

$$\mathbf{a}_A = \mathbf{a}_B + \mathbf{a}_{A/B} \tag{5.45}$$

$$\mathbf{v}_A = \mathbf{v}_B + \boldsymbol{\omega} \times \mathbf{r}_{A/B} \tag{5.46}$$

$$\mathbf{a}_A = \mathbf{a}_B + \dot{\boldsymbol{\omega}} \times \mathbf{r}_{A/B} + \boldsymbol{\omega} \times (\boldsymbol{\omega} \times \mathbf{r}_{A/B}) \tag{5.47}$$

The relative position vector employed is $\mathbf{r}_{A/B}$, not $\mathbf{r}_{B/A}$. On the other hand, the *angular velocity $\boldsymbol{\omega}$ and the angular acceleration $\dot{\boldsymbol{\omega}}$ of the body are independent of the arbitrary reference rotation point used* in describing general motion. For example, consider two lines scribed on a rigid body at two different points. If the body rotates, the line scribed at one point must rotate through the same angle that the line scribed at another point does since the body is rigid. $\boldsymbol{\omega}$ and $\dot{\boldsymbol{\omega}}$

110 REVIEW OF NEWTONIAN MECHANICS

belong then to the class of vectors called *free vectors* since they are not referenced to any specific point in the body and may be moved at will to any reference axis of rotation. Another example of a free vector is a couple; a couple's effect on a rigid body is also independent of its point of application.

5.5 GENERAL PLANE MOTION

A. General Plane Motion: Velocity Analysis

The previous section described a method of analyzing the general motion of a rigid body. Most engineering problems concern plane motion, and therefore, a separate section concerning its analysis is now introduced. The general plane motion of a rigid body has 3 degrees of freedom; 2 degrees of freedom are in translation, and 1 degree of freedom is in rotation. This provides a great deal of simplification. The relative motion of one point on the body to another now becomes, for a rigid body, plane circular motion. For instance, suppose the motion described in Figure 5.8a was planar. The velocity of B relative to A is given by

$$\mathbf{v}_{B/A} = \boldsymbol{\omega} \times \mathbf{r}_{B/A} \tag{5.48}$$

and
$$v_{B/A} = \omega r_{B/A} \tag{5.49}$$

$\mathbf{v}_{B/A}$ lies in the plane of motion perpendicular to $\mathbf{r}_{B/A}$ and is pointed in the direction that $\mathbf{r}_{B/A}$ is turned by ω. With magnitude given by scalar equations such as equation (5.49), and with resultant directions known beforehand, the solution to plane motion kinematic problems can be formed without resort to formal vector cross-product operations.

Most kinematic problems have additional constraints. Take, for instance, the slider crank mechanism shown in Figure 5.9. Crank OA describes plane circular motion about a fixed point with given angular velocity. Connecting rod AP describes general plane motion. Its motion can be broken up into a translation and a rotation. Slider P describes rectilinear translation. For connecting rod AP,

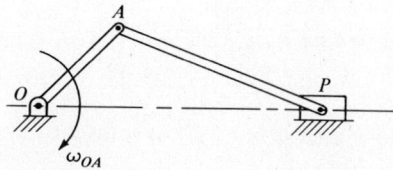

Figure 5.9

$$\mathbf{v}_P = \mathbf{v}_A + \mathbf{v}_{P/A} \tag{a}$$

$$\mathbf{v}_{P/A} = \boldsymbol{\omega}_{AP} \times \mathbf{r}_{P/A} \tag{b}$$

$$v_{P/A} = \omega_{AP} r_{P/A} \quad \text{(perpendicular to } r_{P/A} \text{ and in the direction that } \omega_{AP}, \text{ or an assumed } \omega_{AP}, \text{ turns } r_{P/A}) \quad \text{(c)}$$

For crank OA,

$$\mathbf{v}_A = \boldsymbol{\omega}_{OA} \times \mathbf{r}_A \quad \text{(d)}$$

$$v_A = \omega_{OA} r_A \quad \text{(perpendicular to } r_A \text{ in the direction that } \omega_{OA}, \text{ or an assumed } \omega_{OA}, \text{ turns } r_A) \quad \text{(e)}$$

For slider P,

$$\mathbf{v}_P = v_P \mathbf{i} \quad \text{(f)}$$

The rectilinear motion of point P in a known direction, equation (f), is the additional constraint on the motion of AP that allows the solution of this problem. Otherwise, even with ω_{OA} given, if ω_{AP} was not known, the motion of AP, and consequently P, could not be determined; the motion would be undefined, or unrestricted. With constraint equation (f) given, the component terms of equations (b) and (d) in the x direction can be added and set equal to v_P. Similarly, the y-component terms of equations (b) and (d) can be added and set equal to zero. In the calculations, it is easier to assume all unknown ω's to be in one direction (counterclockwise, for example) and then solve for their actual values. The presence of a negative sign in the answer means an opposite direction of rotation for that ω.

Sample Problem 5.2
Given the slider crank mechanism shown in the figure, find the velocity of the piston P for this instant if the crank OA has an angular velocity of 4000 rpm clockwise.

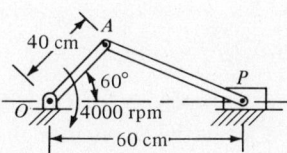

Analysis: Connecting rod AP has general plane motion; i.e., its motion can be broken into translational and rotational parts.
Using equation (5.40),

$$\mathbf{v}_P = \mathbf{v}_A + \mathbf{v}_{P/A}$$

The path of P relative to a translating frame at A is circular; hence equation (5.48) also applies.

$$\mathbf{v}_{P/A} = \boldsymbol{\omega}_{AP} \times \mathbf{r}_{P/A}$$

112   REVIEW OF NEWTONIAN MECHANICS

Point A's absolute path is also circular, or

$$\mathbf{v}_A = \boldsymbol{\omega}_{OA} \times \mathbf{r}_A$$

The solved geometry of the mechanism is shown below.

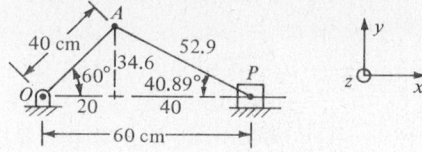

The velocity of P relative to A is

$$\mathbf{v}_{P/A} = \omega_{AP}\mathbf{k} \times (40\mathbf{i} - 34.6\mathbf{j}) = \omega_{AP}(34.6)\mathbf{i} + \omega_{AP}(40)\mathbf{j}$$

The velocity of A is

$$\mathbf{v}_A = 4000 \left(\frac{2\pi}{60}\right)(-\mathbf{k}) \times (20\mathbf{i} + 34.6\mathbf{j}) = 14\,510\mathbf{i} - 8397\mathbf{j}$$

The motion of P is constrained to be horizontal; hence

$$v_P\mathbf{i} = 14\,510\mathbf{i} - 8377\mathbf{j} + \omega_{AP}(34.6)\mathbf{i} + \omega_{AP}(40)\mathbf{j}$$

Equating like components on both sides of the equation yields two scalar equations:

$$v_P = 14\,510 + \omega_{AP}(34.6) \qquad 0 = -8377 + \omega_{AP}(40)$$

Therefore

$$\omega_{AP} = 209.4\mathbf{k} \quad \text{[counterclockwise rotation } (+\mathbf{k}) \text{ that was assumed was correct]}$$

$$\mathbf{v}_P = 217.6\mathbf{i} \quad \text{m/s} \qquad \blacksquare$$

B.   The Method of Instantaneous Centers

Before proceeding to a discussion of acceleration, another method of velocity analysis should be discussed. It is the *method of instantaneous centers*. Its importance lies in the fact that it greatly reduces the amount of calculation necessary for the solution of a plane motion velocity problem.

The *instantaneous center* of a body is defined to be point on the body, or the body imagined extended, that has zero velocity for that instant. The body rotates about the *instantaneous axis of rotation*, which must pass through the body's instantaneous center for that moment. The instantaneous center changes, in general, from moment to moment as the body moves. The locus traced out in space by the instantaneous center is called the *space centrode*. The curve traced out on the body itself by the instantaneous centers is called the *body centrode*. The body centrode appears to roll without slipping on the space centrode.

Some plane motion kinematic examples are depicted in Figures 5.10 to 5.12. In Figure 5.10, the body shown describes plane motion. The velocity of two points A and B on the body are given. The instantaneous center for the body can be found by dropping perpendiculars to the velocities at points A and B. These perpendiculars must intersect at the instantaneous center of the body

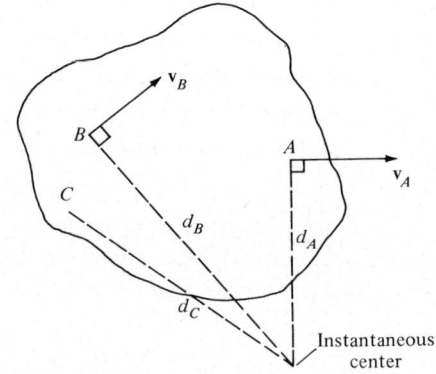

Figure 5.10

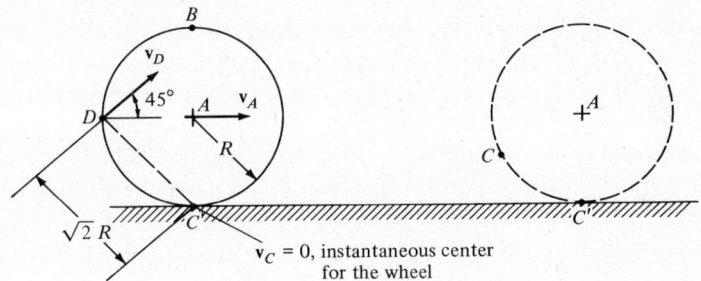

Figure 5.11

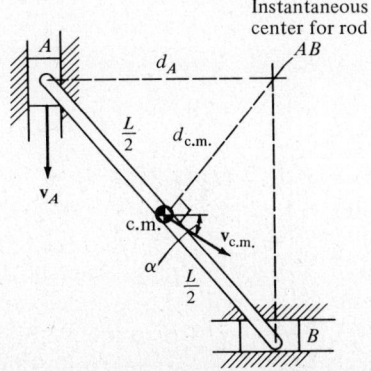

Figure 5.12

if the body is to rotate about that point. This follows from the fact that for a rigid body that rotates about a point having zero velocity, all points in the body have velocities perpendicular to their radii. (See Section 5.3A, Rotation about a Fixed Axis, for example.)

If v_A is known, and d_A and d_B are found using geometry, then one can solve for v_B. Proceeding then,

$$\omega = \frac{v_A}{d_A} \tag{a}$$

From the picture it is seen that the sense of ω must be clockwise for rotation about the instantaneous center (v_A is to the right). So

$$\omega = -\frac{v_A}{d_A}\mathbf{k} \tag{b}$$

$$v_B = \omega d_B = v_A \frac{d_B}{d_A} \tag{c}$$

$\mathbf{v}_B$ must be perpendicular to d_B and point up and to the right to agree with the direction of rotation of ω. The velocity of an arbitrary point C can now be found by first determining the distance d_C and then multiplying d_C by ω of the body. Its direction must be perpendicular to the line d_C, and it must have a direction in agreement with ω. Note that the instantaneous center about which the body rotates for this instant lies off the body, or on the body "extended."

Figure 5.11 shows a wheel that is assumed to be rolling without slipping. because the wheel rolls without slipping, $\mathbf{v}_C = \mathbf{0}$. (The ground is fixed so point C must be the instantaneous center for the wheel.) As the wheel moves to the right, the instantaneous center moves to the right. Its space centrode is a straight line. The body centrode is a circle. By knowing v_A and the radius of the wheel, the angular velocity can be found

$$\omega = -\frac{v_A}{R}\mathbf{k} \tag{d}$$

The velocity of point B then becomes

$$\mathbf{v}_B = 2R\frac{v_A}{R}\mathbf{i} \tag{e}$$

The velocity of point D is

$$v_D = \sqrt{2}R\frac{v_A}{R}(\cos 45°\mathbf{i} + \sin 45°\mathbf{j}) \tag{f}$$

Figure 5.12 illustrates a rod pinned to two sliders that are constrained to move along a vertical and a horizontal track. Since both sliders translate, their instantaneous centers must lie at infinity; i.e., there is no rotation present. By erecting perpendiculars to the velocity at point A and to the direction of the unknown velocity at point B, the instantaneous center for rod AB can be found. Because points A and B lie on the rod, the intersection of these two perpendiculars must be at a point on the body "extended" (imagined) that has an instantaneous velocity equal to zero.

$$\omega = -\frac{v_A}{d_A}\mathbf{k} \tag{g}$$

$$\mathbf{v}_B = d_B \frac{v_a}{d_A}\mathbf{i} \tag{h}$$

$$\mathbf{v}_{\text{c.m.}} = d_{\text{c.m.}} \frac{v_A}{d_A}(\cos\alpha\mathbf{i} - \sin\alpha\mathbf{j}) \tag{i}$$

It is left as an exercise for the reader to find the angle α using geometry.

Sample Problem 5.3
Do previous Sample Problem 5.2 using the method of instantaneous centers.

Analysis: The instantaneous center for connecting rod AP must lie at the intersection of two lines drawn perpendicular to the velocities of two arbitrary points on the rod as shown in the figure. (The instantaneous center for crank OA is located at point O, the instantaneous center for slider P must lie at a point an infinite distance from slider P on a line normal to the horizontal constraint surface.)

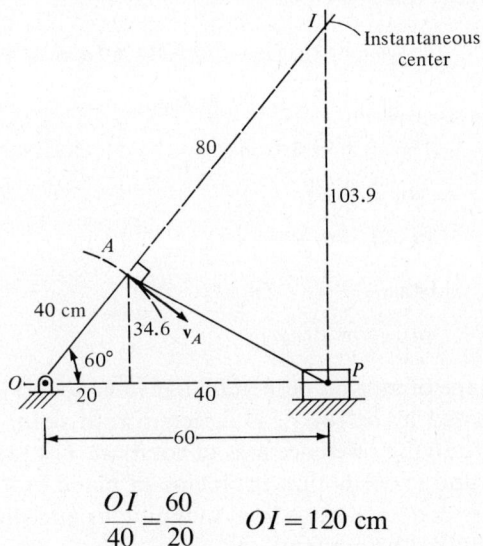

$$\frac{OI}{40} = \frac{60}{20} \qquad OI = 120 \text{ cm}$$

Therefore,
$$IA = 80 \text{ cm} \qquad IP = 103.9 \text{ cm}$$

For circular motion,
$$v = \omega r$$

or

$$v_A = \omega_{OA} r_A = 16\,755 \text{ cm/s}$$

$$\mathbf{v}_A = 16\,755\,(\sin 60°\mathbf{i} - \cos 60°\mathbf{j})$$

Since the instantaneous center by definition is a point on the body, or body extended, that has zero velocity for that instant, the body rotates about that point for the moment.

$$\omega_{AP} = \frac{v_A}{80} = \frac{16\,755}{80} = 209.4$$

$$\omega_{AP} = 209.4\mathbf{k} \quad \text{rad/s}$$

$$v_P = \omega_{AP}(103.9) = 209.4(103.9) = 21\,760 \text{ cm/s}$$

$$\mathbf{v}_P = 217.6\mathbf{i} \quad \text{m/s} \qquad \blacksquare$$

C. *General Plane Motion: Acceleration Analysis*

General plane motion of a rigid body is characterized by 3 degrees of freedom. Two degrees of freedom are in translation, and 1 degree of freedom is in rotation. The relative motion of one point on a rigid body to another point on the same body, as seen by a nonrotating observer, is plane circular motion. The relative acceleration can then be broken up into normal and tangential components (see Sections 2.3 and 5.3A).

$$\mathbf{a}_{B/A} = \dot{\boldsymbol{\omega}} \times \mathbf{r}_{B/A} + \boldsymbol{\omega} \times (\boldsymbol{\omega} \times \mathbf{r}_{B/A}) \tag{5.50}$$

$$\mathbf{a}_{B/A} = (\mathbf{a}_{B/A})_t + (\mathbf{a}_{B/A})_n \tag{5.51}$$

$$(\mathbf{a}_{B/A})_t = \dot{\boldsymbol{\omega}} \times \mathbf{r}_{B/A} \tag{5.52}$$

$$(a_{B/A})_t = \dot{\omega} r_{B/A} \tag{5.53}$$

$$(\mathbf{a}_{B/A})_n = \boldsymbol{\omega} \times (\boldsymbol{\omega} \times \mathbf{r}_{B/A}) \tag{5.54}$$

$$(\mathbf{a}_{B/A})_n = -\omega^2 \mathbf{r}_{B/A} \tag{5.55}$$

$$(a_{B/A})_n = \omega^2 r_{B/A} \tag{5.56}$$

$(\mathbf{a}_{B/A})_t$ acts in the plane of motion perpendicular to $\mathbf{r}_{B/A}$, and points in the direction that $\mathbf{r}_{B/A}$ is turned by $\dot{\boldsymbol{\omega}}$. $(\mathbf{a}_{B/A})_n$ is directed from point B to point A, i.e., central, toward the moving reference axis of rotation. The significance of equations (5.49), (5.53), and (5.56) is that problems in plane motion can be solved without taking cross products since final magnitudes and directions of vectors quantities are recognized beforehand.

Also, bear in mind that since $\mathbf{r}_{B/A} = -\mathbf{r}_{A/B}$,

$$\mathbf{a}_{B/A} = -\mathbf{a}_{A/B} \tag{5.57}$$

$$(\mathbf{a}_{B/A})_t = -(\mathbf{a}_{A/B})_t \tag{5.58}$$

$$(\mathbf{a}_{B/A})_n = -(\mathbf{a}_{A/B})_n \tag{5.59}$$

As mentioned previously, most kinematic problems have additional constraints to consider. Refer again to Figure 5.9. If it is assumed that both the angular velocity and angular acceleration of the crank OA are known, then the acceleration of the piston P can be determined. Crank OA describes plane circular motion about a fixed point O, connecting rod AP describes general plane motion, and slider P describes rectilinear translation. For point P on connecting rod AP,

$$\mathbf{a}_P = \mathbf{a}_A + \mathbf{a}_{P/A} \tag{a}$$

The motion of P relative to A is circular motion; hence,

$$\mathbf{a}_{P/A} = \dot{\boldsymbol{\omega}}_{AP} \times \mathbf{r}_{P/A} + \boldsymbol{\omega}_{AP} \times (\boldsymbol{\omega}_{AP} \times \mathbf{r}_{P/A}) \tag{b}$$

$$\mathbf{a}_{P/A} = (\mathbf{a}_{P/A})_t + (\mathbf{a}_{P/A})_n \tag{c}$$

$$(\mathbf{a}_{P/A})_t = \dot{\boldsymbol{\omega}} \times \mathbf{r}_{P/A} \tag{d}$$

$$(a_{P/A})_t = \dot{\omega}_{AP} r_{P/A} \quad \text{(perpendicular to } \mathbf{r}_{P/A} \text{ and in the direction that } \dot{\omega}_{AP}, \text{ or an assumed } \dot{\omega}_{AP}, \text{ turns } \mathbf{r}_{P/A}) \tag{e}$$

$$(\mathbf{a}_{P/A})_n = \boldsymbol{\omega}_{AP} \times (\boldsymbol{\omega}_{AP} \times \mathbf{r}_{P/A}) \tag{f}$$

$$(\mathbf{a}_{P/A})_n = -\omega_{AP}^2 \mathbf{r}_{P/A} \tag{g}$$

$$(a_{P/A})_n = \omega_{AP}^2 r_{P/A} \quad \text{(directed from point } P \text{ to point } A) \tag{h}$$

$$\mathbf{a}_A = (\mathbf{a}_A)_n + (\mathbf{a}_A)_t \tag{i}$$

$$(\mathbf{a}_A)_n = \boldsymbol{\omega}_{OA} \times (\boldsymbol{\omega}_{OA} \times \mathbf{r}_A) \tag{j}$$

$$(\mathbf{a}_A)_n = -\omega_{OA}^2 \mathbf{r}_A \tag{k}$$

$$(a_A)_n = \omega_{OA}^2 r_A \quad \text{(directed from point } A \text{ to point } O) \tag{l}$$

$$(\mathbf{a}_A)_t = \dot{\boldsymbol{\omega}}_{OA} \times \mathbf{r}_A \tag{m}$$

$$(a_A)_t = \dot{\omega}_{OA} r_A \quad \text{(perpendicular to } \mathbf{r}_A \text{ and in the direction that } \dot{\omega}_{OA}, \text{ or an assumed } \dot{\omega}_{OA} \text{ turns } \mathbf{r}_A) \tag{n}$$

$$\mathbf{a}_P = (a_P)_x \mathbf{i} + (a_P)_y \mathbf{j} \quad \text{(without constraints)} \tag{o}$$

Refer now to equations (a), (b), (d), (f), (j), (m), and (o). With all distances known in the problem, and with ω_{OA} and $\dot{\omega}_{OA}$ given, there are four unknowns to solve for: ω_{AP}, $\dot{\omega}_{AP}$, $(a_P)_x$, $(a_P)_y$. The additional constraint that P must move along a horizontal line reduces the number of unknowns present to three since $(a_P)_y = 0$. Vector equations can be broken up into scalar parts in the x and y directions so that instead of one vector equation, two scalar equations now become available. However, with three unknowns present, this is still not enough. As a general rule, one can state that *in all acceleration problems, one must first perform a velocity analysis to ascertain all angular velocities present that have a bearing on the problem.* With ω_{AP} known there are only two variables left to solve for, and the problem is completed. There can only be two scalar unknowns present in any plane motion velocity problem, and two additional unknowns in any acceleration problem.

A final suggestion may prove helpful. It is recommended that all unknown $\dot{\omega}$'s be assumed to act in one direction, counterclockwise, for example. After solving for their actual values, the presence of a negative sign in the answer would mean an opposite direction of rotation for that $\dot{\omega}$. Note also that the instantaneous center method does not work directly in acceleration analysis. The instantaneous center may accelerate even though it has zero velocity, and it cannot, therefore, be used as a zero reference point.

Sample Problem 5.4

For the four-bar linkage shown (fixed link OC is usually considered as the fourth member), find the angular acceleration of rod *BC* for this instant. Rod *OA* turns at a constant angular velocity of 3 rad/s clockwise.

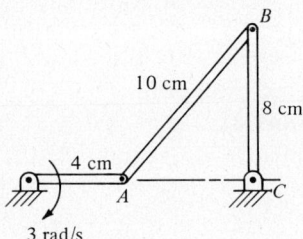

Analysis: This is a typical linear/angular acceleration problem for a mechanism. The solution must start out by finding all the angular velocities present. To do this, the method of instantaneous centers is recommended.

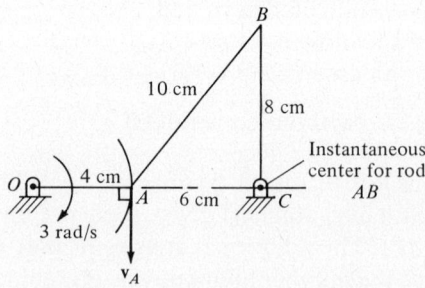

By reviewing the solution of Sample Problem 5.3, it becomes apparent that the instantaneous center of rod *AB* must lie at point *C*.

RIGID BODY DYNAMICS: KINEMATICS 119

$$v = \omega r$$

Then
$$v_A = \omega_{OA} r_A = 3(4) = 12 \qquad \mathbf{v}_A = -12\mathbf{j} \quad \text{cm/s}$$
$$\omega_{AB} = \tfrac{12}{6} = 2 \qquad \boldsymbol{\omega}_{AB} = 2\mathbf{k} \quad \text{rad/s}$$
$$v_B = 2(8) = 16 \qquad \mathbf{v}_B = -16\mathbf{i} \quad \text{cm/s} \qquad (a)$$

However,
$$v_B = \omega_{BC} r_B$$

or
$$\mathbf{v}_B = \omega_{BC} \mathbf{k} \times 8\mathbf{j} = -8\omega_{BC}\mathbf{i} \qquad (b)$$

Comparing equations (a) and (b)
$$\boldsymbol{\omega}_{BC} = 2\mathbf{k} \quad \text{rad/s}$$

The acceleration part of the problem now begins. Employing equation (5.41) for the mechanism,

$$\mathbf{a}_B = \mathbf{a}_A + \mathbf{a}_{B/A}$$

The acceleration of a point traversing a circular path in a plane is given by

$$\mathbf{a} = \boldsymbol{\omega} \times (\boldsymbol{\omega} \times \mathbf{r}) + \boldsymbol{\alpha} \times \mathbf{r}$$

or
$$\mathbf{a} = (\mathbf{a})_n + (\mathbf{a})_t$$

Recalling that for circular motion

$$a_n = r\omega^2 \quad \text{and} \quad a_t = r\alpha = r\dot{\omega}$$

these equations become, for point A, the following:

$$\mathbf{a}_A = (\mathbf{a}_A)_n + (\mathbf{a}_A)_t$$
$$(\mathbf{a}_A)_n = -(4)(3^2)\mathbf{i} = -36\mathbf{i} \quad \text{cm/s}^2$$
$$(\mathbf{a}_A)_t = \alpha_{OA} r_A \mathbf{j} = 0(3)\mathbf{j} = 0$$
$$\mathbf{a}_{B/A} = (\mathbf{a}_{B/A})_n + (\mathbf{a}_{B/A})_t \quad \text{(circular motion also)}$$
$$(\mathbf{a}_{B/A})_n = 2^2(-6\mathbf{i} - 8\mathbf{j})$$
$$= 40(-\cos 53.19°\mathbf{i} - \sin 53.19°\mathbf{j}) \quad \text{cm/s}^2$$
$$(\mathbf{a}_{B/A})_t = \boldsymbol{\alpha}_{AB} \times \mathbf{r}_{B/A}$$
$$(\mathbf{a}_{B/A})_t = \alpha_{AB}(-8\mathbf{i} + 6\mathbf{j}) \quad \text{cm/s}^2$$
$$\mathbf{a}_B = (\mathbf{a}_B)_n + (\mathbf{a}_B)_t$$
$$(\mathbf{a}_B)_n = -8(2^2)\mathbf{j} = -32\mathbf{j} \quad \text{cm/s}^2$$
$$(\mathbf{a}_B)_t = -\alpha_{BC}(8)\mathbf{i}$$

Combining terms for equation (5.41),

$$-\alpha_{BC}(8)\mathbf{i} - 32\mathbf{j} = -36\mathbf{i} + 0 - 2^2(6)\mathbf{i} - 2^2(8)\mathbf{j} - 8\alpha_{AB}\mathbf{i} + 6\alpha_{AB}\mathbf{j}$$

Equating like components on both sides of the result yields

$$-\alpha_{BC}(8) = -36 - 24 - 8\alpha_{AB}$$
$$-32 = -32 + 6\alpha_{AB}$$

Therefore,

$$\alpha_{AB} = 0 \quad \text{and} \quad \alpha_{BC} = 7.5\mathbf{k} \quad \text{rad/s}^2 \quad \blacksquare$$

Sample Problem 5.5
Develop a relation between the input and output crank angles of a four-bar mechanism (*Freudenstein's* equation).

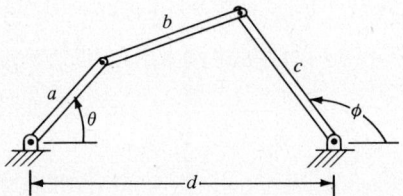

Analysis:

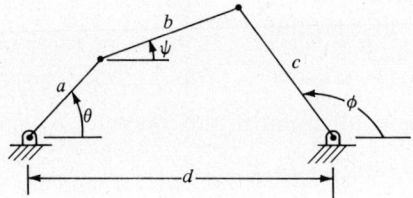

$$a \cos \theta + b \cos \psi - c \cos \phi = d$$
$$a \sin \theta + b \sin \psi - c \sin \phi = 0$$
$$b^2 = (d + c \cos \phi - a \cos \theta)^2 + (c \sin \phi - a \sin \theta)^2$$
$$= d^2 + a^2 + c^2 - 2da \cos \theta + 2dc \cos \phi - 2ac \cos \phi \cos \theta - 2ac \sin \phi \sin \theta$$

Let $R_1 = d/a$, $R_2 = d/c$, and $R_3 = (d^2 + a^2 - b^2 + c^2)/2ca$; then

$$R_3 - R_2 \cos \theta + R_1 \cos \phi - \cos(\phi - \theta) = 0 \quad \blacksquare$$

Sample Problem 5.6
Given the varying input angle θ, write a computer subroutine that will solve for the angle ϕ for one complete revolution of crank a in Sample Problem 5.5. Use the Newton-Raphson method described in Sample Problem 2.11 and the results from Sample Problem 5.5.

Analysis: See Figure 5.13. $\quad \blacksquare$

```
00100 SUBROUTINE FREUD(R1,R2,R3,TH,PH)
00110 F(PH,TH)=R1*COS(PH)-R2*COS(TH)+R3-COS(PH-TH)
00120 FD(PH,TH)=-R1*SIN(PH)+SIN(PH-TH)
00130 PI=3.1415927
00140 N=1
00150 PH=PI/2.
00160 3 PHN=PH-F(PH,TH)/FD(PH,TH)
00170 N=N+1
00180 IF(ABS(PHN-PH).LE.0.00001) GO TO 7
00190 IF(N.GT.100) GO TO 6
00200 PH=PHN
00210 GO TO 8
00220 6 PRINT 1,
00230 1 FORMAT(1X,"   WARNING 100 ITERATIONS HAVE PASSED")
00240 7 PHD=PHN*180./PI
00250 THD=TH*180./PI
00260 PRINT 2,THD,PHD
00270 2 FORMAT(1X,"  THE VALUE OF TH IS  ",F7.3," PH IS",F9.3)
00280 RETURN
```

Figure 5.13

Sample Problem 5.7

Using the subroutine developed in Sample Problem 5.6, develop a program that will solve for $\dot\phi$ and $\ddot\phi$ for the mechanism of Sample Problem 5.4.

Analysis:

$$R_3 - R_2 \cos\theta + R_1 \cos\phi - \cos(\phi-\theta) = 0$$

$$R_2\dot\theta \sin\theta - R_1\dot\phi \sin\phi + (\dot\phi - \dot\theta)\sin(\phi-\theta) = 0$$

$$\dot\phi = \frac{[\sin(\phi-\theta) - R_2 \sin\theta]\dot\theta}{\sin(\phi-\theta) - R_1 \sin\phi} \tag{a}$$

$$R_2\dot\theta^2 \cos\theta - R_1\ddot\phi \sin\phi - R_1\dot\phi^2 \cos\phi + (\dot\phi-\dot\theta)^2 \cos(\phi-\theta) + \ddot\phi \sin(\phi-\theta) = 0$$

$$\ddot\phi = \frac{R_1\dot\phi^2 \cos\phi - R_2\dot\theta^2 \cos\theta - (\dot\phi-\dot\theta)^2 \cos(\phi-\theta)}{\sin(\phi-\theta) - R_1 \sin\phi} \tag{b}$$

See Figures 5.14 and 5.15. ∎

5.6 TIME DERIVATIVE OF A VECTOR REFERENCED TO A ROTATING FRAME

A. Introduction

Many complex kinematic problems, for the general case or for that of plane motion, can be made easier to analyze if a rotating reference system is used

```
00100 PROGRAM LINKAGE(INPUT,OUTPUT)
00110 DATA A/4./,B/10./,C/8./,D/10./,THP/-3./
00120 PI=3.1415927
00130 R1=D/A
00140 R2=D/C
00150 R3=(D**2+A**2+C**2-B**2)/(2.*C*A)
00160 DO 5 I=1,25
00170 THD=(I-1)*15.
00180 TH=PI*THD/180.
00190 CALL FREUD(R1,R2,R3,TH,PH)
00200 PHP=(THP*(-R2*SIN(TH)+SIN(PH-TH)))/(SIN(PH-TH)-
00210+R1*SIN(PH))
00220 PHPP=(R1*PHP*PHP*COS(PH)-R2*THP*THP*COS(TH)-
00230+COS(PH-TH)*(PHP-THP)*(PHP-THP))/(SIN(PH-TH)-
00240+R1*SIN(PH))
00250 PRINT 31,PHP,PHPP
00260 31 FORMAT(4X,"PHIDOT =",E14.7,"RAD/S",6X,"PHIDDOT =",
00270+E14.7,"RAD/S*S")
00280 5 CONTINUE
00290 STOP
00300 END
00310 SUBROUTINE FREUD(R1,R2,R3,TH,PH)
00320 F(PH,TH)=R1*COS(PH)-R2*COS(TH)+R3-COS(PH-TH)
00330 FD(PH,TH)=-R1*SIN(PH)+SIN(PH-TH)
00340 PI=3.1415927
00350 N=1
00360 PH=PI/2.
00370 8 PHN=PH-F(PH,TH)/FD(PH,TH)
00380 N=N+1
00390 IF(ABS(PHN-PH).LE.0.00001) GO TO 7
00400 IF(N.GT.100) GO TO 6
00410 PH=PHN
00420 GO TO 8
00430 6 PRINT 1,
00440 1 FORMAT(1X,"   WARNING 100 ITERATIONS HAVE PASSED")
00450 7 PHD=PHN*180./PI
00460 THD=TH*180./PI
00470 PRINT 2,THD,PHD
00480 2 FORMAT(1X,"   THE VALUE OF TH IS   ",F7.3,"  PH IS",F9.3)
00490 RETURN
00500 END
```

Figure 5.14

```
THE VALUE OF TH IS     0.000  PH IS   90.000
PHIDOT = .2000000E+01RAD/S        PHIDDOT = .7500000E+01RAD/S*S
THE VALUE OF TH IS    15.000  PH IS   81.992
PHIDOT = .1151506E+01RAD/S        PHIDDOT = .1102209E+02RAD/S*S
THE VALUE OF TH IS    30.000  PH IS   78.610
PHIDOT = .2209188E+00RAD/S        PHIDDOT = .9748674E+01RAD/S*S
THE VALUE OF TH IS    45.000  PH IS   79.427
PHIDOT = -.5050222E+00RAD/S       PHIDDOT = .6855812E+01RAD/S*S
THE VALUE OF TH IS    60.000  PH IS   83.251
PHIDOT = -.9882109E+00RAD/S       PHIDDOT = .4337674E+01RAD/S*S

..........................................

THE VALUE OF TH IS   315.000  PH IS  122.475
PHIDOT = .1745197E+01RAD/S        PHIDDOT = -.5251945E+01RAD/S*S
THE VALUE OF TH IS   330.000  PH IS  112.639
PHIDOT = .2173162E+01RAD/S        PHIDDOT = -.4106554E+01RAD/S*S
THE VALUE OF TH IS   345.000  PH IS  101.145
PHIDOT = .2355762E+01RAD/S        PHIDDOT = .5844062E+00RAD/S*S
THE VALUE OF TH IS   360.000  PH IS   90.000
PHIDOT = .2000000E+01RAD/S        PHIDDOT = .7500001E+01RAD/S*S
```

Figure 5.15

(see Section 2.7). Take, for example, the motion depicted in Figure 5.3. The solutions for the velocity and acceleration of point D are given by equations (5.34) and (5.35), respectively. Simplifications would be possible in this problem if the rotations could be divided up; i.e., suppose one could separate the motions so that only the simple circular rotation of the rod relative to the rotating frame $OFGHIA$ was studied. The results for velocity and acceleration would easily be obtained since the rod describes plane circular motion relative to the frame. If some correction could then be added to the terms present to take into account that the frame does rotate, the solution to this complex kinematic problem would have been accomplished. The intent of this section is to show that this is feasible and how other problems like this can be treated. Before proceeding further, an even simpler problem will be discussed to illustrate the usefulness of this concept. In Figure 5.16, a rotating rod OAE bent into the shape of a semicircular hook is depicted. Sliding along the rod is a particle P. All motion occurs in a plane.

For discussion purposes, it will be assumed that ω_{OA} and u are constant. u is the sliding velocity of the particle relative to the rotating hook. The absolute motion of P is complicated. *It describes circular motion on a path that rotates itself.* The analysis would also be simplified here if the simple circular motion on hook OAE were first isolated and studied, and then compensation was made for the hook's actual rotation.

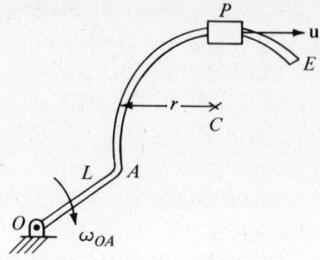

Figure 5.16

Still another example is shown in Figure 5.17. Here an observer at O, fixed to the surface of the earth and rotating with it at an angular rate ω_e, can view and measure the motion of a space object at P. The changes in the relative position vector $\mathbf{r}_P$ that the observer views are not absolute changes. Since the observer rotates, his or her velocity and acceleration calculations must be compensated for.

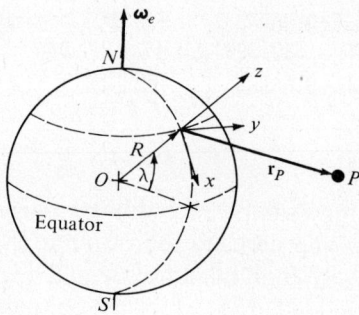

Figure 5.17

B. Velocity Analysis

In Figure 5.18, two coordinate systems are shown. The coordinate axes at point O are fixed and nonrotating, whereas the coordinate set with origin at B has a frame of reference that translates and rotates. It is clear that

$$\mathbf{r}_P = \mathbf{r}_B + \mathbf{r}_{P/B} \tag{5.60}$$

$$\mathbf{v}_P = \mathbf{v}_B + \mathbf{v}_{P/B} \tag{5.61}$$

If $\mathbf{r}_{P/B}$ is expressed in terms of $\mathbf{i}, \mathbf{j}, \mathbf{k}$, then

$$\mathbf{r}_{P/B} = x\mathbf{i} + y\mathbf{j} + z\mathbf{k} \tag{5.62}$$

$$\mathbf{v}_{P/B} = \dot{x}\mathbf{i} + \dot{y}\mathbf{j} + \dot{z}\mathbf{k} + x\dot{\mathbf{i}} + y\dot{\mathbf{j}} + z\dot{\mathbf{k}} \tag{5.63}$$

The last three terms signify that the unit vectors rotate and their derivatives exist. Because a unit vector is of constant length, its derivative has the same

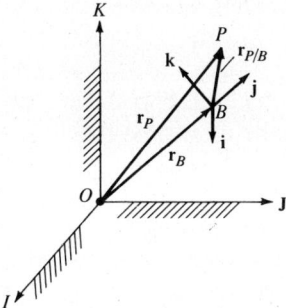

Figure 5.18

form as that established previously for the derivatives of other *constant length* vectors [see equations (5.5) and (5.30)], namely,

$$\dot{\mathbf{i}} = \boldsymbol{\omega} \times \mathbf{i} \tag{5.64}$$

where ω is the rotation rate of $\mathbf{i}$. If now the unit vectors $\mathbf{i}, \mathbf{j}, \mathbf{k}$ forming a coordinate system in a rotating reference frame are considered, then

$$\dot{\mathbf{i}} = \boldsymbol{\Omega} \times \mathbf{i} \tag{5.65}$$
$$\dot{\mathbf{j}} = \boldsymbol{\Omega} \times \mathbf{j} \tag{5.66}$$
$$\dot{\mathbf{k}} = \boldsymbol{\Omega} \times \mathbf{k} \tag{5.67}$$

where $\boldsymbol{\Omega}$ is the angular velocity of the x, y, z frame. With the help of these results, equation (5.63) can be simplified.

$$\mathbf{v}_{P/B} = \dot{x}\mathbf{i} + \dot{y}\mathbf{j} + \dot{z}\mathbf{k} + x(\boldsymbol{\Omega} \times \mathbf{i}) + y(\boldsymbol{\Omega} \times \mathbf{j}) + z(\boldsymbol{\Omega} \times \mathbf{k}) \tag{5.68}$$
$$\mathbf{v}_{P/B} = \dot{x}\mathbf{i} + \dot{y}\mathbf{j} + \dot{z}\mathbf{k} + \boldsymbol{\Omega} \times (x\mathbf{i} + y\mathbf{j} + z\mathbf{k}) \tag{5.69}$$
$$\mathbf{v}_{P/B} = \dot{x}\mathbf{i} + \dot{y}\mathbf{j} + \dot{z}\mathbf{k} + \boldsymbol{\Omega} \times (\mathbf{r}_{P/B}) \tag{5.70}$$

The first three terms on the right-hand side of equation (5.70) have special meaning. They represent the instantaneous velocity of P relative to the local reference frame. They are the rate of change of the components of the relative position vector referred to the local reference coordinates axes that rotate. Equation (5.70) can be rewritten as

$$\mathbf{v}_{P/B} = [\mathbf{v}_{P/B}]_{\text{rot}} + \boldsymbol{\Omega} \times \mathbf{r}_{P/B} \tag{5.71}$$

where now the first term on the right-hand side (in brackets) represents the velocity of P relative to the moving frame at B that rotates with absolute angular velocity $\boldsymbol{\Omega}$. The cross product can be thought of as representing the compensation needed to correct the aforementioned term to attain the "true" relative velocity. The true relative velocity is observed by all *translating* observers. See Section 2.7.

This result is very significant and can be generalized.

$$\frac{d(\ )}{dt} = \left[\frac{d(\ )}{dt}\right]_{rot} + \mathbf{\Omega} \times (\ ) \tag{5.72}$$

Equation (5.72) states that the total time derivative of any vector is equal to the time derivative referenced to a rotating frame plus the cross product of the absolute angular velocity of the reference frame and the vector itself.

Once the results from equation (5.71) are known, they can be added to the velocity of B to find the total velocity of P. The velocity of B can be obtained by a straightforward calculation since $\mathbf{r}_B$ is described in a nonrotating frame [see equation (5.61)]. This may require vector transformations if $\mathbf{v}_B$ is expressed using $\mathbf{I}, \mathbf{J}, \mathbf{K}$ components and $[\mathbf{v}_{P/B}]_{rot}$ is expressed in $\mathbf{i}, \mathbf{j}, \mathbf{k}$ terms. Total solutions, of course, can finally be expressed in either set of coordinates, one set not having any more validity than the other. The important thing to remember is that in adding terms in equation (5.61) *all* the vectors must first be referred to a common base triad.

C. Vector Transformations

Before proceeding to a discussion of acceleration analysis with respect to a rotating reference frame, a development of vector transformation theory will prove useful. Shown in Figure 5.19 are two coordinate sets having a common origin but different orientations.

For one coordinate system,

$$\mathbf{r} = X\mathbf{I} + Y\mathbf{J} + Z\mathbf{K} \tag{5.73}$$

For the other,

$$\mathbf{r} = x\mathbf{i} + y\mathbf{j} + z\mathbf{k} \tag{5.74}$$

Then
$$\begin{aligned} x &= \mathbf{r}\cdot\mathbf{i} = X\mathbf{I}\cdot\mathbf{i} + Y\mathbf{J}\cdot\mathbf{i} + Z\mathbf{K}\cdot\mathbf{i} \\ y &= \mathbf{r}\cdot\mathbf{j} = X\mathbf{I}\cdot\mathbf{j} + Y\mathbf{J}\cdot\mathbf{j} + Z\mathbf{K}\cdot\mathbf{j} \\ z &= \mathbf{r}\cdot\mathbf{k} = X\mathbf{I}\cdot\mathbf{k} + Y\mathbf{J}\cdot\mathbf{k} + Z\mathbf{K}\cdot\mathbf{k} \end{aligned} \tag{5.75}$$

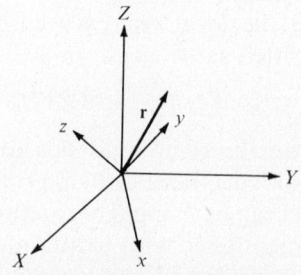

Figure 5.19

Standard notation has

$$\begin{aligned}
\mathbf{i}\cdot\mathbf{I} = \mathbf{I}\cdot\mathbf{i} = l_{xX} & \quad \mathbf{i}\cdot\mathbf{J} = l_{xY} & \quad \mathbf{i}\cdot\mathbf{K} = l_{xZ} \\
\mathbf{j}\cdot\mathbf{I} = l_{yX} & \quad \mathbf{j}\cdot\mathbf{J} = l_{yY} & \quad \mathbf{j}\cdot\mathbf{K} = l_{yZ} \\
\mathbf{k}\cdot\mathbf{I} = l_{zX} & \quad \mathbf{k}\cdot\mathbf{J} = l_{zY} & \quad \mathbf{k}\cdot\mathbf{K} = l_{zZ}
\end{aligned} \quad (5.76)$$

where the l's denote direction cosines. In matrix form one can write

$$\begin{Bmatrix} x \\ y \\ z \end{Bmatrix} = \begin{bmatrix} l_{xX} & l_{xY} & l_{xZ} \\ l_{yX} & l_{yY} & l_{yZ} \\ l_{zX} & l_{zY} & l_{zZ} \end{bmatrix} \begin{Bmatrix} X \\ Y \\ Z \end{Bmatrix} \quad (5.77)$$

or in compact form

$$\{r'\} = [l]\{r\} \quad (5.78)$$

Equation (5.78) formally describes how a vector is transformed from one coordinate set to another. The transpose of $[l]$ helps to reverse the transformation:

$$[l]^t = \begin{bmatrix} l_{xX} & l_{yX} & l_{zX} \\ l_{xY} & l_{yY} & l_{zY} \\ l_{xZ} & l_{yZ} & l_{zZ} \end{bmatrix} \quad (5.79)$$

Through simple trigonometric identities, one can show that

$$[l][l]^t = [1]$$

where $[1]$ is the identity matrix. Then

$$[l]^t = [l]^{-1} \quad (5.80)$$

since

$$[l][l]^{-1} = [l]^{-1}[l] \equiv [1] \quad (5.81)$$

This means that since

$$\{r'\} = [l]\{r\} \quad (5.78)$$

$$\{r\} = [l]^t\{r'\} = [l]^{-1}\{r'\} \quad (5.82)$$

and the reverse transformation is accomplished.

Refer now to Figure 5.20 to clarify this discussion. In this figure two coordinate sets separated by the one angle θ are shown. The transformation takes place in a plane. The vector $\mathbf{r}$ has the following components in the $\mathbf{I}, \mathbf{J}, \mathbf{K}$ system:

$$\mathbf{r} = X\mathbf{I} + Y\mathbf{J} + Z\mathbf{K} \quad (5.83)$$

These components can be projected to the $\mathbf{i}, \mathbf{j}, \mathbf{k}$ coordinate system so that

$$\begin{aligned} x &= X\cos\theta + Y\sin\theta \\ y &= -X\sin\theta + Y\cos\theta \\ z &= Z \end{aligned} \quad (5.84)$$

128   REVIEW OF NEWTONIAN MECHANICS

Then
$$\begin{Bmatrix} x \\ y \\ z \end{Bmatrix} = \begin{bmatrix} \cos\theta & \sin\theta & 0 \\ -\sin\theta & \cos\theta & 0 \\ 0 & 0 & 1 \end{bmatrix} \begin{Bmatrix} X \\ Y \\ Z \end{Bmatrix} \quad (5.85)$$

In compact form
$$\{r'\} = [l]\{r\} \quad (5.78)$$

where now
$$[l] = \begin{bmatrix} \cos\theta & \sin\theta & 0 \\ -\sin\theta & \cos\theta & 0 \\ 0 & 0 & 1 \end{bmatrix} \quad (5.86)$$

and
$$[l]^t = \begin{bmatrix} \cos\theta & -\sin\theta & 0 \\ \sin\theta & \cos\theta & 0 \\ 0 & 0 & 1 \end{bmatrix} \quad (5.87)$$

The reader can verify that
$$[l][l]^t = [l]^t[l] = [1]$$

and thus
$$[l]^t = [l]^{-1} \quad (5.80)$$

Then
$$\begin{Bmatrix} X \\ Y \\ Z \end{Bmatrix} = \begin{bmatrix} \cos\theta & -\sin\theta & 0 \\ \sin\theta & \cos\theta & 0 \\ 0 & 0 & 1 \end{bmatrix} \begin{Bmatrix} x \\ y \\ z \end{Bmatrix} \quad (5.88)$$

or
$$\{r\} = [l]^t\{r'\} \quad (5.82)$$

Expanded, equation (5.88) becomes
$$\begin{aligned} X &= x\cos\theta - y\sin\theta \\ Y &= x\sin\theta + y\cos\theta \\ Z &= z \end{aligned} \quad (5.89)$$

Equation (5.89) can also be derived directly from Figure 5.20.

In all that is said here, one must keep in mind that the vector **r** does not

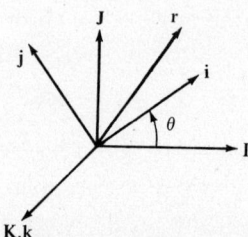

Figure 5.20

change itself in this transformation (due to a coordinate system rotation), only its components change, that is,

$$\mathbf{r} = X\mathbf{I} + Y\mathbf{J} + Z\mathbf{K} = x\mathbf{i} + y\mathbf{j} + z\mathbf{k} \tag{5.90}$$

A linear transformation, such as the one discussed here, where a vector does not change in magnitude is called an *orthogonal transformation*. If $\mathbf{r}$ does not change in magnitude,

$$X^2 + Y^2 + Z^2 = x^2 + y^2 + z^2 \tag{5.91}$$

In matrix form this can be stated as

$$\begin{aligned} \{r'\}^t \{r'\} &= \{r\}^t \{r\} \\ \{r'\}^t [l] \{r\} &= \{r\}^t \{r\} \\ \{r\}^t [l]^t [l] \{r\} &= \{r\}^t \{r\} \end{aligned} \tag{5.92}$$

The last equation requires that

$$[l]^t [l] = [1] = [l][l]^t \tag{5.93}$$

a result established earlier. Equation (5.93) represents the *orthogonality conditions*. The matrix $[l]$ is an *orthogonal matrix*. Equation (5.78) represents an *orthogonal transformation*.

Even though the discussion of orthogonality has been limited to the direction cosine matrix $[l]$, it holds true for the general case, that is,

$$\{r'\} = [A]\{r\} \tag{5.94}$$

For this to be an orthogonal transformation,

$$[A]^t [A] = [A][A]^t = [1] \tag{5.95}$$

If orthogonal, the general linear transformation in equation (5.94) can represent the relationship between the components of the vector in two coordinate systems that are separated by a simple rotation as discussed before, or $[A]$ can represent the pure rotation of the vector itself. Equation (5.94) would then describe the new components of the vector $\mathbf{r}$ in the *same* coordinate system. The magnitude of $\mathbf{r}$ would remain constant. For motion in a plane, the angle θ in Figure 5.20 was shown counterclockwise to represent coordinate rotation. In Figure 5.21 the same angle θ is now shown to act clockwise for actual vector rotation.

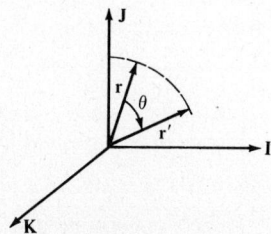

Figure 5.21

One final comment: In order for a general linear transformation to represent a rotation, the necessary orthogonality conditions established in equation (5.93) must be met. In addition, it will be shown in Section 6.3 that the determinant of the transformation matrix must also have the value $+1$. A determinant with the value -1 would represent an inversion of the coordinate axes.

Returning now to the simple transformation described by equation (5.85), the orthogonality conditions are

$$[l][l]^t = [1] \qquad (5.96)$$

and in expanded form they become simply

$$\begin{aligned} \cos^2 \theta + \sin^2 \theta &= 1 \\ -\sin \theta \cos \theta + \cos \theta \sin \theta &= 0 \\ \sin^2 \theta + \cos^2 \theta &= 1 \end{aligned} \qquad (5.97)$$

where zero terms and redundant equations have been ignored. Here

$$\begin{aligned} l_{zX} &= l_{zY} = l_{xZ} = l_{yZ} = 0 \\ l_{zZ} &= 1 \\ l_{xX} &= \cos \theta, \quad l_{yX} = -\sin \theta, \quad l_{xY} = \sin \theta, \quad l_{yY} = \cos \theta \end{aligned} \qquad (5.98)$$

Note that the four direction cosines represented by the last of equations (5.98) are not independent. The three orthogonality conditions represent constraints that reduce the degrees of freedom from 4 to 1. The one remaining degree of freedom represents a simple rotation. In the general case with nine direction cosines, there will be six orthogonality conditions; equation (5.96) represents nine equations, but three are redundant. These six constraints reduce the number of degrees of freedom to 3. These 3 general degrees of freedom characterize the rotational freedom of a rigid body. What is perhaps significant here is the inappropriateness of using direction cosines as coordinates in describing rigid body motion. The direction cosines are not all independent and will prove very cumbersome to work with.

Some other interesting results can also be established. The determinant of $[l]$ in this example is

$$\text{Det } [l] = \cos^2 \theta + \sin^2 \theta = +1 \qquad (5.99)$$

which was predicted by the general theory. Also note that if the vector *itself* *rotates* clockwise through a finite angle θ, the new components of the vector in the same coordinates system, x', y', z' become

$$\begin{Bmatrix} x' \\ y' \\ z' \end{Bmatrix} = \begin{bmatrix} \cos \theta & \sin \theta & 0 \\ -\sin \theta & \cos \theta & 0 \\ 0 & 0 & 1 \end{bmatrix} \begin{Bmatrix} x \\ y \\ z \end{Bmatrix} \qquad (5.100)$$

or

$$\{r_n\} = [l]\{r\}$$

For an *infinitesimal rotation* $d\theta_z$,

$$\begin{Bmatrix} x' \\ y' \\ z' \end{Bmatrix} \simeq \begin{bmatrix} 1 & d\theta_z & 0 \\ -d\theta_z & 1 & 0 \\ 0 & 0 & 1 \end{bmatrix} \begin{Bmatrix} x \\ y \\ z \end{Bmatrix} \tag{5.101}$$

where $[l]$ can be reduced to

$$[l] \simeq \begin{bmatrix} 1 & d\theta_z & 0 \\ -d\theta_z & 1 & 0 \\ 0 & 0 & 1 \end{bmatrix} \tag{5.102}$$

Finite rotations have matrices which do not commute, and the order of their application becomes critical; i.e., finite rotations do not add as vectors since the resultant depends on the rotation sequence. Infinitesimal rotation matrices do commute, and therefore, the rotations themselves act as vectors. Given one infinitesimal rotation $d\theta_z$,

$$[l_1] = \begin{bmatrix} 1 & d\theta_z & 0 \\ -d\theta_z & 1 & 0 \\ 0 & 0 & 1 \end{bmatrix} = [1] + [d\theta_z] \tag{5.103}$$

followed by another rotation $d\theta_y$,

$$[l_2] = \begin{bmatrix} 1 & 0 & -d\theta_y \\ 0 & 1 & 0 \\ d\theta_y & 0 & 1 \end{bmatrix} = [1] + [d\theta_y] \tag{5.104}$$

the total resultant rotation matrix is independent of the rotation order. This can easily be shown since

$$\{r_n\} = [l_2][l_1]\{r\} = ([1] + [d\theta_z] + [d\theta_y])\{r\} \tag{5.105}$$

is identical to

$$\{r_n\} = [l_1][l_2]\{r\} = ([1] + [d\theta_y] + [d\theta_z])\{r\} \tag{5.106}$$

when second-order terms are neglected. Further

$$\{r_n\} - \{r\} = \{dr\} = ([l] - [1])\{r\} \tag{5.107}$$

For the one rotation $d\theta_z$,

$$\begin{Bmatrix} dx \\ dy \\ dz \end{Bmatrix} = \begin{bmatrix} 0 & d\theta_z & 0 \\ -d\theta_z & 0 & 0 \\ 0 & 0 & 0 \end{bmatrix} \begin{Bmatrix} x \\ y \\ z \end{Bmatrix} \tag{5.108}$$

$$\left\{\begin{array}{c}\dfrac{dx}{dt}\\ \dfrac{dy}{dt}\\ \dfrac{dz}{dt}\end{array}\right\}=\begin{bmatrix} 0 & \dfrac{d\theta_z}{dt} & 0 \\ -\dfrac{d\theta_z}{dt} & 0 & 0 \\ 0 & 0 & 0 \end{bmatrix}\left\{\begin{array}{c}x\\ y\\ z\end{array}\right\} \quad (5.109)$$

By changing the sign of $d\theta_z/dt$ to correspond to a positive counterclockwise rotation of the vector **r** and defining $d\theta_z/dt$ to be ω_z, we obtain

$$\left\{\begin{array}{c}v_x\\ v_y\\ v_z\end{array}\right\}=\begin{bmatrix} 0 & -\omega_z & 0 \\ \omega_z & 0 & 0 \\ 0 & 0 & 0 \end{bmatrix}\left\{\begin{array}{c}x\\ y\\ z\end{array}\right\} \quad (5.110)$$

This is the matrix form of the vector expression

$$\mathbf{v}=\omega_z \mathbf{k} \times \mathbf{r} \quad (5.111)$$

In the general case

$$\left\{\begin{array}{c}v_x\\ v_y\\ v_z\end{array}\right\}=\begin{bmatrix} 0 & -\omega_z & \omega_y \\ \omega_z & 0 & -\omega_x \\ -\omega_y & \omega_z & 0 \end{bmatrix}\left\{\begin{array}{c}x\\ y\\ z\end{array}\right\} \quad (5.112)$$

or

$$\mathbf{v}=\boldsymbol{\omega} \times \mathbf{r} \quad (5.113)$$

D. Acceleration Analysis

In Section 5.6B, a velocity expression developed for analysis using a rotating reference frame was derived. That expression, repeated here, is

$$\mathbf{v}_{P/B}=[\mathbf{v}_{P/B}]_{\text{rot}}+\boldsymbol{\Omega}\times(\mathbf{r}_{P/B}) \quad (5.71)$$

Furthermore, this approach was generalized so that the total time derivative referred to a nonrotating reference frame of any arbitrary vector could be obtained from derivative calculations made in a rotating reference frame. The expression that provided the transformation was

$$\frac{d(\ )}{dt}=\left[\frac{d(\ )}{dt}\right]_{\text{rot}}+\boldsymbol{\Omega}\times(\ ) \quad (5.72)$$

It is the purpose of this section to develop an acceleration equation that can be used when analysis with reference to a rotating frame is preferred. The time derivative of equation (5.71) will be taken with the aid of equation (5.72).

$$\frac{d\mathbf{v}_{P/B}}{dt}=\frac{d}{dt}[\mathbf{v}_{P/B}]_{\text{rot}}+\frac{d\boldsymbol{\Omega}}{dt}\times(\mathbf{r}_{P/B})+\boldsymbol{\Omega}\times\mathbf{v}_{P/B}$$

$$\mathbf{a}_{P/B}=\frac{d}{dt}[\mathbf{v}_{P/B}]_{\text{rot}}+\dot{\boldsymbol{\Omega}}\times\mathbf{r}_{P/B}+\boldsymbol{\Omega}\times([\mathbf{v}_{P/B}]_{\text{rot}}+\boldsymbol{\Omega}\times\mathbf{r}_{P/B})$$

RIGID BODY DYNAMICS: KINEMATICS   133

The "arbitrary" vector in equation (5.72) will be $[\mathbf{v}_{P/B}]_{\text{rot}}$.

$$\mathbf{a}_{P/B} = \left[\frac{d}{dt}[\mathbf{v}_{P/B}]_{\text{rot}}\right]_{\text{rot}} + \mathbf{\Omega} \times [\mathbf{v}_{P/B}]_{\text{rot}} + \dot{\mathbf{\Omega}} \times \mathbf{r}_{P/B} + \mathbf{\Omega} \times ([\mathbf{v}_{P/B}]_{\text{rot}} + \mathbf{\Omega} \times \mathbf{r}_{P/B})$$

$$\mathbf{a}_{P/B} = [\mathbf{a}_{P/B}]_{\text{rot}} + 2\mathbf{\Omega} \times [\mathbf{v}_{P/B}]_{\text{rot}} + \dot{\mathbf{\Omega}} \times \mathbf{r}_{P/B} + \mathbf{\Omega} \times (\mathbf{\Omega} \times \mathbf{r}_{P/B}) \quad (5.114)$$

Adding the motion of B to the velocity and acceleration expressions yields

$$\mathbf{v}_P = \mathbf{v}_B + [\mathbf{v}_{P/B}]_{\text{rot}} + \mathbf{\Omega} \times \mathbf{r}_{P/B} \quad (5.115)$$

$$\mathbf{a}_P = \mathbf{a}_B + [\mathbf{a}_{P/B}]_{\text{rot}} + 2\mathbf{\Omega} \times [\mathbf{v}_{P/B}]_{\text{rot}} + \dot{\mathbf{\Omega}} \times \mathbf{r}_{P/B} + \mathbf{\Omega} \times (\mathbf{\Omega} \times \mathbf{r}_{P/B}) \quad (5.116)$$

A summary of the terms in equation (5.116) follows:

$\mathbf{a}_P$ = absolute instantaneous acceleration of point P (Absolute acceleration is defined with respect to a nonrotating, non-accelerating reference frame.)

$\mathbf{a}_B$ = absolute instantaneous acceleration of origin B of moving "local" reference axes

$[\mathbf{a}_{P/B}]_{\text{rot}}$ = instantaneous acceleration of point P measured by any observer in local reference frame (The term represents the component changes with respect to time of the relative velocity terms referred to the local reference coordinate system. It is not the total relative acceleration because of the coordinate axes rotation. It is the acceleration of P in the local reference frame at B that rotates.)

$2\mathbf{\Omega} \times [\mathbf{v}_{P/B}]_{\text{rot}} = \mathbf{a}_{\text{Coriolis}}$ = Coriolis term, named after De Coriolis (1792–1843) (This term is needed to couple the rotating reference frame to that which is nonrotating if motion exists in the local frame.)

$\mathbf{\Omega}, \dot{\mathbf{\Omega}}$ = absolute instantaneous angular velocity, angular acceleration of rotating reference frame

$[\mathbf{v}_{P/B}]_{\text{rot}}$ = instantaneous velocity of point P relative to local reference frame (It is similar to $[\mathbf{a}_{P/B}]_{\text{rot}}$ in perspective; it is the velocity seen by an observer in the local reference frame at B that rotates.)

$\mathbf{r}_{P/B}$ = relative position vector from origin of local coordinate set at B to point P under study

$\mathbf{\Omega} \times \mathbf{r}_{P/B}$ = instantaneous velocity of a coincident point at P, *fixed* to a rotating frame, as seen by a *translating* observer having velocity $\mathbf{v}_B$

$\mathbf{\Omega} \times (\mathbf{\Omega} \times \mathbf{r}_{P/B})$ = instantaneous acceleration of the coincident point at P,
$+ \dot{\mathbf{\Omega}} \times \mathbf{r}_{P/B}$ *fixed* to the rotating frame, as seen by a *translating* observer having the acceleration $\mathbf{a}_B$

As before, all terms must have their components referred to the same coordinate system if they are to be added together. The absolute vector representing the acceleration of point B will still remain the absolute acceleration of B even when its *components* are ultimately found in the local reference coordinate system. What determines whether a derivative is absolute or not has to do with the position vector used and whether the time derivative was taken in a non-rotating frame. It has nothing to do with which of the infinite possible coordinate systems the vector is resolved in.

The terms $[\mathbf{a}_{P/B}]_{\text{rot}}$ and $[\mathbf{v}_{P/B}]_{\text{rot}}$ are easier to understand if one tries to visualize a relative path fixed in the local frame (thus rotating) along which the point P moves. Refer again to the wire shown in Figure 5.16 and redrawn in Figure 5.22. If the local frame of reference is attached to the wire and rotates with it, the relative path APE fixed to this frame is a half-circle. The term $[\mathbf{v}_{P/B}]_{\text{rot}}$ represents the velocity of particle P as it moves along the relative path: a half-circle. Its relative velocity is given as $\mathbf{u}$, and it must be *tangent to the relative path*. The acceleration $[\mathbf{a}_{P/B}]_{\text{rot}}$ can be viewed in the same manner. Since the relative path is a half-circle, the acceleration $[\mathbf{a}_{P/B}]_{\text{rot}}$ would have to reflect a normal acceleration term and a possible tangential term.

$$[\mathbf{a}_{P/B}]_{\text{rot}} = [\mathbf{a}_{P/B}]_{\text{rot}}^n + [\mathbf{a}_{P/B}]_{\text{rot}}^t \quad (5.117)$$

$$[a_{P/B}]_{\text{rot}}^n = \frac{u^2}{r} \quad \text{(toward } C\text{)} \quad (5.118)$$

$$[a_{P/B}]_{\text{rot}}^t = \dot{u} \quad \text{(tangent to the path)} \quad (5.119)$$

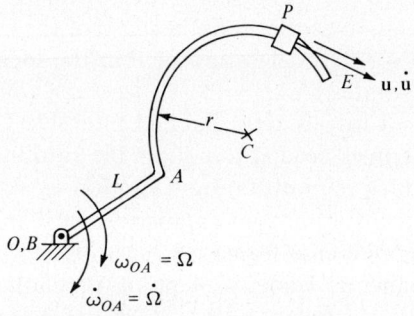

Figure 5.22

The last term, tangent to the relative path, represents the change in speed relative to the path. The normal acceleration term, equation (5.118), has the magnitude u^2/r. u represents the relative speed along the path. The local radius of curvature of the path is r; *it is not the distance* from B to point P shown in Figure 5.22. The *normal acceleration term points toward the local center of curvature of the relative path*. In Figure 5.22, it should point from P to C. If the relative path had been a straight line, as shown in Figure 5.23, $[a_{P/B}]_{\text{rot}}^n$ would be zero.

The last two expressions in equation (5.116) have the following significance for this plane motion problem:

1. $\dot{\boldsymbol{\Omega}} \times \mathbf{r}_{P/B}$ is equal to the tangential acceleration of the coincident point at P fixed to the rotating frame. This acceleration term is perpendicular to the vector drawn from O to P.

RIGID BODY DYNAMICS: KINEMATICS   135

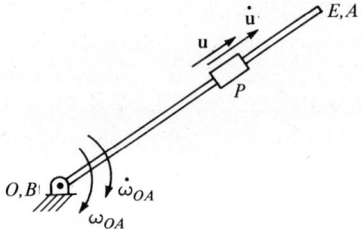

Figure 5.23

2. $\Omega \times (\Omega \times r_{P/B})$ is equal to the normal acceleration of the coincident point at P fixed to the rotating frame. This acceleration points from P toward O.

Sample Problem 5.8
Shown in the figure is a Scotch-yoke mechanism. The disk of radius $R = 12$ cm rotates at a constant angular velocity of 800 rpm counterclockwise. The slider at P is pinned to the disk; it is constrained to move in the slot of the yoke. Find the velocity and acceleration of the yoke for this instant.

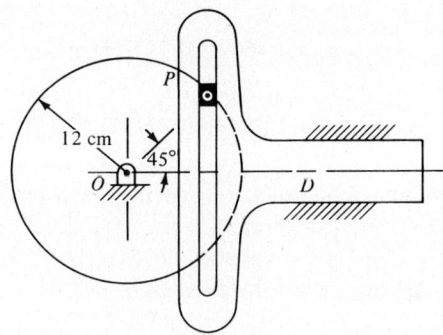

Analysis: Equations (5.115) and (5.116) may be applied; namely,

$$\mathbf{v}_P = \mathbf{v}_D + [\mathbf{v}_{P/D}]_{rot} + \Omega \times \mathbf{r}_{P/D}$$
$$\mathbf{a}_P = \mathbf{a}_D + [\mathbf{a}_{P/D}]_{rot} + \Omega \times (\Omega \times \mathbf{r}_{P/D}) + \dot{\Omega} \times \mathbf{r}_{P/D} + 2\Omega \times [\mathbf{v}_{P/D}]_{rot}$$

For this problem with a nonrotating yoke, and with the reference frame on the yoke,

$$\mathbf{v}_P = \mathbf{v}_D + [\mathbf{v}_{P/D}]_{rot} \qquad \text{(a)}$$

$$v_P = \omega_{OP} r_P = 800 \left(\frac{2\pi}{60}\right)(12) = 1005.4 \quad \text{cm/s}$$

$$\mathbf{v}_P = 1005.4(-\sin 45°\mathbf{i} + \cos 45°\mathbf{j}) = 710(-\mathbf{i} + \mathbf{j})$$

or
$$\mathbf{v}_P = \omega_{OP} \times \mathbf{r}_P = 83.8\mathbf{k} \times 12(\cos 45°\mathbf{i} + \sin 45°\mathbf{j}) = 710(-\mathbf{i} + \mathbf{j})$$

136 REVIEW OF NEWTONIAN MECHANICS

Combining terms, with $[\mathbf{v}_{P/D}]_{\text{rot}} = u\mathbf{j}$,

$$\mathbf{v}_P = -710\mathbf{i} + 710\mathbf{j} = v_D\mathbf{i} + u\mathbf{j}$$

$$\mathbf{v}_D = -710\mathbf{i} \quad \text{cm/s}$$

$$[\mathbf{v}_{P/D}]_{\text{rot}} = \mathbf{u} = 710\mathbf{j} \quad \text{cm/s}$$

For the accelerations,

$$\mathbf{a}_P = \mathbf{a}_D + [\mathbf{a}_{P/D}]_{\text{rot}} \tag{b}$$

$$\mathbf{a}_P = (\mathbf{a}_P)_n + (\mathbf{a}_P)_t \quad \text{(circular motion)}$$

$$(a_P)_n = \omega_{OP}^2 r_P$$

$$(a_P)_n = \left[800\left(\frac{2\pi}{60}\right)\right]^2 (12) = 59\,552(-\mathbf{i} - \mathbf{j}) \quad \text{cm/s}^2$$

$$(a_P)_t = \alpha_{OP} r_P = 0$$

$$\mathbf{a}_D = a_D\mathbf{i}$$

$$[\mathbf{a}_{P/D}]_{\text{rot}} = \dot{u}\mathbf{j}$$

Then from equation (b),

$$\mathbf{a}_D = -59\,552\mathbf{i} \quad \text{cm/s}^2$$

$$[\mathbf{a}_{P/D}]_{\text{rot}} = \dot{\mathbf{u}} = 59\,552\mathbf{j} \quad \text{cm/s}^2 \quad \blacksquare$$

Sample Problem 5.9
Shown below is a Whitworth quick-return mechanism. Crank OP of 12-cm length rotates at 1000 rpm counterclockwise. The slider at P is pinned to the crank but is free to rotate; it slides over rod AB. For the instant shown, determine the angular velocity and acceleration of rod AB.

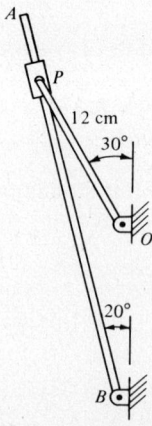

RIGID BODY DYNAMICS: KINEMATICS

Analysis: First the geometry of the problem must be worked out. Using the law of sines,

$$\frac{12}{\sin 20°} = \frac{BP}{\sin 150°} \qquad BP = 17.54 \text{ cm}$$

Then with $\mathbf{v}_B = \mathbf{a}_B = \mathbf{0}$,

$$\mathbf{v}_P = [\mathbf{v}_{P/B}]_{\text{rot}} + \mathbf{\Omega} \times \mathbf{r}_P \tag{a}$$

$$\mathbf{a}_P = [\mathbf{a}_{P/B}]_{\text{rot}} + \mathbf{\Omega} \times (\mathbf{\Omega} \times \mathbf{r}_P) + \dot{\mathbf{\Omega}} \times \mathbf{r}_P + \mathbf{a}_{\text{Coriolis}} \tag{b}$$

$$\mathbf{a}_{\text{Coriolis}} = 2\mathbf{\Omega} \times [\mathbf{v}_{P/B}]_{\text{rot}}$$

Starting with the velocity analysis,

$$\mathbf{v}_P = \boldsymbol{\omega}_{OP} \times \mathbf{r}_P \qquad v_P = \omega_{OP} r_P$$

$$v_P = 1000 \left(\frac{2\pi}{60}\right)(12) = 1256.6$$

$$\mathbf{v}_P = -1088.2\mathbf{i} - 628.4\mathbf{j} \quad \text{cm/s}$$

$$[\mathbf{v}_{P/B}]_{\text{rot}} = u(-\sin 20°\mathbf{i} + \cos 20°\mathbf{j}) = -0.34u\mathbf{i} + 0.94u\mathbf{j}$$

$$\mathbf{\Omega} \times \mathbf{r}_P = \boldsymbol{\omega}_{AB} \times \mathbf{r}_P = \omega_{AB}(-16.48\mathbf{i} - 6\mathbf{j})$$

Combining terms in equation (a),

$$-1088.2\mathbf{i} - 628.4\mathbf{j} = -0.34u\mathbf{i} + 0.94u\mathbf{j} - \omega_{AB}(16.48)\mathbf{i} - \omega_{AB}(6)\mathbf{j}$$

leads to two scalar equations:

$$-1088.2 = -0.34u - \omega_{AB}(16.48)$$

$$-628.4 = 0.94u - \omega_{AB}(6)$$

$$\omega_{AB} = \frac{0.94(-1088.2) - 0.34(628.4)}{-16.48(0.94) - 6(0.34)}$$

$$\omega_{AB} = 70.5\mathbf{k} \quad \text{rad/s}$$

$$u = \frac{6(1088.2) - 16.48(628.4)}{0.34(6) + 0.94(16.48)}$$

$$\mathbf{u} = 218.2(\sin 20°\mathbf{i} - \cos 20°\mathbf{j}) = 74.6\mathbf{i} - 205\mathbf{j} \quad \text{cm/s}$$

In the first part of this problem, it became obvious that the vertical and horizontal coordinate axes used were not the most appropriate to describe the motion. *New coordinate axes* fixed at *B* in the local frame but oriented *along AB* will be used instead for the acceleration analysis.

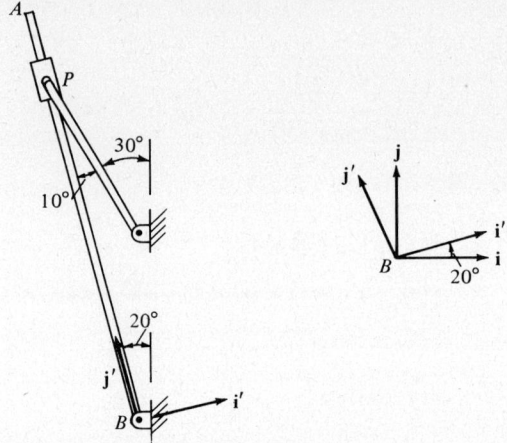

$$\mathbf{a}_P = [\mathbf{a}_{P/B}]_{\text{rot}} + \mathbf{\Omega} \times (\mathbf{\Omega} \times \mathbf{r}_P) + \dot{\mathbf{\Omega}} \times \mathbf{r}_P + \mathbf{a}_{\text{Coriolis}} \tag{b}$$

$$\mathbf{a}_P = (\mathbf{a}_P)_n + (\mathbf{a}_P)_t$$

$$(a_P)_n = \omega_{OP}^2 r_P = 104.7^2(0.12)$$

$$(\mathbf{a}_P)_n = 1315.5(\sin 10°\mathbf{i}' - \cos 10°\mathbf{j}') = 228.4\mathbf{i}' - 1295.4\mathbf{j}' \quad \text{m/s}^2$$

and
$$(a_P)_t = \dot{\omega}_{OP} r_P = \alpha_{OP} r_P$$

$$(\mathbf{a}_P)_t = \mathbf{0} \quad (\omega_{OP} = \text{constant})$$

Furthermore,

$$\mathbf{\Omega} \times (\mathbf{\Omega} \times \mathbf{r}_P) = \boldsymbol{\omega}_{AB} \times (\boldsymbol{\omega}_{AB} \times \mathbf{r}_P) = \omega_{AB}^2 \mathbf{r}_P$$

$$= -(70.5^2)(0.175)\mathbf{j}' = -871.8\mathbf{j}' \quad \text{m/s}^2$$

$$\dot{\mathbf{\Omega}} \times \mathbf{r}_P = \dot{\boldsymbol{\omega}}_{AB} \times \mathbf{r}_P = -\alpha_{AB}(0.175)\mathbf{i}'$$

$$[\mathbf{a}_{P/B}]_{\text{rot}} = \dot{u}\mathbf{j}'$$

$$\mathbf{a}_{\text{Coriolis}} = 2\mathbf{\Omega} \times [\mathbf{v}_{P/B}]_{\text{rot}} = 2\omega_{AB}\mathbf{k}' \times 2.18(-\mathbf{j}')$$

$$= 2(70.5)(2.18)\mathbf{i}' = 307.4\mathbf{i}' \quad \text{m/s}^2$$

Then substituting into equation (b),

$$-1295.4\mathbf{j}' + 228.4\mathbf{i}' = \dot{u}\mathbf{j}' - 871.8\mathbf{j}' - \alpha_{AB}(0.175)\mathbf{i}' + 307.4\mathbf{i}'$$

$$\dot{u} = -1295.4 + 871.8 = -423.6$$

$$[\mathbf{a}_{P/B}]_{\text{rot}} = -423.6\mathbf{j}' \quad \text{m/s}^2$$

$$\alpha_{AB} = \dot{\omega}_{AB} = \frac{228.4 - 307.4}{-0.175} = 451.4$$

$$\boldsymbol{\alpha}_{AB} = 451.4\mathbf{k}' \quad \text{rad/s}^2$$

∎

Sample Problem 5.10

Plate $OAFBE$ shown in the figure rotates at an angular velocity ω and an angular acceleration $\dot{\omega}$ about a vertical axis. On this plate is scribed a circular slot in which a particle is free to slide. When the particle is at point D, it has a speed u and acceleration $\dot{u}$ relative to the plate as shown. Find the instantaneous absolute velocity and acceleration of the particle when it is at point D.

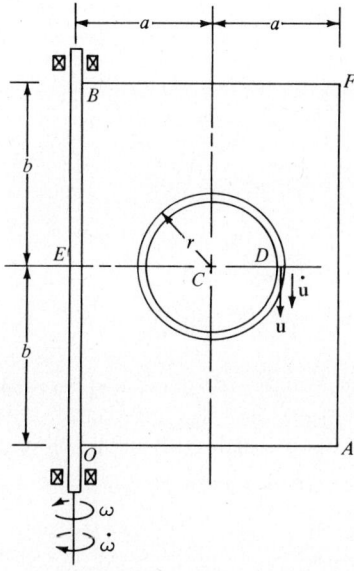

Analysis: The motion is quite complex. The particle's absolute path is on the surface of a torus. Instead of using the absolute path, the motion can be broken into simpler parts. If a reference frame that rotates with the plate is used, the motion of the particle in that frame is circular. This motion can be studied separately, and then corrections can be made for the frame's rotation.

In any reference frame one has the option of placing the coordinate system's origin at several locations. In the first attempt at this problem let the coordinate axes be fixed to the plate at point E and rotate with it. Select the local z axis straight up, the x axis perpendicular to the plate, and the y axis to the right. These last choices are quite arbitrary and unimportant to the problem solution. Employing equations (5.115) and (5.116), and analyzing each term separately, one obtains

$$\Omega = \omega \mathbf{k} \qquad \dot{\Omega} = -\dot{\omega}\mathbf{k}$$

$$\mathbf{r}_{D/E} = (a+r)\mathbf{j} \qquad \mathbf{v}_E = \mathbf{0} \qquad \mathbf{a}_E = \mathbf{0}$$

$$[\mathbf{v}_{D/E}]_{\text{rot}} = -u\mathbf{k}$$

$$[\mathbf{a}_{D/E}]_{\text{rot}} = [\mathbf{a}_{D/E}]^n_{\text{rot}} + [\mathbf{a}_{D/E}]^t_{\text{rot}}$$

$$[\mathbf{a}_{D/E}]_{\text{rot}} = -\frac{u^2}{r}\mathbf{j} - \dot{u}\mathbf{k}$$

$$\mathbf{v}_D = 0 + (-u\mathbf{k}) + \omega\mathbf{k} \times (a+r)\mathbf{j} = -u\mathbf{k} - \omega(a+r)\mathbf{i} \tag{a}$$

$$\mathbf{a}_D = 0 + \left(-\frac{u^2}{r}\mathbf{j} - \dot{u}\mathbf{k}\right) + 2[\omega\mathbf{k} \times (-u\mathbf{k})]$$
$$+ [-\dot{\omega}\mathbf{k} \times (a+r)\mathbf{j}] + \omega\mathbf{k} \times [\omega\mathbf{k} \times (a+r)\mathbf{j}]$$
$$= \left[-\frac{u^2}{r} - \omega^2(a+r)\right]\mathbf{j} - \dot{u}\mathbf{k} + \dot{\omega}(a+r)\mathbf{i} \tag{b}$$

If instead, the arbitrary selection of the coordinate system's origin was at C, then

$$\boldsymbol{\Omega} = \omega\mathbf{k} \qquad \dot{\boldsymbol{\Omega}} = -\dot{\omega}\mathbf{k} \qquad \mathbf{r}_{D/C} = r\mathbf{j}$$

$$\mathbf{v}_C = \omega\mathbf{k} \times a\mathbf{j} = -\omega a\mathbf{i}$$

$$\mathbf{a}_C = (\mathbf{a}_C)_n + (\mathbf{a}_C)_t = \omega\mathbf{k} \times (\omega\mathbf{k} \times a\mathbf{j}) - \dot{\omega}\mathbf{k} \times a\mathbf{j}$$
$$= -\omega^2 a\mathbf{j} + \dot{\omega} a\mathbf{i}$$

$$[\mathbf{v}_{D/C}]_{\text{rot}} = -u\mathbf{k} \qquad [\mathbf{a}_{D/C}]_{\text{rot}} = -\frac{u^2}{r}\mathbf{j} - \dot{u}\mathbf{k}$$

$$\mathbf{v}_D = -\omega a\mathbf{i} - u\mathbf{k} + (\omega\mathbf{k} \times r\mathbf{j}) = -u\mathbf{k} - \omega(a+r)\mathbf{i} \tag{c}$$

$$\mathbf{a}_D = -\omega^2 a\mathbf{j} + \dot{\omega} a\mathbf{i} - \frac{u^2}{r}\mathbf{j} - \dot{u}\mathbf{k} + 2(0) + \omega\mathbf{k} \times (\omega\mathbf{k} \times r\mathbf{j}) + (-\dot{\omega}\mathbf{k} \times r\mathbf{j})$$
$$= \dot{\omega}_1(a+r)\mathbf{i} - \dot{u}\mathbf{k} + \left[-\frac{u^2}{r} - \omega^2(a+r)\right]\mathbf{j} \tag{d}$$

These are the same results that were found in equations (a) and (b). ∎

Sample Problem 5.11
The disk of radius R shown in the figure spins about its own axis located at C

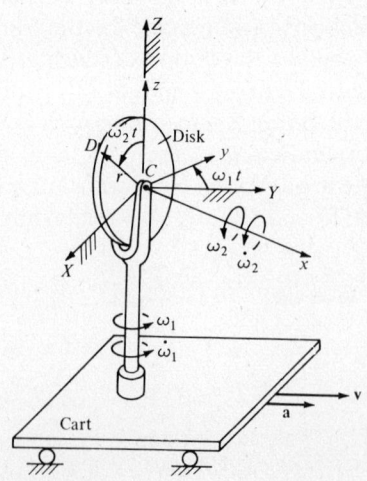

RIGID BODY DYNAMICS: KINEMATICS   141

with an angular speed and acceleration ω_2 and $\dot{\omega}_2$, respectively. The disk's spin axle itself rotates with an angular speed and acceleration ω_1 and $\dot{\omega}_1$ about the vertical direction. The disk, the axle, and the vertical shaft all move with a rolling cart that for this moment has a velocity and acceleration **v** and **a** in the Y direction. For the moment when $\omega_1 t = 40°$ and $\omega_2 t = 30°$, find the absolute instantaneous velocity and acceleration of point D. Express your answer in X, Y, Z components.

Analysis: There are several alternative approaches to this problem. For comparison's sake, the solution will first be developed without using a rotating reference frame. Recalling the work done in Sections 5.3B and 5.4,

$$\mathbf{v}_D = \mathbf{v}_C + \mathbf{v}_{D/C} \tag{a}$$

$$\mathbf{v}_D = \mathbf{v}_C + \boldsymbol{\omega}_{DC} \times \mathbf{r}_{D/C} \tag{b}$$

$$\mathbf{a}_D = \mathbf{a}_C + \mathbf{a}_{D/C} \tag{c}$$

$$\mathbf{a}_D = \mathbf{a}_C + \dot{\boldsymbol{\omega}}_{DC} \times \mathbf{r}_{D/C} + \boldsymbol{\omega}_{DC} \times (\boldsymbol{\omega}_{DC} \times \mathbf{r}_{D/C}) \tag{d}$$

Using X, Y, Z components (although this is not necessary, as will be shown later),

$\mathbf{v}_C = v\mathbf{J} \qquad \mathbf{a}_C = a\mathbf{J}$

$\mathbf{r}_{D/C} = r \sin 30°(\sin 40°\mathbf{I} - \cos 40°\mathbf{J}) + r \cos 30°\mathbf{K}$

$\boldsymbol{\omega}_{DC} = \boldsymbol{\omega}_1 + \boldsymbol{\omega}_2 = \omega_1 \mathbf{K} + \omega_2(\cos 40°\mathbf{I} + \sin 40°\mathbf{J})$

$\dot{\boldsymbol{\omega}}_{DC} = \dot{\omega}_1 \mathbf{K} + \dot{\omega}_2(\cos 40°\mathbf{I} + \sin 40°\mathbf{J}) + \omega_1 \mathbf{K} \times \omega_2(\cos 40°\mathbf{I} + \sin 40°\mathbf{J})$

$\dot{\boldsymbol{\omega}}_{DC} = \dot{\omega}_1 \mathbf{K} + \dot{\omega}_2(\cos 40°\mathbf{I} + \sin 40°\mathbf{J}) + \omega_1 \omega_2 \cos 40°\mathbf{J} - \omega_1 \omega_2 \sin 40°\mathbf{I}$

The solution can now be obtained by substituting these values in equations (5.42) and (5.43) and carrying out the cross products.

If instead, a *rotating reference frame* that rotates with the axle at an angular velocity ω_1 and angular acceleration $\dot{\omega}_1$ is used, the motion of D becomes circular in that frame. Recalling equations (5.115) and (5.116), and changing subscripts,

$$\mathbf{v}_D = \mathbf{v}_C + [\mathbf{v}_{D/C}]_{\text{rot}} + \boldsymbol{\Omega} \times \mathbf{r}_{D/C} \tag{e}$$

$$\mathbf{a}_D = \mathbf{a}_C + [\mathbf{a}_{D/C}]_{\text{rot}} + 2\boldsymbol{\Omega} \times [\mathbf{v}_{D/C}]_{\text{rot}} + \dot{\boldsymbol{\Omega}} \times \mathbf{r}_{D/C} + \boldsymbol{\Omega} \times (\boldsymbol{\Omega} \times \mathbf{r}_{D/C}) \tag{f}$$

It is easier here to use x, y, z components to begin with and then later transform the solution vector to the X, Y, Z coordinates. In terms of x, y, z components,

$\mathbf{v}_C = v(\sin 40°\mathbf{i} + \cos 40°\mathbf{j})$

$\mathbf{a}_C = a(\sin 40°\mathbf{i} + \cos 40°\mathbf{j})$

$\boldsymbol{\Omega} = \omega_1 \mathbf{k} \qquad \dot{\boldsymbol{\Omega}} = \dot{\omega}_1 \mathbf{k}$

$\mathbf{r}_{D/C} = r(-\sin 30°\mathbf{j} + \cos 30°\mathbf{k})$

$$[\mathbf{v}_{D/C}]_{\text{rot}} = \boldsymbol{\omega}_2 \times \mathbf{r}$$
$$= \omega_2 \mathbf{i} \times r(-\sin 30°\mathbf{j} + \cos 30°\mathbf{k})$$
$$= \omega_2 r(-\cos 30°\mathbf{j} - \sin 30°\mathbf{k})$$
$$[\mathbf{a}_{D/C}]_{\text{rot}} = \dot{\boldsymbol{\omega}}_2 \times \mathbf{r} + \boldsymbol{\omega}_2 \times (\boldsymbol{\omega}_2 \times \mathbf{r})$$
$$= \dot{\omega}_2 r(-\cos 30°\mathbf{j} - \sin 30°\mathbf{k}) - \omega_2^2 \mathbf{r}$$
$$= -r(\dot{\omega}_2 \cos 30° - \omega_2^2 \sin 30°)\mathbf{j} - r(\dot{\omega}_2 \sin 30° + \omega_2^2 \cos 30°)\mathbf{k}$$

Substitution of these terms in equations (e) and (f) yields the results for velocity and acceleration expressed in the x, y, z coordinate system. Transformation to X, Y, Z coordinates is now needed.

$$\mathbf{v}_D = v(\sin 40°\mathbf{i} + \cos 40°\mathbf{j}) + \omega_2 r(-\cos 30°\mathbf{j} - \sin 30°\mathbf{k})$$
$$+ \omega_1 \mathbf{k} \times r(-\sin 30°\mathbf{j} + \cos 30°\mathbf{k})$$
$$= (v \sin 40° + \omega_1 r \sin 30°)\mathbf{i} + (v \cos 40° - \omega_2 r \cos 30°)\mathbf{j} - \omega_2 r \sin 30°\mathbf{k}$$

$$[l] = \begin{bmatrix} \cos 40° & \sin 40° & 0 \\ -\sin 40° & \cos 40° & 0 \\ 0 & 0 & 1 \end{bmatrix}$$

and

$$\begin{Bmatrix} x \\ y \\ z \end{Bmatrix} = [l] \begin{Bmatrix} X \\ Y \\ Z \end{Bmatrix}$$

$$[l]^t = \begin{bmatrix} \cos 40° & -\sin 40° & 0 \\ \sin 40° & \cos 40° & 0 \\ 0 & 0 & 1 \end{bmatrix}$$

$$\begin{Bmatrix} v_X \\ v_Y \\ v_Z \end{Bmatrix} = [l]^t \begin{Bmatrix} v_x \\ v_y \\ v_z \end{Bmatrix}$$

$$\begin{Bmatrix} v_X \\ v_Y \\ v_Z \end{Bmatrix} = \begin{bmatrix} \cos 40° & -\sin 40° & 0 \\ \sin 40° & \cos 40° & 0 \\ 0 & 0 & 1 \end{bmatrix} \begin{Bmatrix} v \sin 40° + \omega_1 r \sin 30° \\ v \cos 40° - \omega_2 r \cos 30° \\ -\omega_2 r \sin 30° \end{Bmatrix}$$

$$= \begin{Bmatrix} \omega_1 r \sin 30° \cos 40° + \omega_2 r \cos 30° \sin 40° \\ v + \omega_1 r \sin 30° \sin 40° - \omega_2 r \cos 30° \cos 40° \\ -\omega_2 r \sin 30° \end{Bmatrix}$$

$$\mathbf{v}_D = (\omega_1 r \sin 30° \cos 40° + \omega_2 r \cos 30° \sin 40°)\mathbf{I}$$
$$+ (v + \omega_1 r \sin 30° \sin 40° - \omega_2 r \cos 30° \cos 40°)\mathbf{J}$$
$$+ (-\omega_2 r \sin 30°)\mathbf{K}$$

The solution from equation (b) is

$$\mathbf{v}_D = v\mathbf{J} + \begin{vmatrix} \mathbf{I} & \mathbf{J} & \mathbf{K} \\ \omega_2 \cos 40° & \omega_2 \sin 40° & \omega_1 \\ r \sin 30° \sin 40° & -r \sin 30° \cos 40° & r \cos 30° \end{vmatrix}$$

$$= (\omega_1 r \sin 30° \cos 40° + \omega_2 r \sin 40° \cos 30°)\mathbf{I}$$
$$+ (v + \omega_1 r \sin 30° \sin 40° - \omega_2 r \cos 40° \cos 30°)\mathbf{J} + (-\omega_2 r \sin 30°)\mathbf{K}$$

which checks the velocity results. The reader can now also check the acceleration results. ∎

5.7 MOTION RELATIVE TO THE ROTATING EARTH

The previous section introduced methods that can be used to analyze motion with reference to a rotating frame. The equations developed there will now be employed to describe motion relative to an observer fixed to the rotating earth. The center of the earth will be assumed to have zero acceleration. Refer to Figure 5.24. The observer in the figure, located at O, has coordinate axes that are fixed to the earth and selected so that $\mathbf{k}$ is the local vertical, $\mathbf{j}$ points due east on the local latitude circle, and $\mathbf{i}$ points due south along the local meridian. The $\mathbf{ij}$ plane is the local horizon plane. The radius of the earth is R, the local latitude angle is λ, and the angular velocity of the earth is given as $\boldsymbol{\omega}_e$. For the center of the earth C, $\mathbf{a}_C \simeq 0$, and for the earth's rotation, $\dot{\boldsymbol{\omega}}_e \simeq 0$. The absolute instantaneous acceleration of point D then becomes

$$\mathbf{a}_D = \mathbf{a}_O + [\mathbf{a}_{D/O}]_{rot} + 2\boldsymbol{\omega}_e \times [\mathbf{v}_{D/O}]_{rot} + \boldsymbol{\omega}_e \times (\boldsymbol{\omega}_e \times \mathbf{r}_{D/O}) \quad (5.120)$$

$$\boldsymbol{\omega}_e = \omega_e(-\cos \lambda \mathbf{i} + \sin \lambda \mathbf{k}) \quad (5.121)$$

$$\mathbf{a}_O = \boldsymbol{\omega}_e \times (\boldsymbol{\omega}_e \times R\mathbf{k}) \quad (5.122)$$

$$= -\omega_e^2 R \cos \lambda (\sin \lambda \mathbf{i} + \cos \lambda \mathbf{k}) \quad (5.123)$$

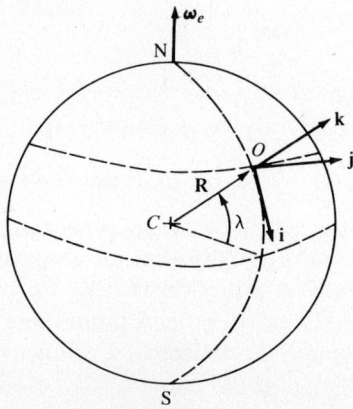

Figure 5.24

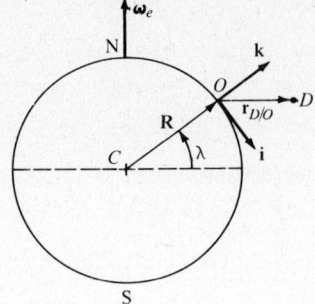

Figure 5.25

The motion an observer on the earth perceives and measures (Figure 5.25) is

$$[\mathbf{a}_{D/O}]_{rot} = \mathbf{a}_D - 2\boldsymbol{\omega}_e \times [\mathbf{v}_{D/O}]_{rot} - \boldsymbol{\omega}_e \times (\boldsymbol{\omega}_e \times \mathbf{r}_{D/O}) + \omega_e^2 R \cos \lambda (\sin \lambda \mathbf{i} + \cos \lambda \mathbf{k}) \quad (5.124)$$

For motion on or near the earth's surface $r_{D/O} \ll R$ and with

$$\omega_e \simeq \frac{1 \text{ rev}}{24 \text{ h}} = 7.27 \times 10^{-5} \text{ rad/s}$$

$\boldsymbol{\omega}_e \times (\boldsymbol{\omega}_e \times \mathbf{r}_{D/O})$ can be neglected. Equation (5.124) then becomes

$$[\mathbf{a}_{D/O}]_{rot} = \mathbf{a}_D - 2\boldsymbol{\omega}_e \times [\mathbf{v}_{D/O}]_{rot} + \omega_e^2 R \cos \lambda (\sin \lambda \mathbf{i} + \cos \lambda \mathbf{k}) \quad (5.125)$$

Some interesting cases follow. For a freely falling body with "little" velocity, close to the surface of the earth, the Coriolis term can be dropped. According to Newton's law of gravitation,

$$\mathbf{F} = -\frac{GM_e m}{r_D^3} \mathbf{r}_D = \text{gravitational force} \quad (5.126)$$

Close to the earth, $\mathbf{r}_D \simeq \mathbf{R} = R\mathbf{k}$; so

$$\mathbf{F} \simeq -\frac{GM_e m}{R^2} \mathbf{k} \quad (5.127)$$

$$\mathbf{a}_D = \frac{\mathbf{F}}{m} = \mathbf{G}_1 \simeq -G_1 \mathbf{k} = -9.824 \mathbf{k} \quad \text{m/s}^2 \quad (5.128)$$

$$[\mathbf{a}_{D/O}]_{rot} = \mathbf{g}_{local} = -G_1 \mathbf{k} + \omega_e^2 R \cos \lambda (\sin \lambda \mathbf{i} + \cos \lambda \mathbf{k}) \quad (5.129)$$

Equation (5.129) states that for a spherical, homogeneous, nonaccelerating earth, the local acceleration of gravity that an observer sees is affected by the earth's rotation. According to equation (5.129), an object appears to weigh less at the equator than at the north pole. A plumb line at O would not point to the earth's center but would be deflected. Calling the local acceleration of gravity g,

$$g = -G_1 + \omega_e^2 R \cos^2 \lambda \quad (5.130)$$

The angle the plumb line makes with the vertical is then

$$\tan \phi = \frac{R\omega_e^2 \cos \lambda \sin \lambda}{g} \tag{5.131}$$

$$= \frac{R\omega_e^2 \sin 2\lambda}{2g} \tag{5.132}$$

The common value of g used in engineering calculation is 9.81 m/s², or 32.2 ft/s², at 45° latitude. At that latitude the angle ϕ becomes

$$\tan \phi = \frac{6371 \times 10^3 \text{ m} \times (7.27 \times 10^{-5})^2 \text{ rad/s}^2}{2(9.81) \text{ m/s}^2}$$

$$= 1.72 \times 10^{-3} \text{ rad}$$

$$\phi = 0.1°$$

For a free-coasting rocket, the velocity relative to the earth's rotating surface can be very large. The Coriolis term then becomes bigger than the centrifugal acceleration term, the last term in equation (5.125). Setting $\mathbf{a}_D \simeq -g\mathbf{k}$,

$$[\mathbf{a}_{D/O}]_{\text{rot}} \simeq \mathbf{a}_D - 2\boldsymbol{\omega}_e \times [\mathbf{v}_{D/O}]_{\text{rot}} \tag{5.133}$$

$$[\mathbf{a}_{D/O}]_{\text{rot}} = (2\dot{y}\omega_e \sin \lambda)\mathbf{i} - (2\dot{x}\omega_e \sin \lambda + 2\dot{z}\omega_e \cos \lambda)\mathbf{j} + (-g + 2\dot{y}\omega_e \cos \lambda)\mathbf{k} \tag{5.134}$$

where

$$[\mathbf{v}_{D/O}]_{\text{rot}} = \dot{x}\mathbf{i} + \dot{y}\mathbf{j} + \dot{z}\mathbf{k} \tag{5.135}$$

$$[\mathbf{a}_{D/O}]_{\text{rot}} = \ddot{x}\mathbf{i} + \ddot{y}\mathbf{j} + \ddot{z}\mathbf{k} \tag{5.136}$$

Then

$$\ddot{x} = 2\dot{y}\omega_e \sin \lambda \tag{5.137a}$$

$$\ddot{y} = -(2\dot{x}\omega_e \sin \lambda + 2\dot{z}\omega_e \cos \lambda) \tag{5.137b}$$

$$\ddot{z} = -g + 2\dot{y}\omega_e \cos \lambda \tag{5.137c}$$

When integrated with the help of a computer, equations (5.137a) to (5.137c) will show that a rocket launched vertically from the earth's surface in the northern hemisphere will land not only west of its launch site, as expected, but will also come down slightly north of its initial launch point.

With the aid of equation (5.133) one can also predict, for instance, the forces produced by the earth's rotation on objects moving along the earth's surface. Take, as an example, a rocket sled traveling due east at 2600 m/s at a latitude of 45°.

$$\ddot{x} = 2\dot{y}\omega_e \sin \lambda$$
$$= 2(2600)(7.27 \times 10^{-5})(0.707) = 0.267 \text{ m/s}^2$$

For a 500-kg test vehicle, this means a Coriolis force

$$F_x = m\ddot{x} = 500(0.267) = 134 \text{ N} \qquad \text{(pointing southward)}$$

acts on the track.

146 REVIEW OF NEWTONIAN MECHANICS

Coriolis forces affect even the wind patterns in the atmosphere. The circulation of wind in a cyclone is counterclockwise in the northern hemisphere but clockwise in the southern hemisphere due to the Coriolis effect.

Sample Problem 5.12

A rocket is fired vertically upward at a latitude of 50°N and a launch speed of 600 m/s. Neglecting air friction and assuming $g = 9.81$ m/s^2 and is constant, write a Fortran program to determine the position of the landing site of the rocket relative to the launch point.

```
00100 PROGRAM RLAND(INPUT,OUTPUT)
00110 DIMENSION Z(6),PZ(9)
00120 DATA OM,XDO,YDO,ZDO,XO,YO,ZO/7.3E-5,0.,0.,600.,0.,0.,0./
00130 DATA T,DT,RLAMB/0.,2.,50./
00140 A=COS(RLAMB/57.3)
00150 B=SIN(RLAMB/57.3)
00160 C=2*OM
00170 Z(1)=XO
00180 Z(2)=YO
00190 Z(3)=ZO
00200 Z(4)=XDO
00210 Z(5)=YDO
00220 Z(6)=ZDO
00230 PRINT 13,
00240 13 FORMAT(12X,'T',16X,'X',16X,'Y',16X,'Z')
00250 5 Z(7)=C*Z(5)*B
00260 Z(8)=-C*(Z(4)*B+Z(6)*A)
00270 Z(9)=-9.81+C*Z(5)*A
00280 DO 7 I=4,6
00290 PZ(I)=Z(I)+Z(I+3)*DT
00300 7 CONTINUE
00310 PZ(7)=C*PZ(5)*B
00320 PZ(8)=-C*(PZ(4)*B+PZ(6)*A)
00330 PZ(9)=-9.81+C*PZ(5)*A
00340 DO 9 I=1,6
00350 Z(I)=Z(I)+DT/2.*(PZ(I+3)+Z(I+3))
00360 9 CONTINUE
00370 T=T+DT
00380 PRINT 8,T,Z(1),Z(2),Z(3)
00390 8 FORMAT(3X,E10.3,3(3X,E14.7))
00400 IF (Z(3).LT.0.) GO TO 10
00410 GO TO 5
00420 10 STOP
00430 END
```

Figure 5.26

RIGID BODY DYNAMICS: KINEMATICS   147

T	X	Y	Z
.200E+01	0.	-.1126250E+00	.1180380E+04
.400E+01	-.4997045E-04	-.4431344E+00	.2321520E+04
.600E+01	-.1978224E-03	-.9841624E+00	.3423420E+04
.800E+01	-.4898195E-03	-.1728343E+01	.4486080E+04
.100E+02	-.9705777E-03	-.2668311E+01	.5509499E+04
.600E+02	-.1711056E+00	-.6825328E+02	.1834186E+05
.620E+02	-.1867571E+00	-.7169830E+02	.1834502E+05
.640E+02	-.2031791E+00	-.7514023E+02	.1830895E+05
.112E+03	-.7999248E+00	-.1376674E+03	.5671009E+04
.114E+03	-.8308363E+00	-.1386381E+03	.4653923E+04
.116E+03	-.8619436E+00	-.1394142E+03	.3597597E+04
.118E+03	-.8932022E+00	-.1399883E+03	.2502030E+04
.120E+03	-.9243663E+00	-.1403531E+03	.1367224E+04
.122E+03	-.9559882E+00	-.1405012E+03	.1931777E+03
.124E+03	-.9874185E+00	-.1404252E+03	-.1020109E+04

Figure 5.27

Analysis: See Figures 5.26 and 5.27.   ■

PROBLEMS

5.1 Rod *BC* of length *L* has end *B* drawn along the floor at constant speed v_B while it remains in contact with a step of height *S*. How many degrees of freedom does the rod have? Find an expression for $\dot\theta$ and $\ddot\theta$ in terms of *S* and v_B.

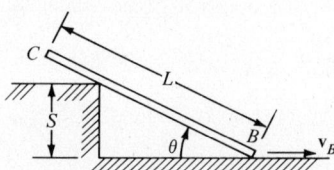

5.2 The wire shown connected to a point mass at *B* unwinds from a spool of radius *R* at a constant rate $\dot\phi = \omega$. How many degrees of freedom does the mass have? Find an expression for $\mathbf{v}_B$ and $\mathbf{a}_B$.

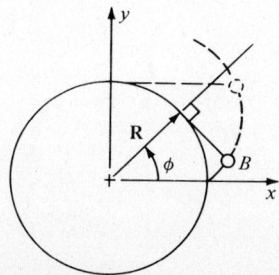

148 REVIEW OF NEWTONIAN MECHANICS

5.3 The point mass shown at B is attached to a rope of thickness b. The rope unwinds from a conical drum that is turning at constant angular speed ω. Find an expression for $\mathbf{v}_B$ and $\mathbf{a}_B$. Assume that at $t=0$, the rope fills the drum.

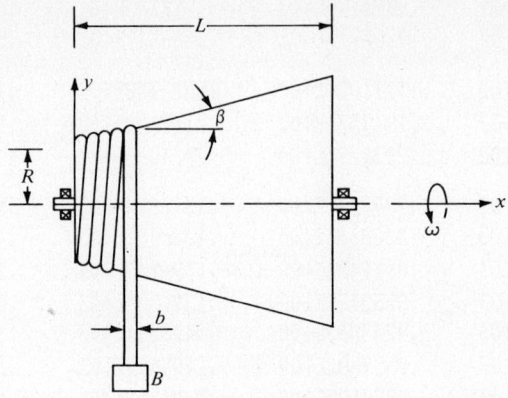

5.4 A simple pendulum consisting of a light rod of length L and a pendulum bob of mass m is free to swing in a vertical plane while its pivot point at A is allowed to move vertically. How many degrees of freedom does the mass have? Derive an expression for its velocity and acceleration.

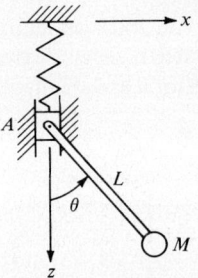

5.5 Mass m is free to move in an inclined slot on a plate. The plate itself is driven at a constant angular speed ω about a fixed horizontal axis. How many degrees of freedom does the mass have? Find an expression for $\mathbf{v}_m$ and $\mathbf{a}_m$.

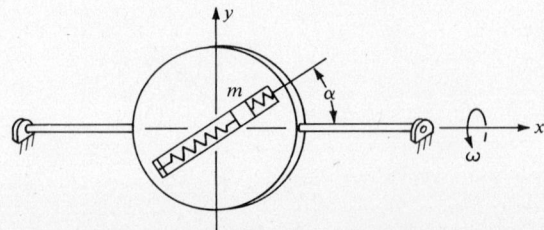

5.6 The disk shown of mass *m* and radius *R* rolls without slipping on the concave surface of the cart. The cart itself moves freely along the ground in a straight line as the disk rolls. How many degrees of freedom does the disk have? Find an expression for the velocity and acceleration of the center of the wheel? The coordinate *x* is an absolute measurement taken from the unstretched position of the spring.

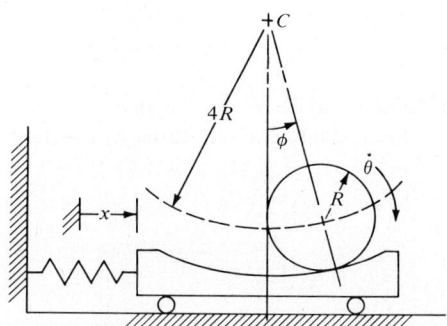

5.7 The double pendulum shown is made up of two light rods of length *L*. Rod *OA* is driven at a constant angular speed ω. How many degrees of freedom exist for the system? Find an expression for the velocity and acceleration of each mass.

5.8 Repeat problem 5.7 using the given angles shown. The top rod is no longer constrained.

5.9 Draw the space and body centrodes for bar *AB* shown in Figure 5.12.

5.10 For the given slider-crank mechanism, determine the velocities and accelerations of the piston when $\theta = 0°$ and when $\theta = 90°$. Crank *OA* is of length *a*, and connecting rod *AB* is of length *b*. The crank turns counterclockwise at constant angular speed ω_0.

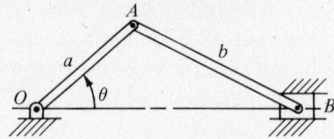

*5.11 Repeat problem 5.10, but derive a general expression for the acceleration of the piston valid for all θ. Write a computer program that will output a_B for various θ given ω_0, a, and b.

5.12 For the mechanism given, determine the angular velocity and angular acceleration of bar BC for this instant. Bar OA turns at constant counterclockwise angular velocity ω_0.

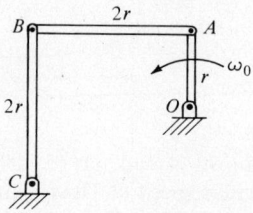

*5.13 Write a computer program for problem 5.12 that will give the x, y coordinates of point B during one complete revolution of arm OA.

5.14 For the mechanism shown, determine the velocity and acceleration of point B for this instant. Arm OA turns at a constant counterclockwise rate ω_0.

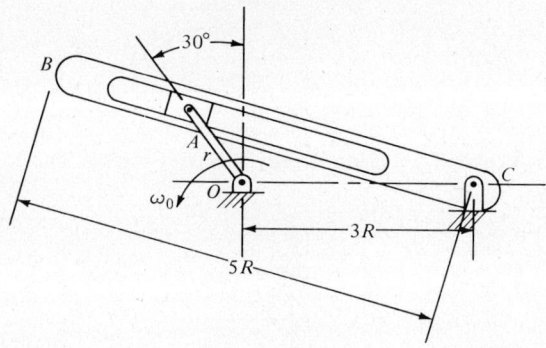

*5.15 For problem 5.14, write a computer program that will give the acceleration of point B during one complete revolution of arm OA.

5.16 Repeat problem 5.5 for the instant when the plate is vertical, as shown, but let the plate with its supports and axle also turn with a varying angular rate $\dot{\phi}$ about the vertical.

5.17 Repeat problem 5.7 allowing the plane of the double pendulum to swing freely about the vertical at a varying angular rate $\dot{\phi}$.

5.18 Repeat problem 5.8 allowing the plane of the double pendulum to swing freely about the vertical at a varying angular rate $\dot{\phi}$.

5.19 The bent rod ACB is free to turn about the vertical at a varying rate $\dot{\phi}$. In addition, rod BD of length L, pinned to ACB, can swing in the vertical plane formed by $ACBD$ at a varying rate $\dot{\theta}$. The distance between the pin at B and the vertical portion of the bent rod is d. How many degrees of freedom are present? Write an expression for $\mathbf{v}_D$ and $\mathbf{a}_D$.

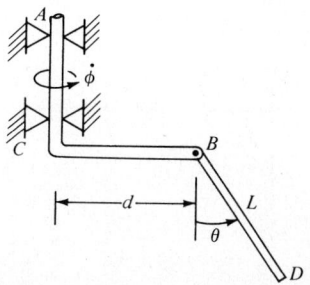

5.20 The ring shown of radius R is free to rotate about the vertical direction at a varying rate $\dot{\phi}$. At the same time, a particle located at B slides about the circumference. How many degrees of freedom are present? Write an expression for $\mathbf{v}_B$ and $\mathbf{a}_B$.

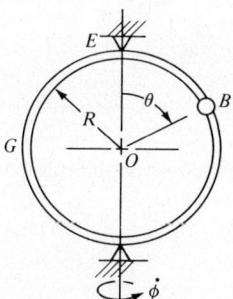

5.21 Repeat problem 5.20 for the case where the vertical axis of rotation is offset and passes through point G instead.

5.22 The communication satellite shown spins at a varying angular rate $\dot{\phi}$ about a fixed direction DE. At the same time, it is known that two solar panel arms rotate relative to the craft with an increasing rate $\dot{\theta}$. For the instant described, find $\mathbf{v}_{B/O}$ and $\mathbf{a}_{B/O}$.

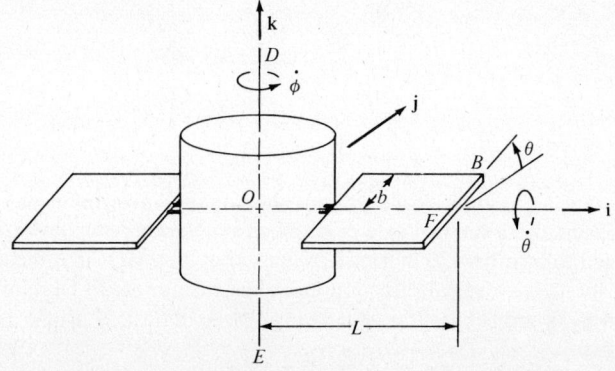

5.23 If the satellite in problem 5.22 was traveling with a period of 24 h in a circular orbit, in the equatorial plane of the earth, what is its altitude? Find velocity and acceleration expressions for $\mathbf{v}_B$ and $\mathbf{a}_B$. Axis DE points toward the center of the earth in the plane of the orbit. Assume the center of the earth is fixed ($GM_e = 3.988 \times 10^{14}$ m³/s², $R = 6371$ km). If $b = 1$ m, $L = 4$ m, $\ddot{\phi} = \ddot{\theta} = 0$, $\dot{\phi} = 3$ rad/s, and $\dot{\theta} = 10$ rad/s, find $\mathbf{v}_B$ and $\mathbf{a}_B$ when $\theta = 0°$.

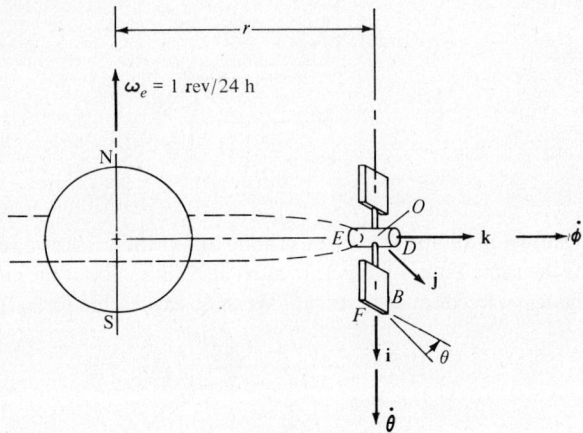

5.24 Particle B travels in a plane on an elliptic path. If the plane itself rotates with an increasing angular rate $\dot{\phi}$ about an axis parallel to the ellipse's semiminor axis and passing through its focus, develop expressions for $\mathbf{v}_B$ and $\mathbf{a}_B$ in terms of r, θ, ϕ, and their derivatives.

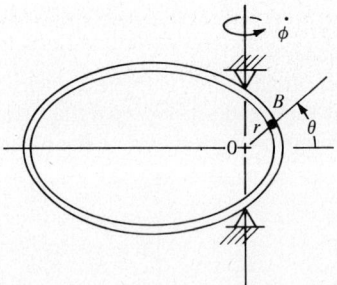

5.25 A thin wheel of radius r rolls without slipping on a horizontal surface. The wheel is free to spin about its own axle of length L at a varying angular rate $\dot{\phi}$. The axle DE is pinned to member CD in such a manner that it remains in a vertical plane while it turns about the vertical direction at an increasing rate $\dot{\theta}$. Develop expressions for $\mathbf{v}_B$ and $\mathbf{a}_B$ in terms of r, L, d, α, $\dot{\phi}$, and $\ddot{\phi}$. Point B for this instant lies in plane CDE as shown.

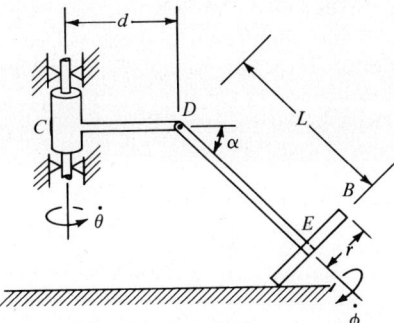

5.26 The right circular cone of radius r and height $\sqrt{3}r$ rolls without slipping on an identical fixed vertical cone. If the plane containing points $ABCDE$ rotates about the vertical at a constant rate, how many degrees of freedom are present? What do $\mathbf{v}_B$ and $\mathbf{a}_B$ equal?

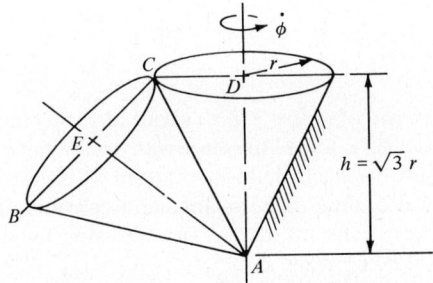

5.27 The thin disk of radius r spins about its own axle at an angular rate $\dot{\phi}$ as shown. The axle of length L is constrained to swing in a vertical plane at an angular speed $\dot{\theta}$. The plane itself turns about the vertical direction at a rate $\dot{\psi}$. All angular rates vary. How many degrees of freedom are present? What is α of the disk? Find $\mathbf{a}_B$ for the special case when $\dot{\psi}, \dot{\phi},$ and $\dot{\theta}$ are all constant. Point B lies in the vertical plane for this instant.

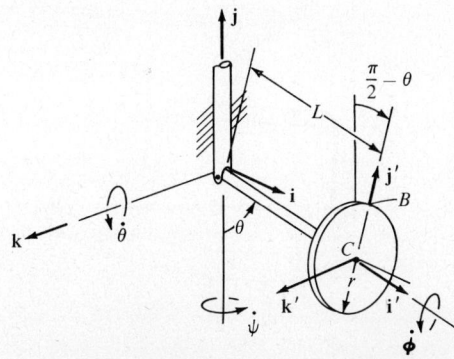

5.28 A child's tricycle moves forward on a flat surface with each wheel of radius r independently rolling without slipping on the ground. There are only 2 degrees of freedom present for this motion. If the spin rate of the front wheel is given by $\dot{\theta}_1$ and its constant angle of attack defined by the angle β, find relations that describe the spin rate $\dot{\theta}_2$ and $\dot{\theta}_3$ of each back wheel, the turning rate $\dot{\phi}$ of the cycle, and the direction of the velocity of the center of mass of the frame in terms of b, $\dot{\theta}_1$, r, c, d, and L.

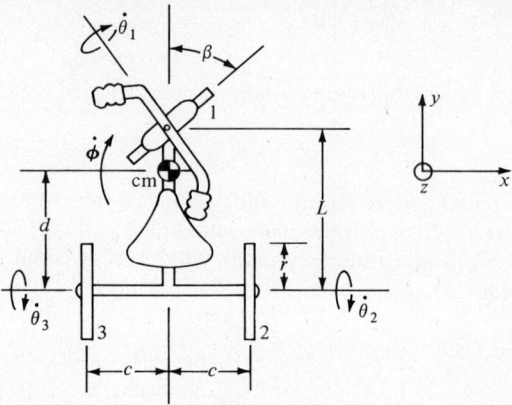

5.29 The thin disk shown of radius r spins about its own axle at A at a constant angular rate $\dot{\psi}$. The axle itself is forced to rotate with arm OA at an increasing rate $\dot{\theta}$. All this occurs while the table to which the mechanism is firmly attached rotates at a decreasing angular speed $\dot{\phi}$. Find the total angular acceleration of the disk. If all the above angular rates were constant, derive an expression for $\mathbf{a}_B$ when point B lies at its uppermost absolute position.

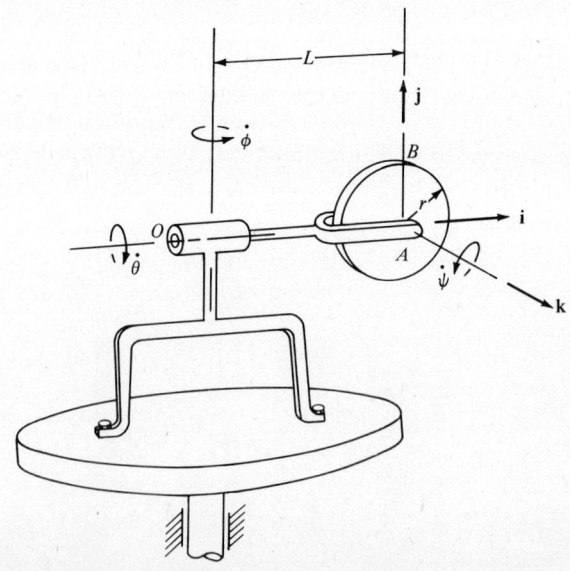

5.30 The thin disk shown of radius r spins about a pin at A with constant angular rate $\dot{\psi}$. The pin itself turns with shaft OA at uniform speed $\dot{\theta}$. What is the total angular acceleration of the disk if arm CBO starts to rotate with increasing angular speed $\dot{\phi}$ about the vertical direction? What is the total acceleration of point E when it lies at its uppermost position?

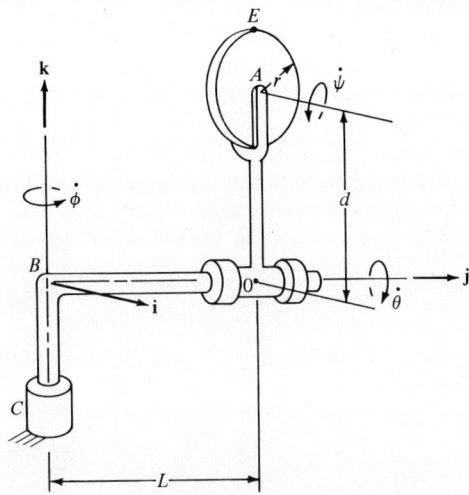

5.31 For the top shown in Figure 5.7, find the components of the angular acceleration along the x, y, z axes.

5.32 The gyro shown has a wheel of radius r that spins about its own axis at a constant rate

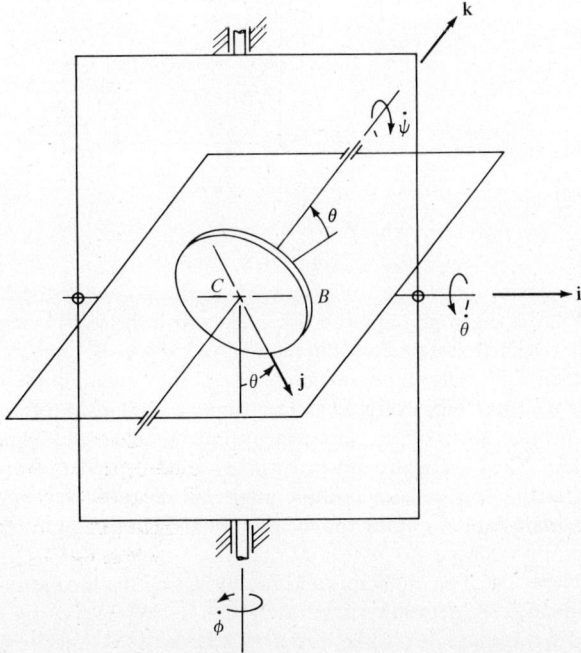

$\dot{\psi}$. For this example, the inner gimbal that supports the wheel is known to "nod" at a varying rate $\dot{\theta}$ while the outer gimbal turns about the vertical direction at a nonuniform rate $\dot{\phi}$. Find the angular acceleration of the wheel. Steady precession for gyroscopic motion occurs when $\ddot{\theta}=\dot{\theta}=0$ and $\dot{\phi}=$ a constant. Assuming now that steady precession exists, find $\mathbf{a}_B$ for the instant when point B on the wheel lies in the plane of the inner gimbal.

5.33 Describe the motion of fluids traveling horizontally from a high-pressure region to one that is low in the northern hemisphere. Neglect frictional effects.

5.34 Taking into account the earth's rotation, derive the equations of motion for a long simple pendulum of length L supported in a frictionless spherical bearing. Such a pendulum was used by Foucault (1819–1868) in 1851 at the Panthéon in Paris in experiments attempting to demonstrate the earth's rotation relative to an "inertial" reference frame. Assume $\mathbf{g}=-g\mathbf{k}$, $\omega_e^2 \sim 0$, and $z \ll L$. Show that when the negligible terms $\dot{z}, \ddot{z}, z, x\dot{y}$, and $y\dot{y}$ are omitted, the following equations result

$$\ddot{x} = \frac{-gx}{L} + 2\omega_e \dot{y} \sin \lambda \qquad \ddot{y} = \frac{-gy}{L} - 2\omega_e \dot{x} \sin \lambda$$

If at $t=0$, $\dot{x}=\dot{y}=x=0$ and $y=A$, show that a solution to these two equations can be written in the form

$$\mathbf{r} = A \cos \sqrt{\frac{g}{L}}\, t\, [\sin (\omega_e \sin \lambda)t\mathbf{i} + \cos (\omega_e \sin \lambda)t\mathbf{j}]$$

The pendulum thus swings in a stationary plane that turns slowly clockwise relative to the rotating earth at a rate $\omega_e \sin \lambda$. [For other initial conditions, elliptic motion results. The semimajor axis of the resulting ellipse precesses at the rate $2\pi/(\omega_e \sin \lambda)$.]

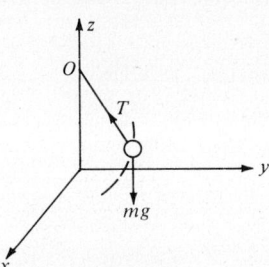

5.35 What is the local acceleration of gravity on the surface of the earth at a latitude of 30°? What is the angle a plumb line makes with the local vertical at that latitude? Assume the earth is a perfect sphere ($GM_e = 3.988 \times 10^{14}$ m³/s², $R = 6.371 \times 10^6$ m).

5.36 In problem 5.35, determine the magnitude of the tension in the string. Let $m=1$ kg.

5.37 What is the total force exerted by the earth on a 1500-kg automobile which is located at 30° latitude north of the equator traveling northeast at 90 km/h.

5.38 A rocket is fired vertically upward with an initial speed v_0 from a launching point on the earth's surface at a latitude angle Nλ degrees. Neglecting friction, find an approximate expression for the location of the landing point due to Coriolis deviation.

5.39 In problem 5.38, if $v_0 = 600$ m/s and $\lambda = 40°$N, find the location of the landing point.

5.40 In problem 5.38, determine separately the Coriolis deviation during the rocket's upward voyage and also when it returns. Check to see that the sum of the two parts equals the result found for problem 5.38.

RIGID BODY PLANE MOTION 6

The discussion in Chapter 4 concerned the dynamic principles governing the motion of a system of particles. If the system of particles becomes a continuum and the measured distances between points in the system remain constant, the system is said to be a rigid body. The same laws of motion that influence a system of particles must also govern the motion of a rigid body. The difference is that with a continuum present, the summation of physical quantities for discrete particles now becomes an integration over the whole volume.

The dynamic principles for a rigid body will now be developed. The subject is so broad that it will be covered in this and a following chapter (Chapter 7). This chapter will concentrate solely on the plane motion of a rigid body. Plane motion of a rigid body allows for 3 degrees of freedom: two in translation, one in rotation. Chapter 7 will study three-dimensional motion when a maximum of 6 degrees of freedom are allowed for each body.

6.1 EQUATIONS OF MOTION

Equation (4.21) for a system of particles states that

$$\mathbf{F}_{ext} = M\mathbf{a}_{c.m.} \tag{6.1}$$

This equation holds true when the body is rigid as well. For plane motion, equation (6.1) becomes, in cartesian coordinates,

$$F_x = M(a_{c.m.})_x \tag{6.2}$$

$$F_y = M(a_{c.m.})_y \tag{6.3}$$

Similar expressions can be formed for the normal and tangential coordinate set,

$$F_n = M(a_{c.m.})_n = \frac{Mv_{c.m.}^2}{\rho} \tag{6.4}$$

$$F_t = M(a_{c.m.})_t = M\dot{v}_{c.m.} \tag{6.5}$$

and for the radial and transverse coordinates,

$$F_r = M(a_{c.m.})_r = M(\ddot{R}_{c.m.} - R_{c.m.}\dot{\theta}^2) \tag{6.6}$$

$$F_\theta = M(a_{c.m.}) = M(R_{c.m.}\ddot{\theta} + 2\dot{R}_{c.m.}\dot{\theta}) \tag{6.7}$$

The moment equations derived to govern rotation for a system of particles were

158 REVIEW OF NEWTONIAN MECHANICS

equations (4.29) and (4.37):

$$\mathbf{M}_O = \frac{d\mathbf{H}_O}{dt} \tag{4.29}$$

$$\mathbf{M}_O = \frac{d\mathbf{H}_O}{dt} \tag{4.37}$$

$\mathbf{H}_O$ and $\mathbf{H}_{c.m.}$ were defined to be

$$\mathbf{H}_O = \sum_{i=1}^{n} \mathbf{r}_i \times m_i \mathbf{v}_i \tag{4.30}$$

$$\mathbf{H}_{c.m.} = \sum_{i=1}^{n} \boldsymbol{\rho}_i \times m_i \dot{\boldsymbol{\rho}}_i \tag{4.34}$$

For a rigid body, the summation process can be replaced by an integration. Equations (4.30) and (4.34) would then become

$$\mathbf{H}_O = \int_{\text{mass}} \mathbf{r} \times \mathbf{v} \, dm \tag{6.8}$$

$$\mathbf{H}_{c.m.} = \int_{\text{mass}} \boldsymbol{\rho} \times \dot{\boldsymbol{\rho}} \, dm \tag{6.9}$$

integrated over the entire mass of the body. For rigid body rotation about a fixed point, the velocity of an arbitrary point fixed to the body is

$$\mathbf{v} = \boldsymbol{\omega} \times \mathbf{r} \tag{6.10}$$

For a rigid body having general motion,

$$\dot{\boldsymbol{\rho}} = \boldsymbol{\omega} \times \boldsymbol{\rho} \tag{6.11}$$

where $\boldsymbol{\omega}$ is the absolute angular velocity of the body. Substitution into equations (6.8) and (6.9) yields

$$\mathbf{H}_O = \int_{\text{mass}} \mathbf{r} \times (\boldsymbol{\omega} \times \mathbf{r}) \, dm \tag{6.12}$$

$$\mathbf{H}_{c.m.} = \int_{\text{mass}} \boldsymbol{\rho} \times (\boldsymbol{\omega} \times \boldsymbol{\rho}) \, dm \tag{6.13}$$

Using vector algebra, equation (6.12) can be simplified to

$$\mathbf{H}_O = \boldsymbol{\omega} \int_{\text{mass}} \mathbf{r} \cdot \mathbf{r} \, dm - \int_{\text{mass}} \mathbf{r}(\mathbf{r} \cdot \boldsymbol{\omega}) \, dm$$

$$= (H_O)_x \mathbf{i} + (H_O)_y \mathbf{j} + (H_O)_z \mathbf{k} \tag{6.14}$$

$$(H_O)_x = \omega_x \int_{\text{mass}} (y^2 + z^2) \, dm - \omega_y \int_{\text{mass}} xy \, dm - \omega_z \int_{\text{mass}} xz \, dm \tag{6.15}$$

$$(H_O)_y = \omega_y \int_{\text{mass}} (x^2 + z^2) \, dm - \omega_x \int_{\text{mass}} yx \, dm - \omega_z \int_{\text{mass}} yz \, dm \tag{6.16}$$

$$(H_O)_z = \omega_z \int_{\text{mass}} (x^2 + y^2)\, dm - \omega_x \int_{\text{mass}} zx\, dm - \omega_y \int_{\text{mass}} zy\, dm \qquad (6.17)$$

In a similar derivation to find the angular momentum of the body about its center of mass, the results will *appear* identical to equations (6.14) to (6.17). The distances x, y, z would be measured from the center of mass of the body instead of from the fixed point O. The integrals in equations (6.15) to (6.17) have a special significance. Their definitions follow.

The *mass moments of inertia* are defined to be

$$I_{xx} = \int_{\text{mass}} (y^2 + z^2)\, dm \qquad \text{kg} \cdot \text{m}^2 \qquad (6.18)$$

$$I_{yy} = \int_{\text{mass}} (x^2 + z^2)\, dm \qquad (6.19)$$

$$I_{zz} = \int_{\text{mass}} (x^2 + y^2)\, dm \qquad (6.20)$$

The *mass products of inertia* are

$$I_{xy} = I_{yx} = -\int_{\text{mass}} xy\, dm \qquad \text{kg} \cdot \text{m}^2 \qquad (6.21)$$

$$I_{xz} = I_{zx} = -\int_{\text{mass}} zx\, dm \qquad (6.22)$$

$$I_{yz} = I_{zy} = -\int_{\text{mass}} zy\, dm \qquad (6.23)$$

The mass product of inertia referenced to two axes is zero when either of the axes is perpendicular to a plane of symmetry.

The mass moment of inertia is the second moment of mass distribution about an axis. The first moment of mass distribution about an axis led to the definition of the center of mass, that is,

$$\mathbf{R}_{\text{c.m.}} \equiv \frac{\sum_{i=1}^{n} m_i \mathbf{r}_i}{\sum_{i=1}^{n} m_i} \qquad (4.1)$$

or
$$\mathbf{R}_{\text{c.m.}} = \frac{\int_{\text{mass}} \mathbf{r}\, dm}{\int_{\text{mass}} dm} \qquad (6.24)$$

In a similar manner one can define the *radius of gyration* of a body with respect to an axis b as

$$k_b^2 \equiv \frac{I_{bb}}{M} \qquad (6.25)$$

For a system of particles this becomes

$$k_b^2 \equiv \frac{\sum_{i=1}^{n} m_i r_i^2}{\sum_{i=1}^{n} m_i} \qquad (6.26)$$

and for a rigid body,

$$k_b^2 = \frac{\int_{mass} r^2\, dm}{\int_{mass} dm} \qquad (6.27)$$

In equations (6.26) and (6.27), r is the perpendicular distance from the reference axis to the mass element in question. The radius of gyration has units of meters. Formulas for the values of mass moments and products of inertia of common shapes are given in Appendix B.

Returning now to equations (6.15) to (6.17), one can write

$$(H_O)_x = \omega_x I_{xx} + \omega_y I_{xy} + \omega_z I_{xz} \qquad (6.28)$$

$$(H_O)_y = \omega_y I_{yy} + \omega_x I_{yx} + \omega_z I_{yz} \qquad (6.29)$$

$$(H_O)_z = \omega_z I_{zz} + \omega_x I_{zx} + \omega_y I_{zy} \qquad (6.30)$$

where it is recognized that the *mass moments and products are referred to the fixed point O.*

In matrix form, equations (6.28) to (6.30) become

$$\begin{Bmatrix}(H_O)_x \\ (H_O)_y \\ (H_O)_z\end{Bmatrix} = \begin{bmatrix} I_{xx} & I_{xy} & I_{xz} \\ I_{yx} & I_{yy} & I_{yz} \\ I_{zx} & I_{zy} & I_{zz}\end{bmatrix}\begin{Bmatrix}\omega_x \\ \omega_y \\ \omega_z\end{Bmatrix} \qquad (6.31)$$

With the center of mass as the reference point, equation (6.31) would be rewritten as

$$\begin{Bmatrix}(H_{c.m.})_x \\ (H_{c.m.})_y \\ (H_{c.m.})_z\end{Bmatrix} = \begin{bmatrix} I_{xx} & I_{xy} & I_{xz} \\ I_{yx} & I_{yy} & I_{yz} \\ I_{zx} & I_{zy} & I_{zz}\end{bmatrix}\begin{Bmatrix}\omega_x \\ \omega_y \\ \omega_z\end{Bmatrix} \qquad (6.32)$$

where it is understood that *mass moments and products of inertia are referred now to the center of mass.*

Equations (6.31) and (6.32) can usually be simplified. For a rigid body in plane motion $\omega_x = \omega_y = 0$. If in turn the body has a plane of symmetry perpendicular to the z axis, the mass products of inertia I_{zx} and I_{zy} vanish. Equations (6.31) and (6.32) then simply become

$$(H_O)_z = I_{zz}\omega_z \qquad (6.33)$$

$$(H_{c.m.})_z = \hat{I}_{zz}\omega_z \qquad (6.34)$$

The circumflex has been added in equation (6.34) to emphasize that the mass moment of inertia is referenced to the center of mass. In equation (6.33), the mass moment of inertia is referenced to the fixed point O.

For the general case I_{xz} and I_{yz} are not zero. As will be shown in a later section, however, every rigid body contains principal axes, or three mutually orthogonal directions in the body about which the mass products of inertia vanish. The mass moments of inertia for these axes are called *principal moments of inertia*. For these axes

$$\mathbf{H} = H_x\mathbf{i} + H_y\mathbf{j} + H_z\mathbf{k} \tag{6.35}$$

$$H_x = I_1\omega_x \tag{6.36}$$

$$H_y = I_2\omega_y \tag{6.37}$$

$$H_z = I_3\omega_z \tag{6.38}$$

where H and I must refer to the same point in equations (6.36) to (6.38), e.g., the center of mass or a fixed point. In matrix form this becomes

$$\begin{Bmatrix} H_x \\ H_y \\ H_z \end{Bmatrix} = \begin{bmatrix} I_1 & 0 & 0 \\ 0 & I_2 & 0 \\ 0 & 0 & I_3 \end{bmatrix} \begin{Bmatrix} \omega_x \\ \omega_y \\ \omega_z \end{Bmatrix} \tag{6.39}$$

where the x, y, and z directions are principal directions.

In general,

$$\{H\} = [I]\{\omega\} \tag{6.40}$$

and the vector $\mathbf{H}$ does not lie along the same direction as ω. The inertia matrix $[I]$ represents a second-order tensor that operates on ω to produce $\mathbf{H}$. Only when the angular velocity acts along a principal direction does $\mathbf{H}$ and ω become collinear. Then one can simply write

$$\mathbf{H} = I\omega \tag{6.41}$$

where I is defined to be the principal moment of inertia about the axis of rotation. As will be seen in Section 6.3, this offers the theoretical basis for calculating the principal moment of inertia values.

When applying the moment equations, equation (4.29) or (4.37), it is important that the coordinate system used to describe the motion be fixed to the body. With such coordinate axes, the body's relative geometry with respect to the coordinate system does not change with time; all mass moment and product of inertia terms are constant.

Since this chapter will concentrate only on the plane motion of rigid bodies having a plane of symmetry perpendicular to the z axis, equations (4.29) and (4.37) become simply

$$(M_O)_z = I_{zz}\dot{\omega}_z \tag{6.42}$$

$$(M_{c.m.})_z = \hat{I}_{zz}\dot{\omega}_z \tag{6.43}$$

To summarize, for *plane motion* constrained to the xy plane with rotation solely about the principal axis z of the body, the three governing equations for the 3 degrees of freedom become

$$F_x = M(a_{c.m.})_x \tag{6.2}$$

$$F_y = M(a_{c.m.})_y \tag{6.3}$$

$$(M_O)_z = I_{zz}\dot{\omega}_z \tag{6.42}$$

or

$$(M_{c.m.})_z = \hat{I}_{zz}\dot{\omega}_z \tag{6.43}$$

Sample Problem 6.1

A 45-kg, 60-cm radius eccentric wheel moves down an inclined plane as shown. What is its angular acceleration for this instant when the velocity of the center of the wheel is 1.2 m/s? The coefficient of static friction is 0.2; that of sliding friction is 0.1. The radius of gyration for the wheel about its center of mass is given as $\hat{k} = 50$ cm.

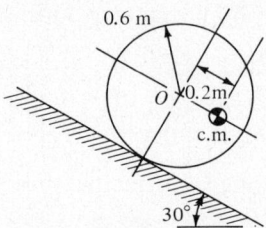

Analysis: The analysis begins with a free-body diagram drawn of the system. The body will first be assumed not to slide. Friction forces necessary for this motion will then be solved for and compared with the maximum allowable by employing the coefficient of static friction. If this value was exceeded, the body is known to slide. A new analysis is called for with a calculated friction force; the friction force would be obtained using the coefficient of sliding friction. The acceleration of the center of the wheel, however, would no longer equal $R\alpha$. If the friction force is less than the maximum, the initial assumption was correct, and the problem can proceed. When writing the equations of motion for the system, proper attention must be given to the *acceleration of the center of mass*. The center of mass does not lie at the center of the wheel here.

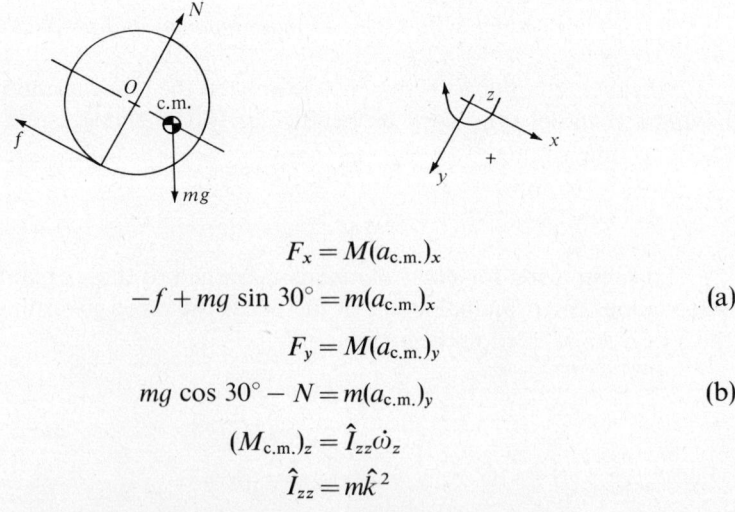

$$F_x = M(a_{c.m.})_x$$
$$-f + mg \sin 30° = m(a_{c.m.})_x \tag{a}$$
$$F_y = M(a_{c.m.})_y$$
$$mg \cos 30° - N = m(a_{c.m.})_y \tag{b}$$
$$(M_{c.m.})_z = \hat{I}_{zz} \dot{\omega}_z$$
$$\hat{I}_{zz} = m\hat{k}^2$$
$$0.6f + 0.2N = m(0.50)^2 \alpha \tag{c}$$

Since rolling without sliding was assumed,

$$v_O = (v_O)_x = R\omega = 1.2 \text{ m/s}$$
$$a_O = (a_O)_x = R\alpha = R\dot{\omega}_z$$
$$\mathbf{a}_{c.m.} = \mathbf{a}_O + \mathbf{a}_{c.m./O}$$
$$\mathbf{a}_{c.m.} = \mathbf{a}_O + (\mathbf{a}_{c.m./O})_n + (\mathbf{a}_{c.m./O})_t$$
$$(\mathbf{a}_{c.m./O})_n = -0.2\omega^2 \mathbf{i} \qquad (\mathbf{a}_{c.m./O})_t = 0.2\alpha \mathbf{j}$$
$$\mathbf{a}_O = 0.6\alpha \mathbf{i}$$

However,

$$\omega = \frac{1.2}{0.6} = 2\mathbf{k} \quad \text{rad/s}$$

therefore,

$$\mathbf{a}_{c.m.} = [0.6\alpha - 0.2(2^2)]\mathbf{i} + 0.2\alpha \mathbf{j}$$

Rewriting equations (a) and (b),

$$-f + mg \sin 30° = m[0.6\alpha - 0.2(2^2)] \qquad (d)$$
$$mg \cos 30° - N = m(0.2\alpha) \qquad (e)$$

and combining equations (c) to (e) yields for the friction force

$$f = 43.96 N$$

The normal force is

$$N = 311.4 \text{ N} \qquad \mu_{\text{required}} = \frac{f}{N} = \frac{43.96}{311.4} = 0.14$$

This is below the value of the static coefficient of friction. The original assumption was therefore correct, and sliding does not occur. Finally, solving for α,

$$\alpha = 7.88\mathbf{k} \quad \text{rad/s}^2$$

The reader should note that the normal force does not equal the weight of the wheel. ∎

6.2 THE MASS MOMENT OF INERTIA MATRIX: TRANSFORMATIONS

In the previous section the importance of the mass moment of inertia matrix to the body's equations of motion was discussed. It is the intent of this section to describe how the matrix $[I]$ transforms due to coordinate system translation and also how it changes under a coordinate axes rotation.

Shown in Figure 6.1 is a rigid body whose mass moment and product of inertia matrix referenced to the center of mass is known. The terms of the matrix

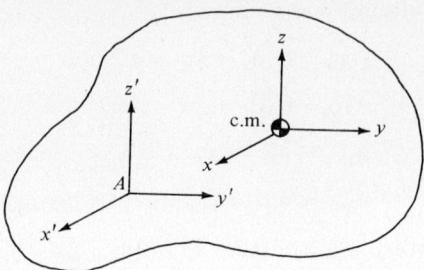

Figure 6.1

will now be transformed from one coordinate set to another set that is *parallel* to the first. For the new coordinate system at point A,

$$I_{x'x'} = \int_{\text{mass}} (y'^2 + z'^2)\, dm$$

$$= \int_{\text{mass}} [(y + y_{\text{c.m.}})^2 + (z + z_{\text{c.m.}})^2]\, dm \tag{6.44}$$

$$I_{x'x'} = \int_{\text{mass}} (y^2 + z^2)\, dm + 2y_{\text{c.m.}} \int_{\text{mass}} y\, dm + 2z_{\text{c.m.}} \int_{\text{mass}} z\, dm$$

$$+ (y_{\text{c.m.}}^2 + z_{\text{c.m.}}^2) \int_{\text{mass}} dm$$

This reduces to equation (6.45) since the second and third integrals vanish.

$$I_{x'x'} = \hat{I}_{xx} + M(y_{\text{c.m.}}^2 + z_{\text{c.m.}}^2) \tag{6.45}$$

Similarly,

$$I_{y'y'} = \hat{I}_{yy} + M(x_{\text{c.m.}}^2 + z_{\text{c.m.}}^2) \tag{6.46}$$

$$I_{z'z'} = \hat{I}_{zz} + M(x_{\text{c.m.}}^2 + y_{\text{c.m.}}^2) \tag{6.47}$$

Equations (6.45) to (6.47) describe the relationship between the mass moments of inertia of a body about an arbitrary point to the moments of inertia about the center of mass of the body referenced to coordinate axes that are parallel. Only equation (6.47) is of interest for plane motion. Equation (6.47) can be rewritten as

$$I_{zz} = \hat{I}_{zz} + Md^2 \tag{6.48}$$

where d is the perpendicular distance between the parallel axes z and z', located at the center of mass and the arbitrary point, respectively. Equation (6.48) is commonly referred to as the *parallel-axis theorem*. Now

$$I_{x'y'} = -\int_{\text{mass}} x'y' \, dm \tag{6.49}$$

$$= -\int_{\text{mass}} (x + x_{\text{c.m.}})(y + y_{\text{c.m.}}) \, dm$$

$$= -\int_{\text{mass}} xy \, dm - x_{\text{c.m.}} \int_{\text{mass}} y \, dm - y_{\text{c.m.}} \int_{\text{mass}} x \, dm - x_{\text{c.m.}} y_{\text{c.m.}} \int_{\text{mass}} dm$$

$$= -\int_{\text{mass}} xy \, dm - x_{\text{c.m.}} y_{\text{c.m.}} \int_{\text{mass}} dm$$

Then
$$I_{x'y'} = \hat{I}_{xy} - M x_{\text{c.m.}} y_{\text{c.m.}} \tag{6.50}$$

$$I_{x'z'} = \hat{I}_{xz} - M x_{\text{c.m.}} z_{\text{c.m.}} \tag{6.51}$$

$$I_{y'z'} = \hat{I}_{yz} - M y_{\text{c.m.}} z_{\text{c.m.}} \tag{6.52}$$

Equations (6.50) to (6.52) transform given mass products of inertia about the center of mass in one cartesian coordinate system to products of inertia referred to an arbitrary point, a distance $x_{\text{c.m.}}$, $y_{\text{c.m.}}$, $z_{\text{c.m.}}$ away, for a parallel coordinate set.

For *coordinate rotation*, when the mass moments and products of inertia are referenced to two sets of axes emanating from a *common point* but having different inclinations, an entirely different problem arises when transformation of the inertia matrix is attempted. To see how this transformation is affected, the angular momentum expression will first be reintroduced. The matrix expression for angular momentum is

$$\{H\} = [I]\{\omega\} \tag{6.40}$$

Since $\{H\}$ and $\{\omega\}$ are vectors, they must transform like vectors do under a coordinate system transformation. That is to say, for a given direction cosine matrix $[l]$,

$$\{H'\} = [l]\{H\} \tag{6.53}$$

$$\{\omega'\} = [l]\{\omega\} \tag{6.54}$$

[see equation (5.78) and Section 5.6C].

Then
$$\{H'\} = [l]\{H\} = [l]([I]\{\omega\})$$

$$\{H'\} = [l][I][l]^{-1}[l]\{\omega\}$$

$$\{H'\} = [l][I][l]^{-1}\{\omega'\}$$

In the new coordinate system
$$\{H'\} = [I']\{\omega'\} \tag{6.55}$$

Then
$$[I'] = [l][I][l]^{-1} \tag{6.56}$$

Equation (6.56) represents a *similarity transformation* and defines how a second-order tensor, such as the mass moment of inertia matrix, transforms under a coordinate axes rotation. Since

$$[l]^t = [l]^{-1} \tag{5.80}$$
$$[I'] = [l][I][l]^t \tag{6.57}$$
$$[I] = [l]^t[I'][l] \tag{6.58}$$

A sample problem illustrating these two types of transformations follows.

Sample Problem 6.2
The homogeneous cylinder shown with an overall length of s and a radius of $s/3$ has a total mass M. Find its mass moments and products of inertia referenced to a coordinate system located at point O and having an inclination of $45°$ with the bottom plane.

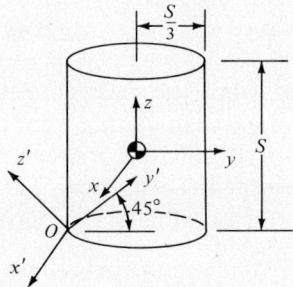

Analysis: Referring to Appendix B,

$$\hat{I}_{xy} = \hat{I}_{xz} = \hat{I}_{yz} = 0$$

$$\hat{I}_{xx} = \hat{I}_{yy} = \frac{M}{12}\left[3\left(\frac{s}{3}\right)^2 + s^2\right] = \frac{Ms^2}{9}$$

$$\hat{I}_{zz} = \frac{M}{2}\left(\frac{s}{3}\right)^2 = \frac{Ms^2}{18}$$

Using the parallel-axis theorem to transfer results to point O,

$$\mathbf{R}_{c.m.} = \frac{s}{3}\mathbf{j} + \frac{s}{2}\mathbf{k}$$

$$I_{xx} = \frac{Ms^2}{9} + M\left[\left(\frac{s}{2}\right)^2 + \left(\frac{s}{3}\right)^2\right] = \frac{17}{36}Ms^2$$

$$I_{yy} = \frac{Ms^2}{9} + M\left(\frac{s}{2}\right)^2 = \frac{13}{36}Ms^2$$

$$I_{zz} = \frac{Ms^2}{18} + M\left(\frac{s}{3}\right)^2 = \frac{Ms^2}{6}$$

$$I_{xy} = I_{xz} = 0 \qquad I_{yz} = 0 - M\left(\frac{s}{2}\right)\left(\frac{s}{3}\right) = -\frac{Ms^2}{6}$$

$$[l] = \begin{bmatrix} 1 & 0 & 0 \\ 0 & \cos 45° & \sin 45° \\ 0 & -\sin 45° & \cos 45° \end{bmatrix}$$

$$[l]^t = \begin{bmatrix} 1 & 0 & 0 \\ 0 & \cos 45° & -\sin 45° \\ 0 & \sin 45° & \cos 45° \end{bmatrix}$$

$$[I] = \frac{Ms^2}{36}\begin{bmatrix} 17 & 0 & 0 \\ 0 & 13 & -6 \\ 0 & -6 & 6 \end{bmatrix}$$

$$[l][I] = \frac{Ms^2}{72}\begin{bmatrix} 34 & 0 & 0 \\ 0 & 7\sqrt{2} & 0 \\ 0 & -19\sqrt{2} & 12\sqrt{2} \end{bmatrix}$$

$$[l][I][l]^t = \frac{Ms^2}{72}\begin{bmatrix} 34 & 0 & 0 \\ 0 & 7 & -7 \\ 0 & -7 & 31 \end{bmatrix} \qquad \blacksquare$$

In this sample problem note that the trace of $[I]$ remains constant during a similarity transformation. This holds true in general for any matrix. The trace of a matrix is the sum of the diagonal terms:

$$\text{Tr}[I] = \sum_{k=1}^{n} I_{kk} \qquad (6.59)$$

It can also be shown that the antisymmetry properties as well as the value of the matrix determinant do not change during a similarity transformation.

6.3 EIGENVALUES AND EIGENVECTORS OF THE INERTIA MATRIX

In the previous section it was shown how the mass inertia matrix transforms under a coordinate rotation. The question now becomes, Is it possible to rotate a given coordinate about a point to find an orientation where *all* the products of inertia vanish? For simplification purposes, it is fortunate that the answer to this question is yes. The found directions are known as the *eigenvectors* or *principal directions*. The mass moments of inertia about these axes are called the *eigenvalues* or *principal mass moments of inertia*.

Recall from Section 6.1 that if a body rotates about a principal direction its

168 REVIEW OF NEWTONIAN MECHANICS

angular momentum vector **H** is parallel to the angular velocity vector ω,

$$\mathbf{H} = I\boldsymbol{\omega} \qquad (6.41)$$

Here I is the principal moment of inertia value about the principal direction. The principal direction is

$$\mathbf{e} = \boldsymbol{\omega}/\omega \qquad (6.60)$$

$$\mathbf{e} = e_1\mathbf{i} + e_2\mathbf{j} + e_3\mathbf{k} \qquad (6.61)$$

$$\mathbf{e} = \{e\} \qquad (6.62)$$

In the **i, j, k** coordinate system

$$\boldsymbol{\omega} = \omega_x\mathbf{i} + \omega_y\mathbf{j} + \omega_z\mathbf{k} \qquad (6.63)$$

and

$$\begin{Bmatrix} H_x \\ H_y \\ H_z \end{Bmatrix} = \begin{bmatrix} I_{xx} & I_{xy} & I_{xz} \\ I_{yx} & I_{yy} & I_{yz} \\ I_{zx} & I_{zy} & I_{zz} \end{bmatrix} \begin{Bmatrix} \omega_x \\ \omega_y \\ \omega_z \end{Bmatrix} \qquad (6.32)$$

or

$$\{H\} = [I]\{\omega\} \qquad (6.40)$$

If the body rotates about a principal direction, then

$$\{H\} = [I]\{\omega\} = I\{\omega\} \qquad (6.64)$$

$$[I]\omega\{e\} = I\omega\{e\}$$

$$[I]\{e\} = I\{e\} \qquad (6.65)$$

Be careful not to confuse the inertia matrix $[I]$ with a principal moment of inertia value I. One can rewrite equation (6.65) as

$$([I] - [1]I)\{e\} = \{0\} \qquad (6.66)$$

where $[1]$ is the identity matrix.

Equation (6.66) written out becomes

$$\begin{bmatrix} I_{xx} - I & I_{xy} & I_{xz} \\ I_{yx} & I_{yy} - I & I_{yz} \\ I_{zx} & I_{zy} & I_{zz} - I \end{bmatrix} \begin{Bmatrix} e_1 \\ e_2 \\ e_3 \end{Bmatrix} = \{0\} \qquad (6.67)$$

Equation (6.67) represents a set of linear homogeneous equations. The solution for the set only becomes nontrivial if the determinant value of the coefficient matrix is zero. That is,

$$|[I] - [1]I| = 0 \qquad (6.68)$$

$$\begin{vmatrix} I_{xx} - I & I_{xy} & I_{xz} \\ I_{yx} & I_{yy} - I & I_{yz} \\ I_{zx} & I_{zy} & I_{zz} - I \end{vmatrix} = 0 \qquad (6.69)$$

For a three-dimensional problem the expanded determinant, also called the *characteristic equation*, must be a cubic

$$aI^3 + bI^2 + cI + d = 0 \tag{6.70}$$

The three roots I_1, I_2, and I_3 are the three *principal moments of inertia*, or *eigenvalues*. The problem started out looking for one value but instead found three. It can be shown that for a *symmetrical matrix with distinct eigenvalues*, the eigenvectors are *orthogonal*. The eigenvalues and eigenvectors are also real for real $[I]$. (Refer to any advanced mathematics textbook on matrix theory for the proof.) When found, these orthogonal eigenvectors, or principal directions, can be conveniently used as the body-fixed coordinates to solve dynamic problems. The angular momentum expression then simplifies to

$$\begin{Bmatrix} H_x \\ H_y \\ H_z \end{Bmatrix} = \begin{bmatrix} I_1 & 0 & 0 \\ 0 & I_2 & 0 \\ 0 & 0 & I_3 \end{bmatrix} \begin{Bmatrix} \omega_x \\ \omega_y \\ \omega_z \end{Bmatrix} \tag{6.71}$$

in the *new* x, y, z coordinate system lying along the principal directions. Returning again to equation (6.67) and substituting the distinct eigenvalue just found, for example, I_1, then

$$\begin{aligned}(I_{xx} - I_1)e_{11} + I_{xy}e_{12} + I_{xz}e_{13} &= 0 \\ I_{yx}e_{11} + (I_{yy} - I_1)e_{12} + I_{yz}e_{13} &= 0 \\ I_{zx}e_{11} + I_{zy}e_{12} + (I_{zz} - I_1)e_{13} &= 0\end{aligned} \tag{6.72}$$

Equation (6.72) represents a set of three homogeneous equations in terms of the unknowns e_{11}, e_{12}, and e_{13}. These unknowns are the components of the unit vector for the first principal direction $\mathbf{e}_1$. In other words, they are the direction cosines of the first principal axis. They cannot be solved for because the set of equations is homogeneous: the rank of the coefficient matrix is less than the number of unknowns present. The best that can be done here is to assume a value for one term and solve for the other two in terms of the first. Since only a *direction* is wanted, the exact magnitudes for e_{11}, e_{12}, e_{13} are inconsequential. Using any two of the subequations from equation (6.72), since a third equation provides no new information, the directions are solved for in terms of one arbitrary variable as follows:

$$\begin{aligned}I_{xy}\frac{e_{12}}{e_{11}} + I_{xz}\frac{e_{13}}{e_{11}} &= -(I_{xx} - I_1) \\ (I_{yy} - I_1)\frac{e_{12}}{e_{11}} + I_{yz}\frac{e_{13}}{e_{11}} &= -I_{yx}\end{aligned} \tag{6.73}$$

The ratios e_{12}/e_{11} and e_{13}/e_{11}, when known, completely describe the direction of the first eigenvalue.

Repeating the calculations for I_2 yields ratios for e_{21}, e_{22}, and e_{23}. Similar calculations for I_3 give results for e_{31}, e_{32}, and e_{33}. Keep in mind that the e's are the direction cosines of the principal directions just found; the $[E]$ matrix is the direction cosine matrix linking the old coordinate axes to the new coordinate axes oriented along the principal directions. This direction cosine matrix can be thought of as operating upon the moment of inertia matrix through a similarity transformation to produce a new moment of inertia matrix for the principal directions *that is diagonal*.

$$[E][I][E]^{-1} = \begin{bmatrix} I_1 & 0 & 0 \\ 0 & I_2 & 0 \\ 0 & 0 & I_3 \end{bmatrix} \quad (6.74)$$

Recall that $[E]^{-1} = [E]^t$ and (6.75)

$$[E] = \begin{bmatrix} e_{11} & e_{12} & e_{13} \\ e_{21} & e_{22} & e_{23} \\ e_{31} & e_{32} & e_{33} \end{bmatrix} \quad (6.76)$$

Sometimes the eigenvalues solved for from equation (6.70) are not all distinct. Arbitrary mutually orthogonal directions in space can then be selected as principal directions for the eigenvalues with repeated roots, since direct solutions would not be feasible. If two eigenvalues are equal, for example, two mutually orthogonal directions perpendicular to the known direction (for the distinct eigenvalue) can be used for eigenvectors. A cylinder represents such a problem, having one distinct eigenvalue along its longitudinal axis. Any two arbitrary perpendicular directions in a transverse plane passing through the center of mass of the cylinder will be suitable for the other two eigenvectors. A sphere has three identical eigenvalues, and any three mutually orthogonal directions can be used for principal axes.

Before proceeding to a sample problem, one further idea should be presented. Equation (6.56) in expanded form for an arbitrary axis passing through a point is

$$I_{x'x'} = l_{x'x}^2 I_{xx} + l_{x'y}^2 I_{yy} + l_{x'z}^2 I_{zz} + 2l_{x'x}l_{x'y}I_{xy} + 2l_{x'x}l_{x'z}I_{xz} + 2l_{x'y}l_{x'z}I_{yz} \quad (6.77)$$

Dividing by the magnitude $I_{x'x'}$ yields

$$1 = I_{xx}a_1^2 + I_{yy}a_2^2 + I_{zz}a_3^2 + 2a_1a_2 I_{xy} + 2a_1a_3 I_{xz} + 2a_2a_3 I_{yz} \quad (6.78)$$

where $\quad a_1 = \dfrac{l_{x'x}}{\sqrt{I_{x'x'}}} \quad a_2 = \dfrac{l_{x'y}}{\sqrt{I_{x'x'}}} \quad a_3 = \dfrac{l_{x'z}}{\sqrt{I_{x'x'}}} \quad (6.79)$

Equation (6.78) is an equation of an ellipsoid; in particular, it is called the *inertia ellipsoid*. Any point on a rigid body has an inertia ellipsoid associated with it. The magnitude of the square of the radius vector, measured from a point on the body to the surface of the inertia ellipsoid along the given direction, is inversely proportional to the moment of inertia for the body, along that direction. If principal axes are used, equation (6.78) becomes

$$1 = I_1 a_1'^2 + I_2 a_2'^2 + I_3 a_3'^2 \quad (6.80)$$

Figure 6.2 shows an arbitrary ellipsoid of inertia for a body about a general point O.

RIGID BODY PLANE MOTION   171

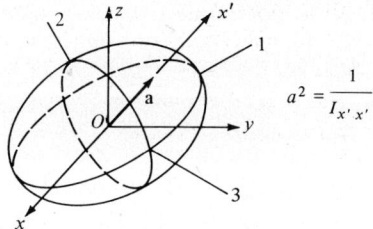

$$a^2 = \frac{1}{I_{x'x'}}$$

Figure 6.2

Sample Problem 6.3

The mass moment of inertia matrix for a body about a given point is known to be

$$[I] = \begin{bmatrix} 900 & 0 & 0 \\ 0 & 400 & 300 \\ 0 & 300 & 1200 \end{bmatrix} \text{ kg} \cdot \text{m}^2$$

Find the principal moments of inertia and the principal directions for this inertia matrix.

Analysis:

$$([I] - [1]I)\{e\} = \{0\}$$

$$|[I] - [1]I| = 0$$

$$\begin{vmatrix} 900 - I & 0 & 0 \\ 0 & 400 - I & 300 \\ 0 & 300 & 1200 - I \end{vmatrix} = 0$$

$$(900 - I)(400 - I)(1200 - I) - (900 - I)(300)(300) = 0$$

$$I_1 = 900 \text{ kg} \cdot \text{m}^2$$

$$(400 - I)(1200 - I) - 300(300) = 0$$

$$(I - 1300)(I - 300) = 0$$

$$I_2 = 1300 \text{ kg} \cdot \text{m}^2 \qquad I_3 = 300 \text{ kg} \cdot \text{m}^2$$

Returning now to the *first eigenvalue,*

$$\begin{bmatrix} 900 - I_1 & 0 & 0 \\ 0 & 400 - I_1 & 300 \\ 0 & 300 & 1200 - I_1 \end{bmatrix} \begin{Bmatrix} e_{11} \\ e_{12} \\ e_{13} \end{Bmatrix} = \{0\}$$

$$(900 - 900)e_{11} = 0 \qquad e_{11} = \text{any value}$$

Take $e_{11} = 1$.

$$(400 - 900)e_{12} + 300e_{13} = 0$$

$$300e_{12} + (1200 - 900)e_{13} = 0$$

172 REVIEW OF NEWTONIAN MECHANICS

With a coefficient determinant not equal to zero, e_{12} and e_{13} must equal to zero.

$$e_{11} = 1 \qquad e_{12} = 0 \qquad e_{13} = 0$$

For the *second eigenvalue*,

$$\begin{bmatrix} 900 - I_2 & 0 & 0 \\ 0 & 400 - I_2 & 300 \\ 0 & 300 & 1200 - I_2 \end{bmatrix} \begin{Bmatrix} e_{21} \\ e_{22} \\ e_{23} \end{Bmatrix} = \{0\}$$

$$(900 - 1300)e_{21} = 0$$

Hence, e_{21} must equal zero.

$$(400 - 1300)e_{22} + 300e_{23} = 0$$

$$300e_{22} + (1200 - 1300)e_{23} = 0$$

The last two equations are proportional. The determinant of the coefficient equals zero. They cannot be solved explicitly. Assuming a value of $e_{22} = 1/\sqrt{10}$, e_{23} becomes $3/\sqrt{10}$. Then $e_{21}^2 + e_{22}^2 + e_{23}^2 = 1 = |\mathbf{e}_2|$.

For the *third eigenvalue*,

$$(900 - 300)e_{31} = 0$$

Thus e_{31} must equal zero.

$$(400 - 300)e_{32} + 300e_{33} = 0$$

$$300e_{32} + (1200 - 300)e_{33} = 0$$

The last two equations are proportional. Taking first $e_{32} = 1$, e_{33} becomes $-\frac{1}{3}$. The normalized solution has $e_{31} = 0$, $e_{32} = 3/\sqrt{10}$, and $e_{33} = -1/\sqrt{10}$, so that $e_{31}^2 + e_{32}^2 + e_{33}^2 = 1 = |\mathbf{e}_3|$. The results are shown in the figure. Note that the eigenvectors, or principal directions, have *no positive or negative sense*. They represent only axes.

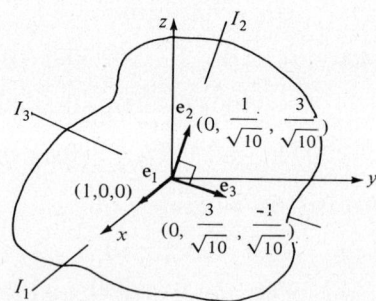

$$[I] = [E] = \begin{bmatrix} 1 & 0 & 0 \\ 0 & \dfrac{1}{\sqrt{10}} & \dfrac{3}{\sqrt{10}} \\ 0 & \dfrac{3}{\sqrt{10}} & -\dfrac{1}{\sqrt{10}} \end{bmatrix}$$

As a check,

$$[E][I][E]^t = \begin{bmatrix} 1 & 0 & 0 \\ 0 & \dfrac{1}{\sqrt{10}} & \dfrac{3}{\sqrt{10}} \\ 0 & \dfrac{3}{\sqrt{10}} & -\dfrac{1}{\sqrt{10}} \end{bmatrix} \begin{bmatrix} 900 & 0 & 0 \\ 0 & 400 & 300 \\ 0 & 300 & 1200 \end{bmatrix} \begin{bmatrix} 1 & 0 & 0 \\ 0 & \dfrac{1}{\sqrt{10}} & \dfrac{3}{\sqrt{10}} \\ 0 & \dfrac{3}{\sqrt{10}} & -\dfrac{1}{\sqrt{10}} \end{bmatrix}$$

$$= \begin{bmatrix} 900 & 0 & 0 \\ 0 & 1300 & 0 \\ 0 & 0 & 300 \end{bmatrix} \text{ kg} \cdot \text{m}^2 \qquad \blacksquare$$

The sample problem given was basically a two-dimensional problem where Mohr's circle construction could have been used for the solution. The third principal direction and moment of inertia is obvious by inspection from the given matrix. For a more difficult matrix, or for vibrational problems involving n dimensions, computer routines are necessary. There are several numerical methods available to solve for the eigenvalues of a n-dimensional real matrix. A few involve polynomial root determination for the characteristic equation, such as Bairstow's method. Others use iteration techniques to "sweep out" known solutions leaving a reduced problem to work with. Still others use iteration techniques to solve for all the eigenvalues and eigenvectors at once, e.g., Jacobi's method.

Sample Problem 6.4
Write a computer program that uses Jacobi's method to solve for the eigenvalues and eigenvectors of a 3 × 3 symmetric matrix.

Analysis: Starting with the symmetric matrix $[A]$ given by

$$[A] = \begin{bmatrix} a_{xx} & a_{xy} & a_{xz} \\ a_{xy} & a_{yy} & a_{yz} \\ a_{xz} & a_{yz} & a_{zz} \end{bmatrix}$$

Jacobi's method attempts to solve for the transformation matrix $[E]$ used in equation (6.74) that operates to diagonalize $[A]$.

$$[E][A][E]^t = [A^d] \qquad \text{(a)}$$

Here $[A^d]$ denotes that the result is a diagonal matrix. The rows of the $[E]$ matrix are the desired eigenvectors; the diagonal elements of $[A^d]$ are the eigenvalues. Jacobi's method uses an iterative approach. Each time, a coordinate rotation is found that reduces the largest absolute value nondiagonal term of the current $[A]$ matrix to zero through a similarity transformation. The succession of these rotation matrices will converge to the $[E]$ matrix. The last transformed $[A]$ matrix is the $[A^d]$ matrix. For example, if a_{xy} is the largest non-diagonal term of the $[A]$ matrix, a coordinate rotation through an angle θ about the z axis transforms $[A]$ as follows:

$$[A'] = [I][A][I]^t$$

$$[A'] = \begin{bmatrix} \cos\theta & \sin\theta & 0 \\ -\sin\theta & \cos\theta & 0 \\ 0 & 0 & 1 \end{bmatrix} \begin{bmatrix} a_{xx} & a_{xy} & a_{xz} \\ a_{xy} & a_{yy} & a_{yz} \\ a_{xz} & a_{yz} & a_{zz} \end{bmatrix} \begin{bmatrix} \cos\theta & -\sin\theta & 0 \\ \sin\theta & \cos\theta & 0 \\ 0 & 0 & 1 \end{bmatrix} \quad (a)$$

On expanding terms, the new a'_{xy} element is found to be

$$a'_{xy} = -(a_{xx} - a_{yy})\sin\theta\cos\theta + a_{xy}(\cos^2\theta - \sin^2\theta)$$

Setting the new a'_{xy} to zero yields a relation for θ,

$$\frac{\cos\theta\sin\theta}{\cos^2\theta - \sin^2\theta} = \frac{a_{xy}}{a_{xx} - a_{yy}} \qquad \theta = \tfrac{1}{2}\tan^{-1}\frac{2a_{xy}}{a_{xx} - a_{yy}} \quad (b)$$

Now knowing θ, the transformation defined by equation (a) will produce $[A']$, with the a'_{xy} terms equal to zero. The new $[A']$ matrix now supplants the $[A]$ matrix, and the procedure is repeated again and again until *all* the nondiagonal terms vanish. A cumulative result of the products of the successively formed rotation matrices will, at the end of the process, yield $[E]$. The elements of the last $[A]$ matrix are the eigenvalues.

In general, if $|a_{ij}|$ is the largest absolute value of an element of the $[A]$ matrix,

$$\tan 2\theta = \frac{2a_{ij}}{a_{ii} - a_{jj}} \quad (c)$$

where rotation is about the third axis. The angle θ lies between $-\pi/2$ and $\pi/2$. Solving for $\tan\theta$, it can be shown that

$$\tan\theta = \frac{\pm 2a_{ij}}{|a_{ii} - a_{jj}| + [(a_{ii} - a_{jj})^2 + 4a_{ij}^2]^{1/2}} \quad (d)$$

where the plus (+) sign is used for $a_{ii} > a_{jj}$. Note that the index i should be less than the index j in this procedure. Further

$$\cos\theta = (1 + \tan^2\theta)^{-1/2} \quad (e)$$

and

$$\sin\theta = \cos\theta\tan\theta \quad (f)$$

The developed program accomplishing these results is shown in Figure 6.3. The program is only applicable for symmetric matrices that are 3×3. ∎

The analytic tools are now in place to finally solve for the axis of rotation and the total angle that a rigid body rotates through about a fixed point (see Euler's theorem and also Section 5.6C). It was shown before that an orthogonal coordinate axes transformation, or a pure vector rotation, can be represented by

$$\{r'\} = [A]\{r\} \quad \text{or} \quad \{r_n\} = [A]\{r\} \quad (5.94)$$

```
00100 PROGRAM JACOBI(INPUT,OUTPUT)
00110C  THIS PROGRAM ONLY WORKS FOR REAL SYMMETRIC MATRICES
00120 DIMENSION A(3,3),B(3,3),E(3,3),E1(3,3),RL(3,3),RLT(3,3)
00130 DO 1 I=1,3
00140 DO 2 J=1,3
00150 E(I,J)=0.
00160 2 CONTINUE
00170 E(I,I)=1.
00180 1 CONTINUE
00190 READ *,((A(I,J),J=1,3),I=1,3)
00200 K1=0
00210C  SEARCH FOR LARGEST OFF-DIAGONAL TERM
00220 30 T=0.
00230 DO 3 I1=1,2
00240 M=I1+1
00250 DO 3 J1=M,3
00260 IF(T.GT.ABS(A(I1,J1))) GO TO 3
00270 T=ABS(A(I1,J1))
00280 I=I1
00290 J=J1
00300 3 CONTINUE
00310C  CHECK TO SEE IF FINISHED
00320 IF (T.LE.1.E-6) GO TO 999
00330 IF (K1.GE.25) GO TO 999
00340 K1=K1+1
00350 K=6-(I+J)
00360C  FIND TRANSFORMATION MATRIX
00370 R=A(I,I)-A(J,J)
00380 TAN =2.*A(I,J)/(ABS(R)+SQRT(R*R+4.*A(I,J)*A(I,J)))
00390 IF (A(J,J).LE.A(I,I)) GO TO 4
00400 TAN=-TAN
00410 4 COS=1./SQRT(1.+TAN*TAN)
00420 SIN=COS*TAN
00430 RL(I,I)=COS
00440 RL(J,J)=COS
00450 RL(I,J)=SIN
00460 RL(J,I)=-SIN
00470 RL(K,K)=1.
00480 DO 5 I1=1,3
00490 IF (I1.EQ.K) GO TO 5
00500 RL(I1,K)=0.
00510 RL(K,I1)=0.
00520 5 CONTINUE
00530C  FIND TRANSPOSE OF THE MATRIX
```

Figure 6.3

```
00540 DO 6 I1=1,3
00550 DO 6 J1=1,3
00560 RLT(I1,J1)=RL(J1,I1)
00570 6 CONTINUE
00580C  PERFORM SIMILARITY TRANSFORMATION
00590 DO 7 I1=1,3
00600 DO 7 J1=1,3
00610 B(I1,J1)=0.
00620 DO 8 M1=1,3
00630 B(I1,J1)=B(I1,J1)+A(I1,M1)*RLT(M1,J1)
00640 8 CONTINUE
00650 7 CONTINUE
00660 DO 9 I1=1,3
00670 DO 9 J1=1,3
00680 A(I1,J1)=0.
00690 DO 10 M1=1,3
00700 A(I1,J1)=A(I1,J1)+RL(I1,M1)*B(M1,J1)
00710 10 CONTINUE
00720 9 CONTINUE
00730C   FIND THE NEW EIGENVECTOR MATRIX
00740 DO 12 I1=1,3
00750 DO 12 J1=1,3
00760 E1(I1,J1)=0.
00770 DO 13 M1=1,3
00780 E1(I1,J1)=E1(I1,J1)+RL(I1,M1)*E(M1,J1)
00790 13 CONTINUE
00800 12 CONTINUE
00810 DO 14 I1=1,3
00820 DO 14 J1=1,3
00830 E(I1,J1)=E1(I1,J1)
00840 14 CONTINUE
00850 GO TO 30
00860 997 PRINT 998,
00870 998 FORMAT(1X,'   WARNING 25 ITERATIONS HAVE PASSED'///)
00880 999 PRINT 996,
00890 996 FORMAT(1X,'   THE DIAGONALIZED MATRIX IS:'/////)
00900 PRINT 995, ((A(I,J),J=1,3),I=1,3)
00910 995 FORMAT(1X,3(E10.3,6X))
00920 PRINT 994,
00930 994 FORMAT(1X,'   THE EIGENVECTOR MATRIX IS:'/////)
00940 PRINT 993,((E(I,J),J=1,3),I=1,3)
00950 993 FORMAT(1X,3(F11.4,6X))
00960 STOP
00970 END
```

Figure 6.3 (continued)

If the rotating vector $\{r\}$ is fixed in the body, $[A]$ also describes rotation of the body. If $\{r\}$ lies along the single axis of rotation of Euler's theory, then

$$[A]\{r\} = \lambda\{r\} \tag{6.81}$$

The value of λ must equal $+1$, and only $+1$, if the vector $\{r\}$ is not to change length during a rotation; this is supported by the mathematical theory developed for eigenvalue determination and similarity transformations.

Rewriting equation (6.81) in a more familiar form yields

$$([A] - [1]\lambda)\{r\} = \{0\} \tag{6.82}$$

since

$$\{e\} = \frac{\{r\}}{r} \tag{6.83}$$

then also

$$([A] - [1]\lambda)\{e\} = \{0\} \tag{6.84}$$

Using the eigenvalue theory just discussed,

$$|[A] - [1]\lambda| = 0 \tag{6.85}$$

or

$$a\lambda^3 + b\lambda^2 + c\lambda + d = 0 \tag{6.86}$$

The matrix $[E]$ transforms $[A]$ by means of a similarity transformation to the diagonal matrix $[\lambda]$.

$$[E][A][E]^t = \begin{bmatrix} \lambda_1 & 0 & 0 \\ 0 & \lambda_2 & 0 \\ 0 & 0 & \lambda_3 \end{bmatrix} = [\lambda] \tag{6.87}$$

For an orthogonal transformation

$$[A][A]^t = [1] \tag{5.95}$$

then the determinant of $[A]$ can be found as follows:

$$\text{Det}([A][A]^t) = \text{Det}[A]\,\text{Det}[A]^t = (\text{Det}[A])^2 = 1 \quad \text{Det}[A] = \pm 1 \tag{6.88}$$

The determinant of a matrix must remain unchanged under a similarity transformation, the determinant of $[\lambda]$ must therefore be

$$\lambda_1 \lambda_2 \lambda_3 = \pm 1 \tag{6.89}$$

The possible solution in which a unique λ could be -1 must be proscribed. This would physically mean that the vector $\{r\}$ changes not by rotation but by sudden inversion through $180°$. Since this is not possible

$$\text{Det}[A] = +1 \tag{6.90}$$

and

$$\lambda_1 \lambda_2 \lambda_3 = +1 \tag{6.91}$$

For equation (6.91) to hold true, except for the trivial case where all λ's are equal, there can be only one unique real eigenvalue λ having the magnitude $+1$. This eigenvalue lies on the single axis of rotation called for by Euler's theory. This

178 REVIEW OF NEWTONIAN MECHANICS

axis of rotation can now be found by substituting $\lambda = +1$ into equation (6.84) and solving for the eigenvector. This eigenvector represents the direction of that single axis of rotation.

If through an orthogonal transformation the original coordinate axes are transformed so that the x' axis (y' or z' will work too) coincides with the eigenvector corresponding to the eigenvalue $\lambda = +1$, then any rotation matrix about this new axis would be of the form

$$[A'] = \begin{bmatrix} 1 & 0 & 0 \\ 0 & \cos\phi & \sin\phi \\ 0 & -\sin\phi & \cos\phi \end{bmatrix} \qquad (6.92)$$

for a positive ϕ. Since the trace of a matrix remains unchanged during a similarity transformation, the trace of the original matrix $[A]$ must be equal to the trace of $[A']$ or

$$\text{Tr}\,[A] = 1 + 2\cos\phi \qquad (6.93)$$

This equation provides a means of calculating the single angle of rotation of the body about its axis of rotation. In summary, the rigid body rotation represented by the matrix $[A]$ has been, as in agreement with Euler's theorem, replaced by a single rotation ϕ about a single instantaneous axis of rotation.

Sample Problem 6.5

A rectangular body fixed at a corner point O first rotates about a fixed Z direction through a positive angle $\pi/4$. It then rotates about a body-fixed axis x, now separated from X by $\pi/4$, through a positive angle $\pi/4$.

1. Find the total transformation matrix $[A]$ for the body.
2. Find a single axis of rotation for the total rotation.
3. Find the single angle of rotation about this axis that will lead to the same transformation results as matrix $[A]$.

Analysis: Figures (a) through (d) show the rotation occurring. Keep in mind that the $[A]$ matrix used for coordinate transformations has positive rotation measured counterclockwise whereas the same matrix used for vector, or body rotation, has angles measured positive clockwise. For example, the relationship between coordinate components in the x, y, z and X, Y, Z directions caused by the first rotation can be expressed by

$$\begin{Bmatrix} x \\ y \\ z \end{Bmatrix} = \begin{bmatrix} \frac{\sqrt{2}}{2} & \frac{\sqrt{2}}{2} & 0 \\ -\frac{\sqrt{2}}{2} & \frac{\sqrt{2}}{2} & 0 \\ 0 & 0 & 1 \end{bmatrix} \begin{Bmatrix} X \\ Y \\ Z \end{Bmatrix} \qquad (a)$$

or

$$\{r'\} = [A]\{r\}$$

RIGID BODY PLANE MOTION   179

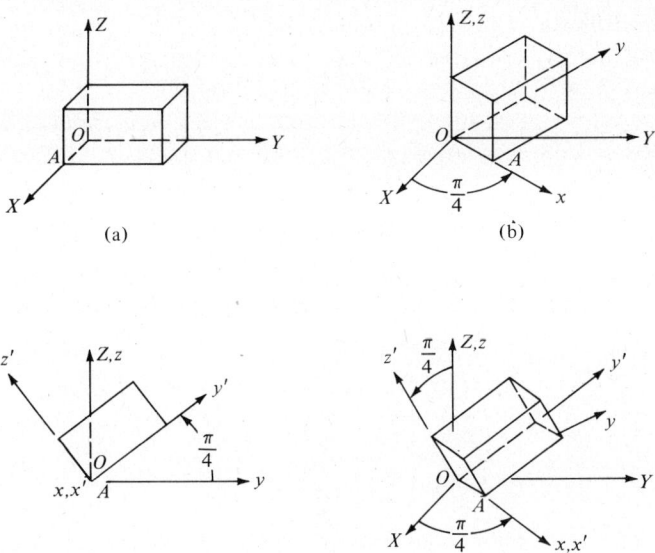

(a)   (b)   (c)   (d)

This last expression relates the components of the same vector $\{r\}$ in *two different coordinate systems*. A rotation of $\{r\}$ itself, and not the coordinate axes, can best be described by holding the vector fixed while turning the coordinate axes in the opposite direction. Then the new vector $\{r_n\}$ is expressed by

$$\{r_n\} = [A_1]\{r\} \tag{b}$$

$$[A_1] = \begin{bmatrix} \frac{\sqrt{2}}{2} & -\frac{\sqrt{2}}{2} & 0 \\ \frac{\sqrt{2}}{2} & \frac{\sqrt{2}}{2} & 0 \\ 0 & 0 & 1 \end{bmatrix} = [A]^t \tag{c}$$

As an example, take a vector $\{r\}$ initially lying along the X axis, $(1, 0, 0)$. By using equations (b) and (c), one can obtain the new vector $\{r_n\}$ caused by the rotation. It is equal to $(\sqrt{2}/2, \sqrt{2}/2, 0)$. The vector $\{r\}$ lies along edge OA of the prism.

The next coordinate system rotation can be expressed by

$$\begin{Bmatrix} x' \\ y' \\ z' \end{Bmatrix} = \begin{bmatrix} 1 & 0 & 0 \\ 0 & \frac{\sqrt{2}}{2} & \frac{\sqrt{2}}{2} \\ 0 & -\frac{\sqrt{2}}{2} & \frac{\sqrt{2}}{2} \end{bmatrix} \begin{Bmatrix} x \\ y \\ z \end{Bmatrix} \tag{d}$$

180 REVIEW OF NEWTONIAN MECHANICS

The total coordinate system rotation is then

$$[A_t] = \begin{bmatrix} \frac{\sqrt{2}}{2} & \frac{\sqrt{2}}{2} & 0 \\ -\frac{1}{2} & \frac{1}{2} & \frac{\sqrt{2}}{2} \\ \frac{1}{2} & -\frac{1}{2} & \frac{\sqrt{2}}{2} \end{bmatrix} \qquad (e)$$

The total vector rotation or *body rotation*, is the transpose of $[A_t]$,

$$\{r_n\} = [A_t]^t [r] \qquad (f)$$

$$\{r_n\} = \begin{bmatrix} \frac{\sqrt{2}}{2} & -\frac{1}{2} & \frac{1}{2} \\ \frac{\sqrt{2}}{2} & \frac{1}{2} & -\frac{1}{2} \\ 0 & \frac{\sqrt{2}}{2} & \frac{\sqrt{2}}{2} \end{bmatrix} \{r\} \qquad (g)$$

Part (2) requires finding the eigenvector of the total rotation matrix. Then by equation (6.84) with $\lambda = +1$,

$$\left(\frac{\sqrt{2}}{2} - 1\right) e_x - \tfrac{1}{2} e_y + \tfrac{1}{2} e_z = 0 \qquad (h)$$

$$\frac{\sqrt{2}}{2} e_x + (\tfrac{1}{2} - 1) e_y - \tfrac{1}{2} e_z = 0 \qquad (i)$$

$$0 e_x + \frac{\sqrt{2}}{2} e_y + \left(\frac{\sqrt{2}}{2} - 1\right) e_z = 0 \qquad (j)$$

To expedite matters, since actual magnitudes are unimportant, assume $e_y = 1$ in equations (h) and (j). Then e_z and e_x take on values

$$e_z = e_x = \frac{1}{\sqrt{2} - 1}$$

The eigenvector lies along the direction just found, or

$$\{e\} = c \begin{Bmatrix} \frac{1}{\sqrt{2}-1} \\ 1 \\ \frac{1}{\sqrt{2}-1} \end{Bmatrix} \qquad (k)$$

where c is an arbitrary constant. This direction is the direction for the axis of rotation in Euler's theorem. The values of e_x, e_y, and e_z just found, that is, $\{e\}$ can be normalized so that $\{e\}$ has a magnitude of one.

$$\{e\} = \begin{Bmatrix} 0.679 \\ 0.281 \\ 0.679 \end{Bmatrix}$$

The angle turned through about the single axis of rotation can be found using the trace of $[A_t]^t$.

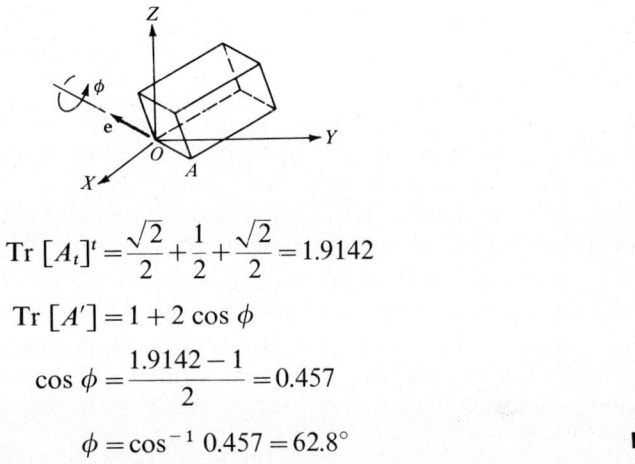

$$\text{Tr}\,[A_t]^t = \frac{\sqrt{2}}{2} + \frac{1}{2} + \frac{\sqrt{2}}{2} = 1.9142$$

$$\text{Tr}\,[A'] = 1 + 2\cos\phi$$

$$\cos\phi = \frac{1.9142 - 1}{2} = 0.457$$

$$\phi = \cos^{-1} 0.457 = 62.8°$$ ∎

6.4 THE MOMENT EQUATION ABOUT AN ARBITRARY POINT

When a body is in static equilibrium, the sum of the moments due to the external forces about any arbitrary point must equal zero. However, for a moving body, the derived equations of motion have different results depending upon which point the moment of external forces are taken about (Figure 6.4). From particle analysis, the angular momentum of a particle about an arbitrary point is defined to be

$$\mathbf{H}_{Ai} = (\mathbf{r}_i - \mathbf{r}_A) \times m_i(\mathbf{v}_i - \mathbf{v}_A) \tag{6.94}$$

For a system of particles

$$\mathbf{H}_A = \sum_{i=1}^{n} \mathbf{r}_i \times m_i \mathbf{v}_i + \mathbf{v}_A \times \sum_{i=1}^{n} m_i \mathbf{r}_i - \mathbf{r}_A \times \sum_{i=1}^{n} m_i \mathbf{v}_i + \mathbf{r}_A \times M\mathbf{v}_A \tag{6.95}$$

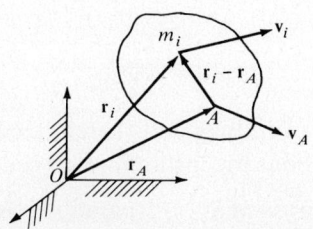

Figure 6.4

182 REVIEW OF NEWTONIAN MECHANICS

Substituting equations (4.2) and (4.7) yields

$$\mathbf{H}_A = \mathbf{H}_O + \mathbf{v}_A \times M\mathbf{R}_{c.m.} - \mathbf{r}_A \times M\mathbf{v}_{c.m.} + \mathbf{r}_A \times M\mathbf{v}_A \tag{6.96}$$

The time derivative of $\mathbf{H}_A$ becomes

$$\dot{\mathbf{H}}_A = \dot{\mathbf{H}}_O + \mathbf{a}_A \times M\mathbf{R}_{c.m.} + \mathbf{v}_A \times M\mathbf{v}_{c.m.} - \mathbf{v}_A \times M\mathbf{v}_{c.m.}$$
$$- \mathbf{r}_A \times M\mathbf{a}_{c.m.} + \mathbf{v}_A \times M\mathbf{v}_A + \mathbf{r}_A \times M\mathbf{a}_A$$

$$= \mathbf{M}_O - \mathbf{r}_A \times \mathbf{F} + (\mathbf{r}_A - \mathbf{R}_{c.m.}) \times M\mathbf{a}_A \tag{6.97}$$

or finally,

$$\mathbf{M}_A = \dot{\mathbf{H}}_A + (\mathbf{R}_{c.m.} - \mathbf{r}_A) \times M\mathbf{a}_A \tag{6.98}$$

The last moment equation has one more term than those previously studied. It is more general in application, but it reduces to the equations for $\mathbf{M}_O$ and $\mathbf{M}_{c.m.}$ when point A is fixed or is the center of mass. Reduced equations are also possible when point A moves with constant velocity or when $(\mathbf{R}_{c.m.} - \mathbf{r}_A)$ is parallel to $\mathbf{a}_A$. This last condition develops for the point of contact of a wheel that is rolling without slipping. (If moments are taken about the wheel's instantaneous center, the point of contact, then mass moments of inertia must be referred to that point as well.) For the plane motion of a rolling wheel having a principal direction along the z axis,

$$(M_c)_z = (I_c)_{zz} \dot{\omega}_z \tag{6.99}$$

where the subscript c denotes the instantaneous center.

Sample Problem 6.6
The compound wheel shown of weight Mg rolls to the right without slipping. Find the minimum coefficient of static friction μ necessary for the motion. The force P is applied to the wheel at a distance r from the center at an angle β with the horizontal. The radius of the wheel is R, and its radius of gyration is $\hat{k}$. For rolling, what is the maximum angle β can assume? Discuss the relationships between $\hat{k}$, β, R, and r that will lead to a friction force acting always to the left.

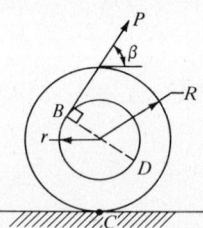

Analysis: This is a case of rolling, and therefore equation (6.99) applies. The mass moment of inertia about the instantaneous center is

$$(I_c)_{zz} = M\hat{k}^2 + MR^2 \quad \text{(parallel-axis theorem)}$$
$$= M(\hat{k}^2 + R^2)$$

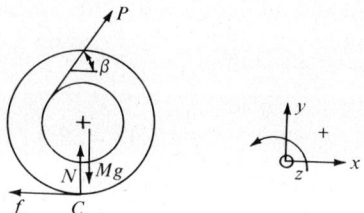

From the free-body diagram, the following force equations can be obtained

$$F_x = M(a_{c.m.})_x \qquad P\cos\beta - f = M(a_{c.m.})_x \tag{a}$$

$$F_y = M(a_{c.m.})_y \qquad P\sin\beta - Mg + N = 0 \qquad N = Mg - P\sin\beta \tag{b}$$

$$(M_c)_z = (I_c)_{zz}\dot\omega_z \qquad -Pr - PR\cos\beta = M(\hat k^2 + R^2)\dot\omega_z \tag{c}$$

From the kinematics for *rolling without sliding*

$$(a_{c.m.})_x = -R\dot\omega_z \tag{d}$$

Note the need for a minus sign in equation (d) to agree with the coordinate sign conventions used. Solving equations (a), (c), and (d) simultaneously leads to

$$f = \frac{P(\hat k^2 \cos\beta - Rr)}{\hat k^2 + R^2} \tag{e}$$

but from statics

$$\mu \geq \frac{|f|}{|N|} \qquad \mu \geq \frac{|P/\hat k^2 \cos\beta - Rr)|}{|(\hat k^2 + R^2)(Mg - P\sin\beta)|} \tag{f}$$

Equation (f) represents the minimum value μ must have for rolling to occur. From equation (b) it is seen that

$$\sin\beta < \frac{Mg}{P} \tag{g}$$

otherwise the wheel loses contact with the ground. From equation (e), for friction always acting to the left,

$$\hat k^2 \cos\beta > Rr \tag{h}$$

Therefore, β cannot be greater than 90°. For β equal to 90°, the friction force and the acceleration must be to the right. If the force is applied on the other side of the wheel at point D, the wheel will always have friction acting to the left for β less than 90°.

If the *wheel slides*, equation (d) is no longer valid since

$$(a_{c.m.})_x \neq -R\dot\omega_z$$

In that case the friction force is known, namely,

$$f = \mu_k N \tag{i}$$

where μ_k is the coefficient of kinetic friction. One can summarize the discussion of friction for rolling, impending sliding, and sliding as follows:

Rolling: $\quad F \leqslant \mu N \quad |(a_{c.m.})_x| = |R\dot{\omega}_z|$ (j)

Impending sliding: $\quad F = \mu N \quad |(a_{c.m.})_x| = |R\dot{\omega}_z|$ (k)

Sliding: $\quad F = \mu_k N \quad |(a_{c.m.})_x| \neq |R\dot{\omega}_z|$ (l)

∎

6.5 PRINCIPLE OF WORK AND KINETIC ENERGY FOR A RIGID BODY

The work-energy principle for a system of particles is, of course, valid for the special case of the rigid body; namely,

$$W_{1 \to 2} = T_2 - T_1 \tag{4.44}$$

For a conservative system,

$$T_1 + V_1 = T_2 + V_2 = E \tag{4.45}$$

The kinetic energy expression for a system of particles is

$$T = \frac{Mv_{c.m.}^2}{2} + \sum_{i=1}^{n} \frac{m_i \dot{\rho}_i^2}{2} \tag{4.47}$$

For a rigid body, the relative velocity of a particle with respect to the center of mass is given by

$$\dot{\rho} = \omega \times \rho$$

and the summation process for n particles becomes an integration over the entire mass or volume. Equation (4.47) then becomes

$$T = \frac{Mv_{c.m.}^2}{2} + \frac{1}{2} \int_{\text{mass}} (\omega \times \rho) \cdot (\omega \times \rho) \, dm \tag{6.100}$$

Upon expansion, this becomes

$$T = \frac{Mv_{c.m.}^2}{2} + \frac{\hat{I}_{xx}\omega_x^2}{2} + \frac{\hat{I}_{yy}\omega_y^2}{2} + \frac{\hat{I}_{zz}\omega_z^2}{2}$$
$$+ \hat{I}_{xy}\omega_x\omega_y + \hat{I}_{xz}\omega_x\omega_z + \hat{I}_{yz}\omega_y\omega_z \tag{6.101}$$

The last six terms represent the *rotational* kinetic energy; the first term is the *translational* energy. For *principal axes*,

$$T = \frac{Mv_{c.m.}^2}{2} + \frac{\hat{I}_{xx}\omega_x^2}{2} + \frac{\hat{I}_{yy}\omega_y^2}{2} + \frac{\hat{I}_{zz}\omega_z^2}{2} \tag{6.102}$$

For *plane motion*,

$$T = \frac{Mv_{c.m.}^2}{2} + \frac{\hat{I}_{zz}\omega_z^2}{2} \tag{6.103}$$

or
$$T = M \frac{[(v_{c.m.})_x^2 + (v_{c.m.})_y^2]}{2} + \frac{\hat{I}_{zz}\omega_z^2}{2} \qquad (6.104)$$

The last two equations are the most significant for work in this chapter. In Chapter 7, where three-dimensional motion is discussed, equations (6.101) and (6.102) will be used.

If there is a fixed point in the body, then a special relation can be developed. Starting from equation (4.43),

$$T = \sum_{i=1}^{n} \frac{m_i v_i^2}{2} \qquad (4.43)$$

and using, for rigid body motion about a fixed point,

$$\mathbf{v} = \boldsymbol{\omega} \times \mathbf{r}$$

equation (4.43) becomes

$$T_O = \int_{\text{mass}} (\boldsymbol{\omega} \times \mathbf{r}) \cdot (\boldsymbol{\omega} \times \mathbf{r}) \, dm \qquad (6.105)$$

$$T_O = \frac{I_{xx}\omega_x^2}{2} + \frac{I_{yy}\omega_y^2}{2} + \frac{I_{zz}\omega_z^2}{2} + I_{xy}\omega_x\omega_y + I_{xz}\omega_x\omega_z + I_{yz}\omega_y\omega_z \qquad (6.106)$$

The subscript on T_O is there to remind the reader that this equation represents the total kinetic energy of a rigid body turning about a fixed point O. The *mass moments and products of inertia must be referred to this point*. For *principal axes*,

$$T_O = \frac{I_{xx}\omega_x^2}{2} + \frac{I_{yy}\omega_y^2}{2} + \frac{I_{zz}\omega_z^2}{2} \qquad (6.107)$$

For *plane motion*,

$$T_O = \frac{I_{zz}\omega_z^2}{2} \qquad (6.108)$$

The total kinetic energy represented by equation (6.108) for motion about a fixed point must be equal to the general expression for kinetic energy given by equation (6.103). This can be shown easily for the plane motion case since for rotation about a fixed point O,

$$v_{c.m.} = \omega_z R_{c.m.}$$

Equation (6.103) then becomes

$$T = \frac{MR_{c.m.}^2 \omega_z^2}{2} + \frac{\hat{I}_{zz}\omega_z^2}{2} = \frac{\hat{I}_{zz} + MR_{c.m.}^2}{2} \omega_z^2$$

Using the parallel-axis theorem, this finally becomes

$$T = \frac{I_{zz}\omega_z^2}{2} = T_O$$

where $\quad \hat{I}_{zz} + MR_{c.m.}^2 = I_{zz} \quad$ (about the fixed point O)

Returning now to the work expression given in Chapter 4, namely,

$$W_{1\to 2} = \int_1^2 \sum_{i=1}^n \mathbf{F}_i \cdot d\boldsymbol{\rho}_i + \int_1^2 \mathbf{F}_{\text{ext}} \cdot d\mathbf{R}_{\text{c.m.}} + \int_1^2 \sum_{\substack{i=1 \\ i\neq j}}^n \sum_{j=1}^n \mathbf{f}_{ij} \cdot d\boldsymbol{\rho}_i \quad (4.41)$$

simplifications can now be affected for rigid body motion. The last term on the right, the internal work done, must be zero for a nonelastic body.

The first term becomes, with s forces acting and $d\boldsymbol{\rho}_i = d\boldsymbol{\theta} \times \boldsymbol{\rho}_i$ for a rigid body,

$$\int_1^2 \sum_{i=1}^n \mathbf{F}_i \cdot d\boldsymbol{\rho}_i = \int_1^2 \sum_{i=1}^s \mathbf{F}_i \cdot (d\boldsymbol{\theta} \times \boldsymbol{\rho}_i)$$

Through vector algebra this reduces to

$$\int_1^2 \sum_{i=1}^n \mathbf{F}_i \cdot d\boldsymbol{\rho}_i = \int_1^2 \sum_{i=1}^s (\boldsymbol{\rho}_i \times \mathbf{F}_i) \cdot d\boldsymbol{\theta} = \int_1^2 \mathbf{M}_{\text{c.m.}} \cdot d\boldsymbol{\theta} \quad (6.109)$$

where $\mathbf{M}_{\text{c.m.}}$ represents the total moment of external forces about the center of mass. This term represents the *rotational work* done upon a rigid body.

$$W_{1\to 2, \text{rot}} = \int_1^2 \mathbf{M}_{\text{c.m.}} \cdot d\boldsymbol{\theta} \quad (6.110)$$

The *translational work* done is the second term:

$$W_{1\to 2, \text{transl}} = \int_1^2 \mathbf{F}_{\text{ext}} \cdot d\mathbf{R}_{\text{c.m.}} \quad (6.111)$$

The *total work* done is

$$W_{1\to 2} = W_{1\to 2, \text{rot}} + W_{1\to 2, \text{transl}} \quad (6.112)$$

The kinetic energy expressions, equations (6.101) and (6.102), can also be broken up into translation and rotational parts. It can then be shown that

$$W_{1\to 2, \text{rot}} = (T_2 - T_1)_{\text{rot}} \quad (6.113)$$

$$W_{1\to 2, \text{transl}} = (T_2 - T_1)_{\text{transl}} \quad (6.114)$$

and the sum, of course, is

$$W_{1\to 2} = T_2 - T_1 \quad (4.44)$$

Rotational and translational terms for power, the time rate at which work is done, can be derived from equations (6.110) and (6.111), namely,

$$P_{\text{transl}} = \mathbf{F}_{\text{ext}} \cdot \mathbf{v}_{\text{c.m.}} \quad (6.115)$$

$$P_{\text{rot}} = \mathbf{M}_{\text{c.m.}} \cdot \boldsymbol{\omega} \quad (6.116)$$

One final note: Since the translational work done on a body is a function of the displacement of the body's center of mass, and since for conservative forces the change in potential energy is equal to minus the work done, the *gravitational potential energy* change for a body must be a function only of the variation in height of the body's center of mass.

Sample Problem 6.7
The gear shown moves on the horizontal fixed rack. The gear has mass m and a radius of gyration of $\hat{k}$ equal to its radius. Rod BC connected to the gear has length l_r and mass $2m$. The slider has negligible mass, and the attached spring has a spring constant k equal to mg/l_r. If the system is released from rest in a vertical position with the spring unstretched, find the velocity of the gear after it has moved a distance $l_r/2$ to the right. Neglect friction and the mass of the spring.

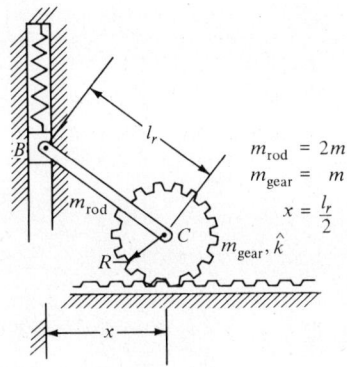

Analysis: Since friction is neglected here, the principle of conservation of mechanical energy can be used.

$$T_1 + V_1 = T_2 + V_2$$
$$T_1 = 0 \qquad V_1 = m_{\text{rod}} g \frac{l_r}{2}$$

$$V = \frac{m_{\text{rod}} g (l_r^2 - x^2)^{1/2}}{2} + \frac{k[l_r - (l_r^2 - x^2)^{1/2}]^2}{2}$$

$$= \frac{m_{\text{rod}} g l_r \sqrt{3}}{4} + \frac{k l_r^2 (2 - \sqrt{3})^2}{8}$$

$$V_1 - V_2 = \frac{m_{\text{rod}} g l_r (2 - \sqrt{3})}{4} - \frac{k l_r^2 (2 - \sqrt{3})^2}{8}$$

Note that in calculating the potential energy of rod BC it is the vertical height to the center of mass that is used.

$$T_2 = \frac{m_{\text{rod}} (v_{\text{c.m.}})_{\text{rod}}^2}{2} + \frac{I_{\text{rod}} \omega_{\text{rod}}^2}{2} + \frac{m_{\text{gear}} (v_{\text{c.m.}})_{\text{gear}}^2}{2} + \frac{I_{\text{gear}} \omega_{\text{gear}}^2}{2}$$

Using the method of instantaneous centers, it can be shown that

$$\omega_{rod} = \frac{2(v_{c.m.})_{gear} \mathbf{k}}{l_r \sqrt{3}}$$

and

$$(\mathbf{v}_{c.m.})_{rod} = \frac{l_r}{2} \omega_{rod}(\cos 30° \mathbf{i} - \sin 30° \mathbf{j})$$

Therefore,

$$(\mathbf{v}_{c.m.})_{rod} = \frac{(v_{c.m.})_{gear}}{\sqrt{3}} (\cos 30° \mathbf{i} - \sin 30° \mathbf{j})$$

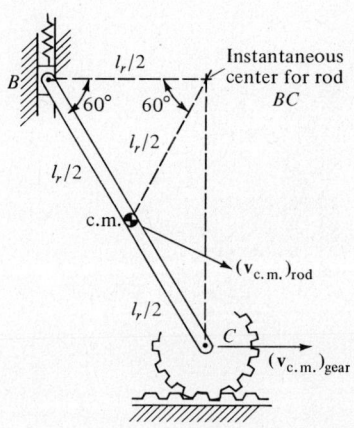

For the gear, with its own instantaneous center at the point of contact

$$\omega_{gear} = -\frac{(v_{c.m.})_{gear}}{R} \mathbf{k}$$

Now

$$I_{rod} = \frac{m_{rod} l_r^2}{12}$$

and

$$I_{gear} = m_{gear} \hat{k}^2$$

Combining terms, T_2 equals

$$T_2 = \frac{(v_{c.m.})_{gear}^2}{2} \left[\frac{m_{rod}}{3} + \frac{m_{rod}}{9} + m_{gear} + m_{gear} \left(\frac{\hat{k}}{R} \right)^2 \right]$$

Then

$$(v_{c.m.})_{gear} = \left[\frac{9}{13m} \left(\frac{mgl_r(2-\sqrt{3})}{2} - \frac{mgl_r(2-\sqrt{3})^2}{8} \right) \right]^{1/2}$$

$$= \left[\frac{9gl_r}{13} (0.134 - 0.009) \right]^{1/2} = 0.294 \sqrt{gl_r} \quad \blacksquare$$

6.6 PRINCIPLE OF IMPULSE AND MOMENTUM FOR A RIGID BODY

It is again apparent that the equations developed in Chapter 4 for a system of particles must now hold for the special case of a rigid body when impulse and momentum are considered.

$$\int_1^2 \mathbf{F}_{\text{ext}} \, dt = \mathbf{P}_2 - \mathbf{P}_1 \tag{4.52}$$

$$\mathbf{P} = M\mathbf{v}_{\text{c.m.}} \tag{4.53}$$

$$\int_1^2 F_x \, dt = (Mv_{\text{c.m.}})_{x,2} - (Mv_{\text{c.m.}})_{x,1}$$

$$\int_1^2 F_y \, dt = (Mv_{\text{c.m.}})_{y,2} - (Mv_{\text{c.m.}})_{y,1} \tag{4.54}$$

$$\int_1^2 F_z \, dt = (Mv_{\text{c.m.}})_{z,2} - (Mv_{\text{c.m.}})_{z,1}$$

For angular impulse and momentum

$$\int_1^2 \mathbf{M}_O \, dt = \mathbf{H}_{O,2} - \mathbf{H}_{O,1} \tag{4.58}$$

$$\int_1^2 \mathbf{M}_{\text{c.m.}} \, dt = \mathbf{H}_{\text{c.m.},2} - \mathbf{H}_{\text{c.m.},1} \tag{4.59}$$

$$\int_1^2 M_x \, dt = H_{x,2} - H_{x,1}$$

$$\int_1^2 M_y \, dt = H_{y,2} - H_{y,1} \tag{4.60}$$

$$\int_1^2 M_z \, dt = H_{z,2} - H_{z,1}$$

In equation (4.60) the reference point used is the center of mass or a fixed point. The angular momentum expressions on the right are further described by equation (6.31) or equation (6.32).

For the *plane motion* of a body which has a plane of symmetry perpendicular to the z axis these equations reduce to the following three:

$$\int_1^2 F_x \, dt = (Mv_{\text{c.m.}})_{x,2} - (Mv_{\text{c.m.}})_{x,1} \tag{6.117}$$

$$\int_1^2 F_y \, dt = (Mv_{\text{c.m.}})_{y,2} - (Mv_{\text{c.m.}})_{y,1} \tag{6.118}$$

$$\int_1^2 (M_{\text{c.m.}})_z \, dt = (\hat{I}_{zz}\omega_z)_2 - (\hat{I}_{zz}\omega_z)_1 \tag{6.119}$$

or with a fixed-point reference,

$$\int_1^2 (M_O)_z \, dt = (I_{zz}\omega_z)_2 - (I_{zz}\omega_z)_1 \tag{6.120}$$

With these equations, it is now possible to complete the discussion of impact begun in Chapter 4. For the case of *oblique impact*, when the line of action of the impact forces does not coincide with the line passing through each body's center of mass, rotation is encountered. The momentum principles, equations (6.117) to (6.120), combined with the expression for the coefficient of restitution are sufficient in themselves for most oblique impact problem denouement. It must be kept in mind, however, that the coefficient of restitution now is written in terms of the components of relative velocity of the *contacting* points before and after impact, along the direction of the line of impact; it is *not* written for the centers of mass.

Sample Problem 6.8

A bullet of mass m_B has an initial velocity $\mathbf{v}_0$ before it impacts the thin square wood box of mass m_A shown. The bullet strikes the box halfway down its side and then becomes immediately embedded at point C. Determine the impulsive reaction at A. Find the velocity and angular velocity of the box immediately after impact.

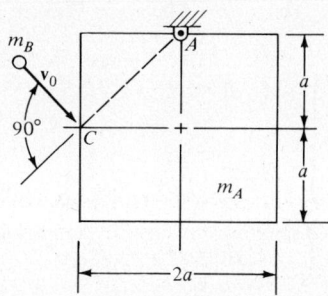

Analysis: This is an example of oblique impact. The system's total angular momentum with reference to fixed point A is conserved, for there are no external moments acting about A. One can write for the system

$$H_{A,1} = H_{A,2}$$

$$a\sqrt{2}m_B v_0 = (I_{A,\text{box}} + m_B 2a^2)\omega \tag{a}$$

$$a\sqrt{2}m_B v_0 = \frac{a^2}{3}(5m_A + 6m_B)\omega$$

$$\omega = \frac{3\sqrt{2}m_B v_0}{a(5m_A + 6m_B)} \qquad v_{\text{c.m.}} = \frac{3\sqrt{2}m_B v_0}{5m_A + 6m_B} \tag{b}$$

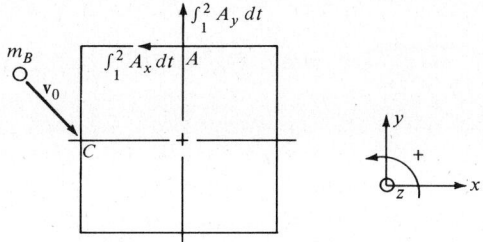

Referring to the free-body diagram shown, and ignoring the effect of weight during the instantaneous impact, the linear momentum equations for the system become for the x direction

$$-\int_1^2 A_x \, dt = \frac{(m_A + m_B)3\sqrt{2}m_B v_0}{5m_A + 6m_B} - \frac{m_B v_0 \sqrt{2}}{2}$$

or

$$\int_1^2 A_x \, dt = \frac{m_B v_0 \sqrt{2}}{2} - \frac{(m_A + m_B)3\sqrt{2}m_B v_0}{5m_A + 6m_B} \qquad (c)$$

For the y direction,

$$\int_1^2 A_y \, dt = -\frac{m_B 3\sqrt{2}m_B v_0}{5m_A + 6m_B} - \left(-m_B v_0 \frac{\sqrt{2}}{2}\right)$$

$$\int_1^2 A_y \, dt = \frac{m_B v_0 \sqrt{2}}{2} - \frac{3m_B^2 \sqrt{2} v_0}{5m_A + 6m_B} \qquad (d)$$

∎

Sample Problem 6.9

The thin rod shown of mass M and length L rests on a smooth horizontal table. If a particle of the same mass impacts with the end of the rod with a velocity v_0 perpendicular to the rod, describe the ensuing motion immediately after impact. The coefficient of restitution ε equals 1 for the impact.

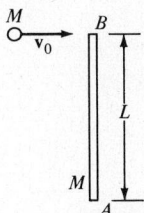

Analysis: Because the line of impact does not coincide with the line joining the centers of mass for both objects, rotation occurs after impact. This is, therefore, an oblique impact problem requiring angular momentum considerations. For the rod, one can write the following momentum equations:

192   REVIEW OF NEWTONIAN MECHANICS

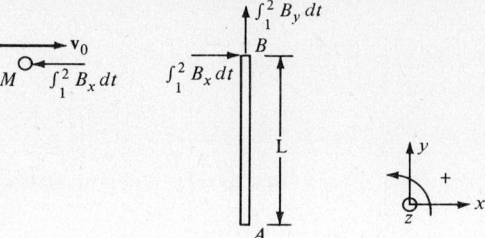

For the x direction,
$$\int_1^2 B_x \, dt = M(v_{\text{c.m.}})_{x,2} - 0 \tag{a}$$

For the y direction,
$$\int_1^2 B_y \, dt = 0 \tag{b}$$

For the angular momentum about the z direction,
$$-\int_1^2 \frac{L}{2} B_x \, dt = \frac{ML^2}{12} \omega_2 - 0 \tag{c}$$

For the impacting particle,
$$-\int_1^2 B_x \, dt = Mv_2 - Mv_0 \tag{d}$$

Also,
$$\varepsilon = -\frac{v_B - v_2}{0 - v_0} = 1 \tag{e}$$

where
$$v_B = (v_{\text{c.m.}})_{x,2} - \omega_2 \frac{L}{2}$$

Then
$$v_0 = (v_{\text{c.m.}})_{x,2} - \omega_2 \frac{L}{2} - v_2 \tag{f}$$

From equations (a) and (d) one obtains
$$M(v_{\text{c.m.}})_{x,2} + Mv_2 - Mv_0 = 0 \tag{g}$$

and from equations (c) and (d),
$$\frac{ML}{2} v_2 - \frac{ML}{2} v_0 = \frac{ML^2}{12} \omega_2 \tag{h}$$

Equations (f) to (h) rewritten are
$$v_0 = (v_{\text{c.m.}})_{x,2} - \omega_2 \frac{L}{2} - v_2$$

$$v_0 = (v_{c.m.})_{x,2} + v_2$$
$$v_0 = -\frac{\omega_2 L}{6} + v_2$$

With solutions

$$v_2 = \tfrac{3}{5}v_0 \qquad (v_{c.m.})_{x,2} = \tfrac{2}{5}v_0 \qquad \omega_2 = -\frac{12v_0}{5L}$$ ∎

PROBLEMS

6.1 Find the moments of inertia of the hollow semicircular cylinder shown with respect to the x, y, z coordinate system.

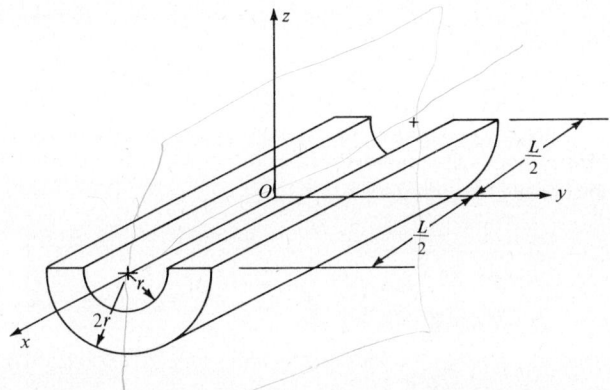

6.2 Find the center of mass of the hollow semicircular cylinder given in problem 6.1, and then find the new moments of inertia with respect to parallel coordinates passing through the center of mass.

6.3 If the coordinates in problem 6.1 are rotated 90° clockwise about an axis passing through the origin and making a 45° angle with the y axis in the yz plane, find the rotation matrix $[l]$. What are the new moments of inertia? Take the total mass to be 4 kg, the inside radius to be 10 cm, and the length to be 50 cm.

6.4 Find the eigenvalues and eigenvectors of the following moment of inertia matrix:

$$[I] = \begin{bmatrix} 320 & 60 & 0 \\ 60 & 160 & 0 \\ 0 & 0 & 500 \end{bmatrix} \text{ kg} \cdot \text{m}^2$$

*6.5 Find the eigenvalues and eigenvectors of the following moment of inertia matrix using a computer.

$$[I] = \begin{bmatrix} 18 & -9 & 0 \\ -9 & 18 & -9 \\ 0 & -9 & 12 \end{bmatrix} \text{ kg} \cdot \text{m}^2$$

6.6 For the right triangular prism shown, find the moments of inertia with respect to the x, y, z coordinates given.

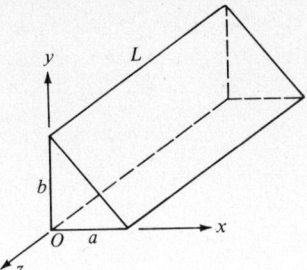

6.7 Using the parallel-axis theorem, transfer the moments of inertia found in problem 6.6 back to the center of mass.

*6.8 If, in problem 6.7, the coordinates passing through the center of mass are positively rotated about the z axis through an angle of 30°, find the new moments of inertia. Take $b=1$ m, $a=1$ m, $L=4$ m, and let the mass equal 1800 kg. Check your results with the computer.

*6.9 Use a computer to find the eigenvalues and eigenvectors for the moment of inertia matrix found in problem 6.7.

*6.10 With the help of a computer, find the principal moments of inertia and the principal directions about the combined center of mass for the aluminum body shown. The specific density of aluminum is 2.7.

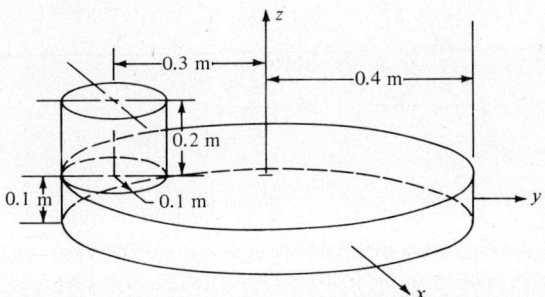

6.11 A clock pendulum consists of a thin homogeneous rod having length 1.4 m and 1 kg mass and a thin circular disk of radius 10 cm with mass equal to that of the rod's. The rod goes through the disk's center. The friction-free motion of the rod and disk lies in a vertical plane. Find the distance x from the fixed pivot point at O to the point where the disk's center must be attached so that the pendulum's period will be 2 s.

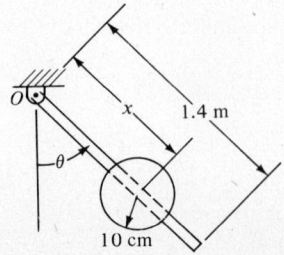

6.12 A sphere of radius R and mass m rolls without sliding inside a fixed cylindrical surface having radius $3R$. If when $\phi = 60°$, the velocity of the center of mass of the sphere is $\sqrt{2gR}$, what is the normal force exerted on the sphere from the fixed surface? What friction acts? What is the angular acceleration of the sphere at this instant?

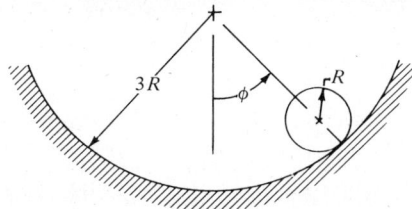

6.13 Suspended inside a thin ring of radius a and mass m_1 is a very narrow plate having mass m_2, length b, and width c. Initially the ring is held fixed while the plate rotates at an angular speed ω_0. After the ring is released, it comes to rotate at a common speed with the plate after t seconds have elapsed. Calculate the internal friction moment, assumed to be constant, that is present. The outside supports do not contribute any moments.

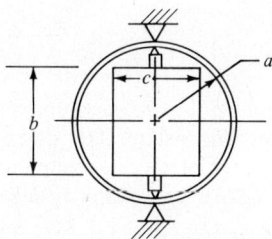

6.14 The thin rod of mass m and length L shown has a pivot at end A free to slide in a frictionless horizontal track. At the same time the rod can rotate. Find for this instant the acceleration of point A if a force P is applied to the lower end of the rod.

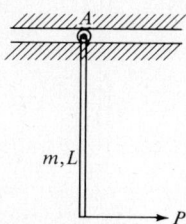

6.15 The automotive dynamometer drum shown has a moment of inertia I_O. If it turns in friction-free bearings, for what size (mass) automobile can the dynamometer load accurately simulate road conditions of vehicle acceleration? The drum's radius is R. Describe what happens if a larger car is tested.

6.16 In order to determine the mass moment of inertia of a body of mass m having the outside shape of a cylinder of radius R, the body is placed on an inclined plane, and the time for it to travel (t_1) a distance (x_1) along the plane starting from rest is recorded. If the center of mass is known to lie at the "cylinder's" center, derive an expression for the mass moment of inertia. No sliding occurs.

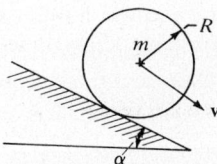

6.17 The Scotch-yoke mechanism shown, operating in a horizontal plane, pulls a load P that acts to the right for this instant. The slider has mass m_2, and the crank OA, which turns at constant angular speed ω counterclockwise, has mass m_1 and length L. What is the necessary driving torque provided to OA for this instant? Neglect all friction.

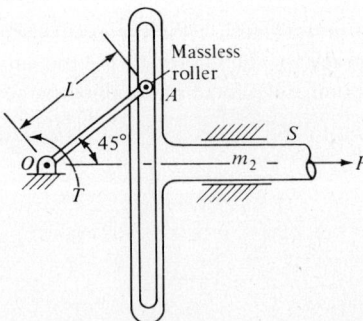

6.18 The thin rod OA of mass m_1 and length L is driven about the vertical at a constant angular speed ω by an applied torque T. A slider of mass m_2 moves outward simultaneously along the rod. It is resisted by a spring having unstretched length $L/2$ and a spring constant k. Write the equations of motion for the system for the instant when $r = \frac{3}{4}L$. Find what torque is necessary. What is the total horizontal restraining force acting on the rod? Neglect friction.

RIGID BODY PLANE MOTION   197

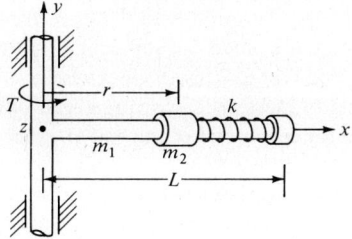

6.19 A thin straight rod of mass *m* and length *L* lies on a smooth horizontal surface. A force *P* is applied perpendicular to the rod at one end. What is the maximum bending moment developed in the rod for the instant when the force is first applied?

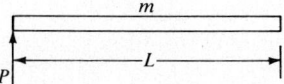

6.20 A thin rod having mass *m* and length *L* is constrained to slide along a smooth horizontal circular track having radius *R*. If at one end a constant force *P* is applied perpendicular to the rod, determine the resultant angular acceleration as a function of *L*.

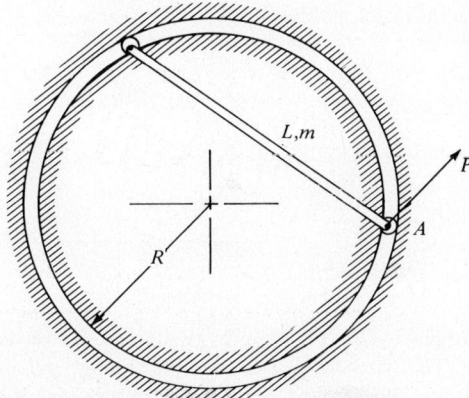

6.21 A thin uniform plate in the form of an equilateral triangle having mass *m* and sides of length *a* is connected to the shaft of a rotating horizontal sun gear. A peg on the

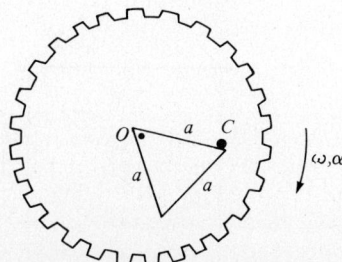

gear at C drives the plate around with the gear. Determine the extra loading on the shaft caused by the plate when the sun gear rotates with an angular acceleration α and an angular speed $\omega = \sqrt{3\alpha}$, both clockwise.

6.22 A thin uniform rod of mass m_1 can swing about point O in a vertical plane. A thin wheel is pinned to the lower end of the rod at point A. It has mass m_2 and radius r, and it rolls without slipping on the inside of a curved surface of radius R. Neglecting bearing friction, calculate the period of the motion for small oscillation.

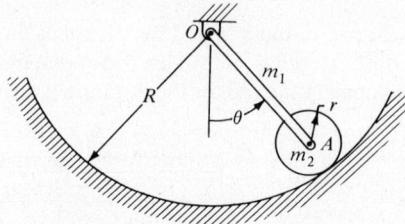

6.23 A wheel of radius R rests on a horizontal table. A spool having radius r about which a cord is wound is rigidly attached to the wheel. If the cord is pulled vertically downward with a constant force P, what must be the minimum value of the friction coefficient to permit rolling without sliding on the table independent of the size of P? Assume the motion of the wheel is in a plane. The wheel and spool have a combined mass m and a radius of gyration C.

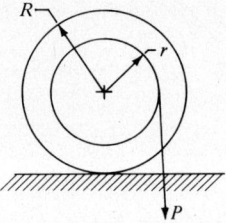

6.24 A thin square plate $ABCD$ is hung by two wires in a vertical plane so that edge AB is horizontal. The wires form a 45° angle with edge AB and are attached to a fixed point. If one wire breaks, the other wire's tension drops suddenly from D_0 to D_1. Find the ratio D_0/D_1.

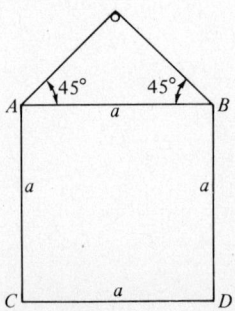

6.25 A thin uniform rod having mass *m* and length *L* is supported in the horizontal position with the aid of two strings. If the lower string is suddenly cut, the force in the other string immediately takes on the value $mg/2$. What does *x* equal?

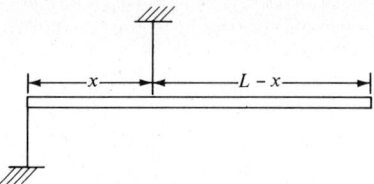

6.26 Two identical rods of length *L* are joined together by a frictionless hinge at *A*. They rest symmetrically upon two friction-free supports having the same height and separated by the distance $1.5L$. Initially, both rods are held in a horizontal position by a string held at the hinge. What is the magnitude of the force on the supports immediately before and after the string is cut?

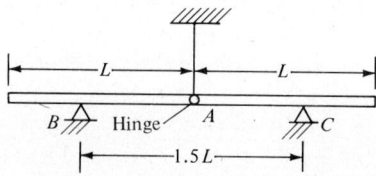

6.27 Find the distance *b* in the figure from the center of mass to the impact point *A* that will reduce the horizontal reaction at the pivot point *O* to zero during the impact of the thin rod with the fixed peg. The mass of the rod is *m* and its length is *L*. Point *A* is called the *center of percussion*.

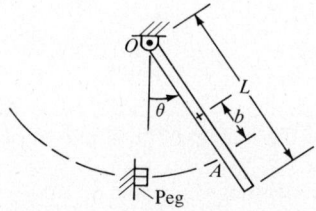

6.28 The square prism shown having mass *m* and sides of length *a* rests on one of its edges before it falls and strikes the horizontal floor. If the edge always remains in contact with the frictionless floor, determine the velocity of the center of mass just before impact.

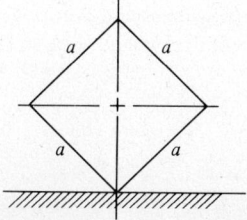

6.29 Two identical rods BA and DA of mass m and length a are connected by a pin at A. The rod ends at B and D are constrained to move along a friction-free horizontal track. If the system is released from rest in a horizontal position, determine the angular velocity of each rod at the moment when the rods form a right angle with each other.

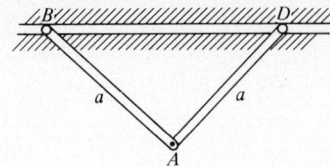

6.30 Two identical rods DA and BA of length a and mass m are attached by a pin at A. End B is constrained to move in a friction-free vertical track. If the system is released from rest in a horizontal position, develop an expression for the velocity of B as a function of y.

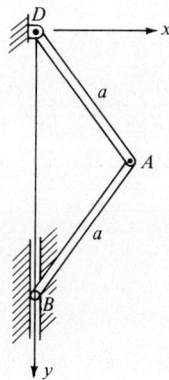

6.31 A thin rod of mass m and length $2a$ falls in a horizontal position, and strikes a fixed knife edge at C. What must the distance S be so that the rod attains the maximum angular velocity after the impact? The impact is plastic.

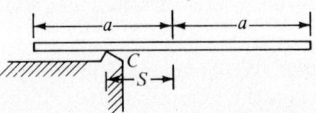

6.32 An equilateral triangular prism having mass m and sides of length a slides on a smooth horizontal floor with speed v_0 before it plastically impacts with a fixed peg at C. How large must the initial speed v_0 be so that the prism turns completely over after the impact?

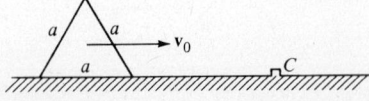

6.33 A circular cylinder of mass m and radius R starts from rest and rolls without slipping down an inclined plane a distance S. It then plastically impacts with a raised step of height $h = 0.5R$, i.e., the contact line with the step comes to rest. What must the minimum distance S be so that the cylinder swings up and over the edge of the step?

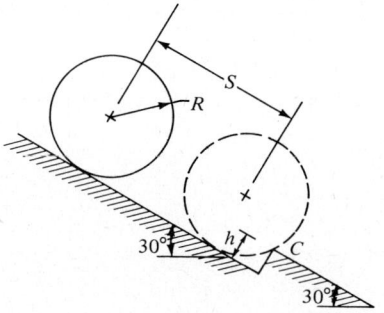

6.34 Two identical rods AB and BC of length a and mass m are hinged at B in a vertical plane. When the system is at rest, rod BC receives a horizontal impact a distance x from B. Immediately after the impact, rod BC translates only. Find the value of x that will produce this result.

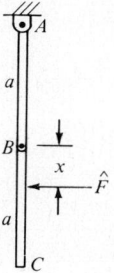

PART TWO: ANALYTICAL MECHANICS

RIGID BODY DYNAMICS 7

7.1 INTRODUCTION

Part One of this book was a detailed review of elementary dynamics. Part Two will build on the subjects already explored, concentrating on more advanced topics. The discussion in Chapter 7, for instance, will no longer be limited to motion in a plane but will explore general space motion. Chapter 8 will introduce Lagrange's equation, a reformulation of the laws of dynamics using generalized energy terms. The extensive contributions of Hamilton will also be discussed. Chapters 9 and 10 will introduce the topics of vibrations, stability analysis, and computer simulation of continuous (time) dynamic systems.

7.2 GENERAL RIGID BODY MOTION

In Chapters 4 and 6, the equations of motion governing the motion of a rigid body in space were derived. They were

$$\mathbf{F}_{ext} = M\mathbf{a}_{c.m.} \tag{4.21}$$

$$\mathbf{M}_O = \dot{\mathbf{H}}_O \tag{4.29}$$

or

$$\mathbf{M}_{c.m.} = \dot{\mathbf{H}}_{c.m.} \tag{4.37}$$

where

$$\{\mathbf{H}\} = [I]\{\omega\} \tag{6.40}$$

In equation (6.40) *the mass moments and products of inertia and the angular momentum are referenced to the same point.* Equation (6.40) written in cartesian coordinates is

$$H_x = \omega_x I_{xx} + \omega_y I_{xy} + \omega_z I_{xz} \tag{7.1}$$

$$H_y = \omega_y I_{yy} + \omega_x I_{yx} + \omega_z I_{yz} \tag{7.2}$$

$$H_z = \omega_z I_{zz} + \omega_x I_{zx} + \omega_y I_{zy} \tag{7.3}$$

In the discussion in Chapter 6 it was stated that unless the coordinate system is fixed to the moving body (an exception being a body of revolution), the relative geometry of the body with respect to the coordinate system will change. The mass moments and products of inertia terms in equation (6.40) will have a time derivative. By fixing a coordinate system (reference frame) to the body, this problem can be removed. However, since the body rotates, the derivative of the angular momentum vector found in the body-fixed reference frame is no longer the total derivative. Because of the reference frame rotation, the derivative of the coordinate components in equations (7.1) to (7.3) are only the changes measured relative to the local frame. Section 5.6 described a general expression for finding the time derivative of any vector in terms of the apparent derivative and the *angular velocity of the local axes* (to be read now as the angular velocity of the body when using a body-fixed reference frame); namely,

$$\frac{d(\ )}{dt} = \left[\frac{d(\ )}{dt}\right]_{rot} + \mathbf{\Omega} \times (\ ) \qquad (5.72)$$

When applied to the angular momentum, this becomes

$$\mathbf{M}_O = \left[\frac{d\mathbf{H}_O}{dt}\right]_{rot} + \mathbf{\Omega} \times \mathbf{H}_O \qquad (7.4)$$

or

$$\mathbf{M}_{c.m.} = \left[\frac{d\mathbf{H}_{c.m.}}{dt}\right]_{rot} + \mathbf{\Omega} \times \mathbf{H}_{c.m.} \qquad (7.5)$$

The term $[d\mathbf{H}_O/dt]_{rot}$, or $[d\mathbf{H}_{c.m.}/dt]_{rot}$, represents the time rate of change of the angular momentum vector observed in a rotating reference frame. In other words, with $\mathbf{H}_O = (H_O)_x \mathbf{i} + (H_O)_y \mathbf{j} + (H_O)_z \mathbf{k}$,

$$\left[\frac{d(\mathbf{H})}{dt}\right]_{rot} = \left[\frac{d(H_O)_x}{dt}\right]_{rot} \mathbf{i} + \left[\frac{d(H_O)_y}{dt}\right]_{rot} \mathbf{j} + \left[\frac{d(H_O)_z}{dt}\right]_{rot} \mathbf{k} \qquad (7.6)$$

Substituting equations (7.1) to (7.3), one obtains

$$\left[\frac{d(H_O)_x}{dt}\right]_{rot} = \left[\frac{d\omega_x}{dt}\right]_{rot} I_{xx} + \left[\frac{d\omega_y}{dt}\right]_{rot} I_{xy} + \left[\frac{d\omega_z}{dt}\right]_{rot} I_{xz} \qquad (7.7)$$

$$\left[\frac{d(H_O)_y}{dt}\right]_{rot} = \left[\frac{d\omega_y}{dt}\right]_{rot} I_{yy} + \left[\frac{d\omega_x}{dt}\right]_{rot} I_{yx} + \left[\frac{d\omega_z}{dt}\right]_{rot} I_{yz} \qquad (7.8)$$

$$\left[\frac{d(H_O)_z}{dt}\right]_{rot} = \left[\frac{d\omega_z}{dt}\right]_{rot} I_{zz} + \left[\frac{d\omega_x}{dt}\right]_{rot} I_{zx} + \left[\frac{d\omega_y}{dt}\right]_{rot} I_{zy} \qquad (7.9)$$

where

$$\left[\frac{d\omega}{dt}\right]_{rot} = \left[\frac{d\omega_x}{dt}\right]_{rot} \mathbf{i} + \left[\frac{d\omega_y}{dt}\right]_{rot} \mathbf{j} + \left[\frac{d\omega_z}{dt}\right]_{rot} \mathbf{k} \qquad (7.10)$$

Equation (7.10) simplifies for *body-fixed axes*. Using equation (5.72) once more,

$$\frac{d\omega}{dt} = \left[\frac{d\omega}{dt}\right]_{rot} + \mathbf{\Omega} \times (\omega)$$

and since $\omega = \mathbf{\Omega}$ for these axes,

$$\frac{d\omega}{dt} = \left[\frac{d\omega}{dt}\right]_{rot} \qquad (7.11)$$

or

$$\frac{d\omega_x}{dt} = \left[\frac{d\omega_x}{dt}\right]_{rot} = \dot{\omega}_x \qquad (7.12)$$

$$\frac{d\omega_y}{dt} = \left[\frac{d\omega_y}{dt}\right]_{rot} = \dot{\omega}_y \qquad (7.13)$$

$$\frac{d\omega_z}{dt} = \left[\frac{d\omega_z}{dt}\right]_{rot} = \dot\omega_z \tag{7.14}$$

With these results, and using equations (7.1) to (7.3) and (7.4) or (7.5), the moment equation in expanded form can be found.

$$M_x = I_{xx}\dot\omega_x + I_{xy}\dot\omega_y + I_{xz}\dot\omega_z + \omega_y\omega_z(I_{zz} - I_{yy}) + \omega_y\omega_x I_{zx} + (\omega_y^2 - \omega_z^2)I_{zy} - \omega_x\omega_z I_{yx} \tag{7.15}$$

$$M_y = I_{yy}\dot\omega_y + I_{yx}\dot\omega_x + I_{yz}\dot\omega_z + \omega_x\omega_z(I_{xx} - I_{zz}) + \omega_y\omega_z I_{xy} + (\omega_z^2 - \omega_x^2)I_{xz} - \omega_y\omega_x I_{zy} \tag{7.16}$$

$$M_z = I_{zz}\dot\omega_z + I_{zx}\dot\omega_x + I_{zy}\dot\omega_y + \omega_x\omega_y(I_{yy} - I_{xx}) + (\omega_x^2 - \omega_y^2)I_{yx} + \omega_x\omega_z I_{yz} - \omega_y\omega_z I_{xz} \tag{7.17}$$

It must be remembered that equations (7.15) to (7.17) are only valid for a coordinate system fixed to the body, that is $\Omega = \omega$, and for a fixed or center-of-mass reference point.

If *principal axes* fixed to the body were used instead, equations (7.15) to (7.17) would reduce to

$$M_1 = I_1\dot\omega_1 + \omega_2\omega_3(I_3 - I_2) \tag{7.18}$$

$$M_2 = I_2\dot\omega_2 + \omega_1\omega_3(I_1 - I_3) \tag{7.19}$$

$$M_3 = I_3\dot\omega_3 + \omega_1\omega_2(I_2 - I_1) \tag{7.20}$$

These last equations are called *Euler's equations for body-fixed principal axes*.

Sample Problem 7.1
A very thin rectangular plate of 12-kg mass is welded to a massless shaft as shown. The shaft is of length 80 cm and is supported by two frictionless bearings. The plate is centered on the shaft, and the shaft and plate rotate together at a constant angular speed of 1000 rpm. For the instant that the plate lies in a vertical plane, find the reactions from the bearings at points A and B.

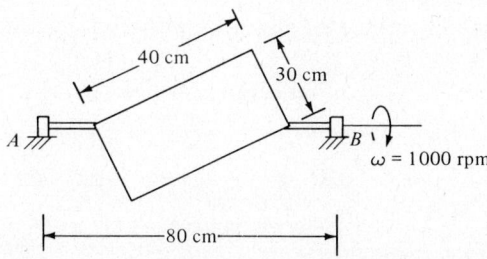

Analysis: This cannot be treated as a plane motion problem since there is no plane of symmetry perpendicular to the axis of rotation. The methods introduced in this chapter are required.

208 ANALYTICAL MECHANICS

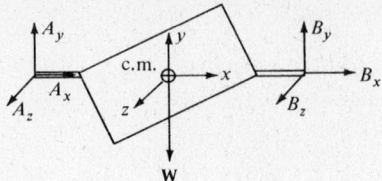

Solution 1. This solution will use a body-fixed coordinate system that is not parallel to principal directions. Since $\mathbf{a}_{c.m.} = \mathbf{0}$,

$$A_y + B_y - W = 0 \tag{a}$$

$$A_x + B_x = 0 \qquad A_x = -B_x \tag{b}$$

$$A_z + B_z = 0 \qquad A_z = -B_z \tag{c}$$

From Appendix B, for a thin rectangular plate,

$$I_1 = \frac{ma^2}{12} = \frac{12(0.3^2)}{12} = 0.09 \text{ kg} \cdot \text{m}^2$$

$$I_2 = \frac{mb^2}{12} = \frac{12(0.4^2)}{12} = 0.16 \text{ kg} \cdot \text{m}^2$$

$$I_3 = \frac{m(a^2 + b^2)}{12} = \frac{12(0.3^2 + 0.4^2)}{12} = 0.25 \text{ kg} \cdot \text{m}^2$$

Rotating the axes as shown will lead to an expression for the mass moments and products of inertia with respect to the x, y, z coordinate axes.

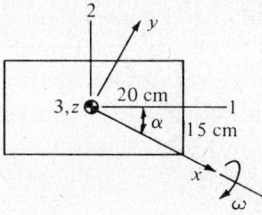

From the figure

$$\alpha = \tan^{-1} \tfrac{3}{4}$$

Then
$$[l] = \begin{bmatrix} 0.8 & -0.6 & 0 \\ 0.6 & 0.8 & 0 \\ 0 & 0 & 1 \end{bmatrix} \qquad [l]^t = \begin{bmatrix} 0.8 & 0.6 & 0 \\ -0.6 & 0.8 & 0 \\ 0 & 0 & 1 \end{bmatrix}$$

$$[I'] = [l][I][l]^t \tag{6.57}$$

$$[I'] = \begin{bmatrix} 0.1152 & -0.0336 & 0 \\ -0.0336 & 0.1348 & 0 \\ 0 & 0 & 0.25 \end{bmatrix} \text{ kg} \cdot \text{m}^2$$

$$I_{xx}=0.1152 \quad I_{yy}=0.1348 \quad I_{zz}=0.25$$
$$I_{xy}=I_{yx}=-0.0336 \quad I_{xz}=I_{zx}=I_{yz}=I_{zy}=0$$

Further,
$$\dot{\omega}_x=\dot{\omega}_y=\dot{\omega}_z=0$$
$$\omega_y=\omega_z=0 \quad \omega_x=1000\frac{2\pi}{60}=104.7 \text{ rad/s}$$

Then equations (7.15) to (7.17) become, with moments taken about the center of mass,

$$M_x=0=0 \tag{d}$$
$$M_y=0.4(A_z-B_z)=0 \tag{e}$$
$$M_z=0.4(B_y-A_y)=\omega_x^2 I_{yx}=104.7^2(-0.0336)=-368.47$$
$$B_y-A_y=-921.2 \tag{f}$$

Combining equations (a) to (f) leads to

$A_x=-B_x$ (undetermined, equal to zero for no external load in x direction)
$A_z=B_z=0$ $B_y=-401.7$ N $A_y=519.46$ N

Solutions 2.   Now, instead of using the x, y, z coordinate system, principal axes will be employed.

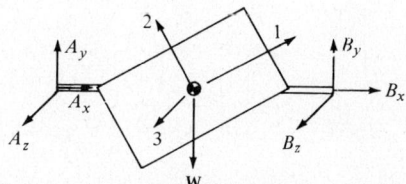

$$I_1=0.09 \text{ kg} \cdot \text{m}^2 \quad I_2=0.16 \text{ kg} \cdot \text{m}^2 \quad I_3=0.25 \text{ kg} \cdot \text{m}^2$$
$$\boldsymbol{\omega}=\omega_x(0.8\mathbf{e}_1-0.6\mathbf{e}_2)$$

or
$$\omega_1=0.8\omega_x=0.8(104.7)=83.78 \text{ rad/s}$$
$$\omega_2=-0.6\omega_x=-62.83 \text{ rad/s}$$
$$\omega_3=0$$
$$\dot{\boldsymbol{\omega}}=\mathbf{0} \quad \dot{\omega}_1=\dot{\omega}_2=\dot{\omega}_3=0$$

Further, as shown in the figure,
$$M_x=0 \quad M_y=0.4(A_z-B_z) \quad M_z=0.4(B_y-A_y)$$

and the components of M_x, M_y, and M_z become, in the principal axes directions,
$$M_1=M_y(0.6)=0.24(A_z-B_z)$$

210 ANALYTICAL MECHANICS

$$M_2 = 0.8 M_y = 0.32(A_z - B_z)$$
$$M_3 = M_z = 0.4(B_y - A_y)$$

Substituting these values into Euler's equation for body-fixed principal axes leads to

$$M_1 = 0.24(A_z - B_z) = 0 \tag{g}$$
$$M_2 = 0.32(A_z - B_z) = 0 \tag{h}$$
$$M_3 = 0.4(B_y - A_y) = (83.78)(-62.83)(0.16 - 0.09)$$
$$B_y - A_y = -921.2 \tag{i}$$

Combining equations (a), (b), (c), (g), (h), and (i) gives the same results as before:

$$A_x = -B_x \qquad A_z = B_z = 0$$
$$B_y = -401.7 \text{ N} \qquad A_y = 519.46 \text{ N} \qquad \blacksquare$$

Sample Problem 7.2

A thin wheel of radius R and mass m rolls without sliding on the inclined surface shown with a constant speed v_0. It is free to turn relative to the massless shaft of length L with negligible friction. The shaft itself rotates about the vertical. Find the support force at A as well as the normal force of the ground on the wheel.

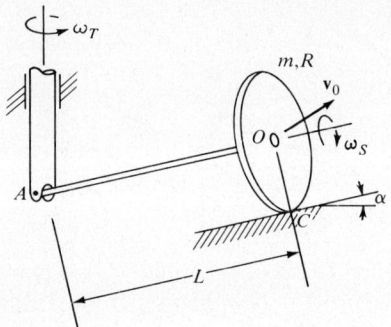

Analysis: Principal axes fixed to the body will be used in Euler's equations referenced to fixed point A. In this example the kinematics are described as follows:

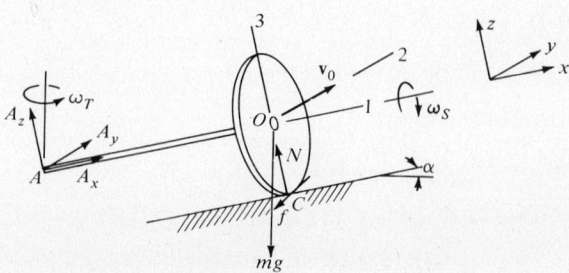

$$\mathbf{v}_C = \mathbf{v}_O + \boldsymbol{\omega} \times \mathbf{R} \tag{a}$$

$$\mathbf{v}_C = 0 \quad \text{(instantaneous center)}$$

$$\mathbf{v}_O = v_O \mathbf{j} = \boldsymbol{\omega}_T \times L\mathbf{i}$$

Here

$$\boldsymbol{\omega} = \omega_T(\sin \alpha \mathbf{i} + \cos \alpha \mathbf{k}) - \omega_s \mathbf{i}$$

Then
$$v_O \mathbf{j} = \omega_T L \cos \alpha \mathbf{j} \tag{b}$$

and from equation (a),

$$v_O \mathbf{j} = -\boldsymbol{\omega} \times \mathbf{R} = (\omega_s - \omega_T \sin \alpha) R \mathbf{j} = \omega_T L \cos \alpha \mathbf{j}$$

Then from equation (b),

$$\omega_T = \frac{v_O}{L \cos \alpha}$$

and
$$\omega_s = \frac{\omega_T(L \cos \alpha + R \sin \alpha)}{R} \tag{c}$$

or
$$\omega_s = \frac{v_O(1 + (R/L)\tan \alpha)}{R}$$

With axes fixed to the body

$$\frac{d\boldsymbol{\omega}}{dt} = \left[\frac{d\boldsymbol{\omega}}{dt}\right]_{\text{rot}}$$

$$\frac{d\boldsymbol{\omega}}{dt} = \boldsymbol{\omega}_T \times \boldsymbol{\omega}_s = -\omega_T \omega_s \cos \alpha \mathbf{j} \tag{d}$$

The cartesian components are then

$$\dot{\omega}_x = \dot{\omega}_z = 0 \qquad \dot{\omega}_y = -\omega_T \omega_s \cos \alpha \tag{e}$$

From Appendix B, the mass moments of inertia for a thin disk about principal axes are

$$I_{xx} = I_1 = \frac{mR^2}{2} \qquad I_{yy} = I_{zz} = I_2 = I_3 = \frac{mR^2}{4} \tag{f}$$

It will provide greater ease in problem solution if the moment equations are written for fixed point A since the three reaction forces at A would then be eliminated. This means that the principal axes must be moved to point A:

$$I_1' = \frac{mR^2}{2} \qquad I_2' = I_3' = \frac{mR^2}{4} + mL^2 = \frac{m(R^2 + 4L^2)}{4}$$

Euler's equations now become

$$M_1' = 0 = -Rf \quad (f = 0 \text{ for this constant speed problem}) \tag{g}$$

$$M_2' = \frac{m(R^2 + 4L^2)}{4}(-\omega_T\omega_s \cos\alpha) + \frac{m(R^2 - 4L^2)}{4}(\omega_T \sin\alpha - \omega_s)\omega_T \cos\alpha$$

$$= L(mg \cos\alpha - N) \tag{h}$$

$$M_3' = 0 = -Lf \tag{i}$$

$$L(mg \cos\alpha - N) = -\frac{mR^2}{2}\omega_T\omega_s \cos\alpha + m\left(\frac{R^2 - 4L^2}{4}\right)\omega_T^2 \sin\alpha \cos\alpha$$

$$N = mg \cos\alpha + \frac{mR^2}{2L}(\omega_T\omega_s \cos\alpha) + m\left(\frac{4L^2 - R^2}{4L}\right)\omega_T^2 \sin\alpha \cos\alpha \tag{j}$$

Because point A was used as a reference point, the normal force was obtained directly from the moment equation (h). Note that for $v_O = 0$, the static case, $N = mg \cos\alpha$. Since the wheel spins and turns at the same time, gyroscopic moments are called into play to affect the motion. In this case the extra normal force needed is a function of the squared angular velocity of the shaft. This term obviously cannot be neglected. The reaction forces at A can now be found with the aid of the force equations.

$$F_x = m(a_{c.m.})_x$$

$$-mg \sin\alpha + A_x = -\frac{mv_O^2}{L} \tag{k}$$

$$F_y = m(a_{c.m.})_y$$

$$A_y = f = 0 \tag{l}$$

$$F_z = m(a_{c.m.})_z$$

$$A_z + N - mg \cos\alpha = \frac{mv_O^2}{L}\tan\alpha \tag{m}$$

The results are

$$A_x = mg \sin\alpha - \frac{mv_O^2}{L} \tag{n}$$

$$A_y = 0 \tag{o}$$

$$A_z = -\frac{mR^2}{2L}(\omega_T\omega_s \cos\alpha) + m\left(\frac{R^2 - 4L^2}{4L}\right)\omega_T^2 \sin\alpha \cos\alpha + \frac{mv_O^2}{L}\tan\alpha \tag{p}$$

For $\alpha = 0°$, the following results are obtained:

$$N = mg + \frac{mR}{2}\left(\frac{v_O}{L}\right)^2 \tag{q}$$

$$A_x = -\frac{mv_O^2}{L} \tag{r}$$

$$A_y = f = 0 \tag{s}$$

$$A_z = -\frac{mR}{2}\left(\frac{v_O}{L}\right)^2 \tag{t}$$

∎

The kinetic energy expressions for a rigid body having general space motion were also derived in Chapters 4 and 6. Rewritten here in matrix form they are

$$T = \frac{M\{v_{c.m.}\}'\{v_{c.m.}\}}{2} + \frac{\{\omega\}'[\hat{I}]\{\omega\}}{2} \quad (7.21)$$

and for a fixed-point reference

$$T_O = \frac{\{\omega\}'[I_O]\{\omega\}}{2} \quad (7.22)$$

The reader is well-advised at this point to review the materials contained in Chapters 4 and 6 concerning the principle of work-energy, conservation of energy, and impulse and momentum before proceeding to the following sample problems.

Sample Problem 7.3
Derive an expression for the total kinetic energy of the wheel described in Sample Problem 7.2.

Analysis: Since the wheel moves about a fixed point, and since the principal moments of inertia about that point were solved for previously, it is convenient to use the expression for T_O, equation (7.22), to solve the problem.

$$T_O = \frac{\{\omega\}'[I_O]\{\omega\}}{2}$$

$$[I_O] = \begin{bmatrix} I_1 & 0 & 0 \\ 0 & I_2 & 0 \\ 0 & 0 & I_3 \end{bmatrix} = \begin{bmatrix} \frac{mR^2}{2} & 0 & 0 \\ 0 & \frac{m(R^2+4L^2)}{4} & 0 \\ 0 & 0 & \frac{m(R^2+4L^2)}{4} \end{bmatrix}$$

$$\{\omega\} = \begin{Bmatrix} \omega_T \sin\alpha - \omega_s \\ 0 \\ \omega_T \cos\alpha \end{Bmatrix}$$

$$T_O = \frac{I_1\omega_1^2}{2} + \frac{I_2\omega_2^2}{2} + \frac{I_3\omega_3^2}{2}$$

$$= \frac{mR^2}{4}(\omega_T \sin\alpha - \omega_s)^2 + \frac{m(R^2+4L^2)}{8}(\omega_T \cos\alpha)^2 \quad \blacksquare$$

Sample Problem 7.4
It is desired to steer a cylindrical space probe with the help of an internal spinning wheel. The wheel initially spins about the y direction shown in the figure with a maintained constant speed ω_{sw} relative to the ship. The probe's

214 ANALYTICAL MECHANICS

cylindrical body turns with speed ω_{SS} about the x axis. If an internal force is applied to turn the wheel so that its spin axis eventually lines up with the cylinder's spin axis, through what angle does the cylinder's axis (space probe) turn? For the wheel, the mass moment of inertia about its axis of symmetry is I_{AW} and about a transverse axis is I_{TW}. For the cylinder, the mass moments of inertia are I_{AS} and I_{TS}, respectively.

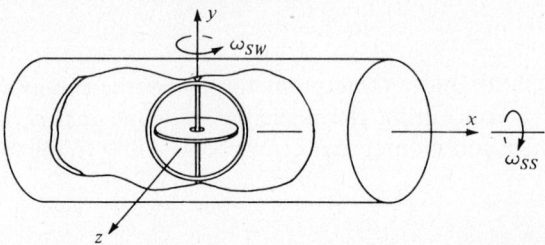

Analysis: Since no external moments are applied to the system, the momentum of the system about the combined center of mass must be conserved.

$$(\mathbf{H}_{c.m.})_{initially} = (\mathbf{H}_{c.m.})_{finally}$$

$$(\mathbf{H}_{c.m.})_{initially} = (I_{TW} + I_{AS})\omega_{SS}\mathbf{i} + (I_{AW})\omega_{SW}\mathbf{j}$$

$$(\mathbf{H}_{c.m.})_{finally} = [I_{AW}\omega_{SW} + (I_{AS} + I_{AW})\omega'_{SS}]\mathbf{i}'$$

The new axis $\mathbf{i}'$ for the space probe must point along the same direction as the initial total angular momentum. This direction is fixed in space for the moment-free body.

This means that

$$\mathbf{i}' = \frac{(I_{TW} + I_{AS})\omega_{SS}\mathbf{i} + (I_{AW})\omega_{SW}\mathbf{j}}{[(I_{TW} + I_{AS})^2\omega_{SS}^2 + I_{AW}^2\omega_{SW}^2]^{1/2}}$$

From the figure, the angle turned through is

$$\theta = \tan^{-1}\frac{I_{AW}\omega_{SW}}{(I_{TW} + I_{AS})\omega_{SS}}$$

∎

Sample Problem 7.5
Shown in the figure is a thin disk of mass m and radius r that rolls with constant

angular spin $\dot{\psi}$. The radius of the circular path that the center of mass of the disk describes is R. If it is also known that the disk maintains a constant angle of inclination θ during its motion, write the applicable equations of motion for the disk. What is the relationship between the precession rate of turning $\dot{\phi}$ and the angle of inclination θ?

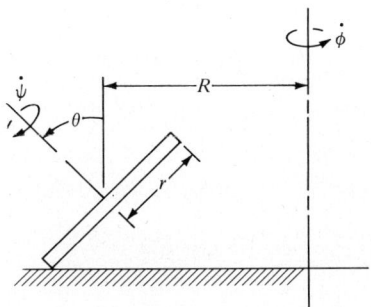

Analysis: The kinematics of the problem are such that if the velocity of the contact point of the wheel on the ground is zero for pure rolling,

$$\dot{\psi} = \frac{R + r \cos \theta}{r} \dot{\phi} \qquad (a)$$

where $\omega = \dot{\psi} + \dot{\phi}$.

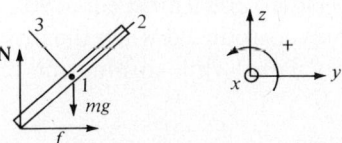

The free-body diagram shown depicts all the forces acting on the disk for assumed constant speed motion; i.e., there is no friction force acting in the x direction. The applicable force equations for the disk are

$$F_y = m(a_{c.m.})_y \qquad f = mR\dot{\phi}^2 \qquad (b)$$
$$F_z = m(a_{c.m.})_z \qquad N - mg = 0 \qquad (c)$$

The moment equations about the center of mass for body-fixed principal axes are

$$M_1 = I_1 \dot{\omega}_1 + (I_3 - I_2)\omega_3 \omega_2 \qquad (d)$$
$$M_2 = I_2 \dot{\omega}_2 + (I_1 - I_3)\omega_1 \omega_3 \qquad (e)$$
$$M_3 = I_3 \dot{\omega}_3 + (I_2 - I_1)\omega_1 \omega_2 \qquad (f)$$

where $I_1 = I_2 = Mr^2/4$ and $I_3 = Mr^2/2$. $\omega = (-\dot{\psi} + \dot{\phi} \cos \theta)\mathbf{i}_3 + \dot{\phi} \sin \theta \mathbf{i}_2$, for the instant shown when axis 2 lies in the vertical plane of N, f, W, and axis 3.

Therefore,
$$\omega_1 = 0 \qquad \omega_2 = \dot{\phi} \sin \theta \qquad \omega_3 = -\dot{\psi} + \dot{\phi} \cos \theta$$

The angular acceleration is
$$\dot{\omega} = \dot{\phi} \times \dot{\psi} = -\dot{\phi}\dot{\psi} \sin \theta \mathbf{i}$$
$$\dot{\omega}_1 = -\dot{\phi}\dot{\psi} \sin \theta \qquad \dot{\omega}_2 = \dot{\omega}_3 = 0$$

Euler's equations reduce to one useful equation about axis 1, namely,
$$rf \sin \theta - rN \cos \theta = \frac{mr^2}{4}(-\dot{\phi}\dot{\psi} \sin \theta) + \frac{mr^2}{4}(-\dot{\psi} + \dot{\phi} \cos \theta)(\dot{\phi} \sin \theta) \qquad (d')$$

To help simplify this, equation (a) may be rewritten as
$$-\dot{\psi} + \dot{\phi} \cos \theta = -\frac{\dot{\phi} R}{r} \qquad (g)$$

Then equation (d') becomes
$$Rmr\dot{\phi}^2 \sin \theta - rmg \cos \theta = \frac{mr^2}{2}\left(-\frac{\dot{\phi} R}{r}\right)(\dot{\phi} \sin \theta) - \frac{mr^2}{4}\dot{\phi}^2 \sin \theta \cos \theta$$
$$4g \cos \theta = 6R\dot{\phi}^2 \sin \theta + r\dot{\phi}^2 \sin \theta \cos \theta$$
$$\dot{\phi}^2 = \frac{4g \cot \theta}{6R + r \cos \theta} \qquad (h)$$

For straight line motion, $\dot{\phi}$ is zero and θ must equal 90°. The gyroscopic moment that causes precession is only introduced when the disk leans over. A similarity exists between the leaning of a disk while turning and the banking of an aircraft. ∎

7.3 MODIFIED EULER EQUATIONS

The restriction imposed during the development of Euler's moment equations for body-fixed principal axes can be relaxed somewhat when the body is axially symmetric. The argument used in deriving equations (7.18) to (7.20) was that the axes had to be fixed to the body to eliminate time derivatives of the moment of inertia terms in the local rotating frame. For an axially symmetric body, if one principal axis is aligned and fixed to the axis of symmetry, the body can rotate with respect to the other two axes without introducing time derivatives of the moments of inertia. Due to the symmetry present, the geometry of the body relative to the coordinate system does not change. The body's angular velocity can now differ from that of the coordinate system by an extra spin term ω_s along the axis of symmetry. Defining the z axis to be the axis of symmetry, and $I_{zz} = I_3$, then

$$\omega_{\text{body}} = \omega_x \mathbf{i} + \omega_y \mathbf{j} + (\omega_z + \omega_s)\mathbf{k} \qquad (7.23)$$
$$\mathbf{\Omega} = \omega_x \mathbf{i} + \omega_y \mathbf{j} + \omega_z \mathbf{k} \qquad (7.24)$$

With $I_{xx} = I_{yy} = I_1$,

$$\mathbf{H} = I_1(\omega_x \mathbf{i} + \omega_y \mathbf{j}) + I_3(\omega_s + \omega_z)\mathbf{k} \tag{7.25}$$

The sum of the moments about the center of mass or about a fixed point in the body is equal to the change in angular momentum referenced to that point. Then

$$\frac{d\mathbf{H}}{dt} = \left[\frac{d\mathbf{H}}{dt}\right]_{\text{rot}} + \boldsymbol{\Omega} \times \mathbf{H} \tag{7.26}$$

and
$$M_1 = I_1[\dot{\omega}_x]_{\text{rot}} + (I_3 - I_1)\omega_z\omega_y + I_3\omega_s\omega_y \tag{7.27}$$
$$M_2 = I_1[\dot{\omega}_y]_{\text{rot}} + (I_1 - I_3)\omega_x\omega_z - I_3\omega_s\omega_x \tag{7.28}$$
$$M_3 = I_3[(\dot{\omega}_s + \dot{\omega}_z)]_{\text{rot}} \tag{7.29}$$

The advantage of these equations is that they allow for the use of a reference frame having one less degree of rotational freedom. This can allow, for instance, for the constant orientation of an axis relative to the ground.

Sample Problem 7.6
Rework Sample Problem 7.5 using the modified Euler equations. Take the z axis to be the axis of symmetry ($I_z = I_3$). Let axis 1 remain parallel to the ground. This means that $\boldsymbol{\Omega} = \dot{\boldsymbol{\phi}}$.

Analysis:

$$\boldsymbol{\Omega} = \dot{\boldsymbol{\phi}} = \dot{\phi}(\sin\theta \mathbf{j} + \cos\theta \mathbf{k}) \tag{a}$$

$$\Omega_x = 0 \qquad \Omega_y = \dot{\phi}\sin\theta \qquad \Omega_z = \dot{\phi}\cos\theta$$

$$\boldsymbol{\omega}_{\text{body}} = \boldsymbol{\Omega} - \dot{\psi}\mathbf{k} \qquad \omega_s = -\dot{\psi}$$

$$[\dot{\boldsymbol{\omega}}_{\text{body}}]_{\text{rot}} = \dot{\boldsymbol{\omega}}_{\text{body}} - \boldsymbol{\Omega} \times \boldsymbol{\omega}_{\text{body}} \tag{b}$$

$$\dot{\boldsymbol{\omega}}_{\text{body}} = \dot{\boldsymbol{\phi}} \times \dot{\boldsymbol{\psi}} \qquad \boldsymbol{\Omega} \times \boldsymbol{\omega}_{\text{body}} = \dot{\boldsymbol{\phi}} \times (\dot{\boldsymbol{\psi}} + \dot{\boldsymbol{\phi}})$$

Therefore, from equation (b),

$$[\dot{\boldsymbol{\omega}}_{\text{body}}]_{\text{rot}} = 0$$
$$[\dot{\omega}_x]_{\text{rot}} = [\dot{\omega}_y]_{\text{rot}} = [\dot{\omega}_s + \dot{\omega}_z]_{\text{rot}} = 0$$

Then
$$M_1 = (I_3 - I_2)\dot{\phi}^2 \sin\theta \cos\theta + I_3(-\dot{\psi})(\dot{\phi}\sin\theta) \tag{c}$$
$$M_2 = M_3 = 0 \tag{d}$$

Using the values of $I_3 = Mr^2/2$ and $I_1 = Mr^2/4$, equation (c) becomes

$$M_1 = \frac{mr^2}{4}\dot{\phi}^2 \sin\theta \cos\theta - \frac{mr^2}{2}\dot{\psi}\dot{\phi}\sin\theta \tag{e}$$

This last equation derived, equation (e), should now be compared with the right-hand side of equation (d′) from the previous solution. ∎

7.4 EULERIAN ANGLES

A rigid body has 3 degrees of rotational freedom. It was found earlier that the direction cosine matrix, though it adequately described three-dimensional rotation of a body, had elements that were not independent of each other. Because of this, they are not suitable choices to use as coordinates. This section concerns itself with an alternate method of describing three-dimensional rotation or body orientation, using eulerian angles. Eulerian angles describe the orientation of one cartesian coordinate system with respect to another. Though various sets of Euler angle systems exist in the literature, they all have in common the capability to describe a rigid body's general orientation in terms of three independent angles of rotation or angular coordinates. Note that the angles of rotation found by integrating the components of angular velocity ω_x, ω_y, ω_z do not represent the body's orientation relative to a non-rotating frame, nor do the angles found by integrating the components ω_X, ω_Y, ω_Z.

The set of Euler angles defined here and their order of application are as follows:

1. First, a positive rotation ϕ of the body about the Z axis is taken to yield an intermediate position where body-fixed coordinates, which previously coincided with fixed directions X, Y, Z, are now the x', y', z' coordinate axes, respectively. The x' axis is called the *line of nodes*. See Figures 7.1 to 7.3.
2. Then a positive rotation θ about the new x' axis, the line of nodes, produces a new intermediate position where the body-fixed axes x', y', z' become the x'', y'', z'' coordinate set.
3. A final positive rotation ψ about the new z'' axis produces the final orientation of the body where the body-fixed axes are now the x, y, z coordinate system.

If the XYZ cartesian coordinate system is selected to have a *fixed* direction in space, then the angular velocity of the body is

$$\boldsymbol{\omega} = \dot{\boldsymbol{\psi}} + \dot{\boldsymbol{\phi}} + \dot{\boldsymbol{\theta}} \tag{7.30}$$

or in terms of rectangular components,

$$\omega_x = \dot{\theta} \cos \psi + \dot{\phi} \sin \theta \sin \psi \tag{7.31}$$

$$\omega_y = -\dot{\theta} \sin \psi + \dot{\phi} \sin \theta \cos \psi \tag{7.32}$$

$$\omega_z = \dot{\psi} + \dot{\phi} \cos \theta \tag{7.33}$$

Similarly,

$$\omega_X = \dot{\psi} \sin \theta \sin \phi + \dot{\theta} \cos \phi \tag{7.34}$$

$$\omega_Y = -\dot{\psi} \sin \theta \cos \phi + \dot{\theta} \sin \phi \tag{7.35}$$

$$\omega_Z = \dot{\psi} \cos \theta + \dot{\phi} \tag{7.36}$$

RIGID BODY DYNAMICS 219

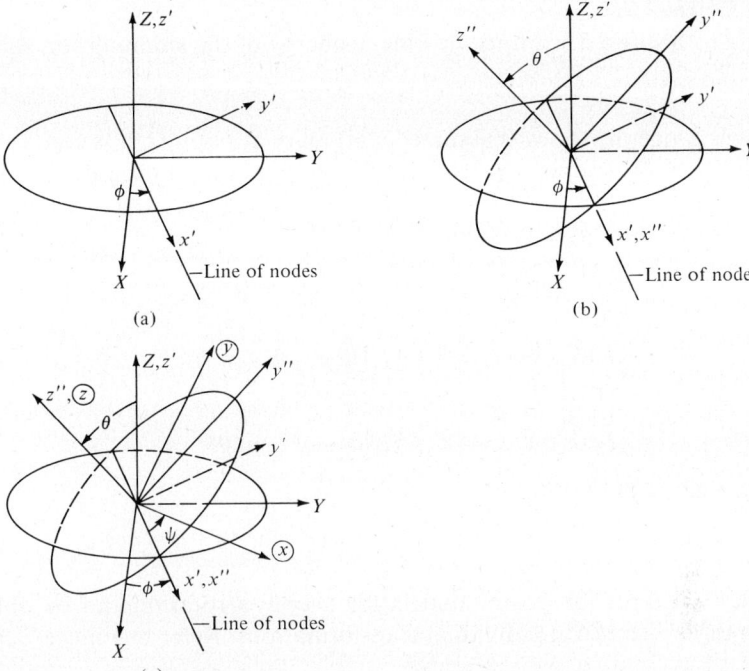

Figure 7.1

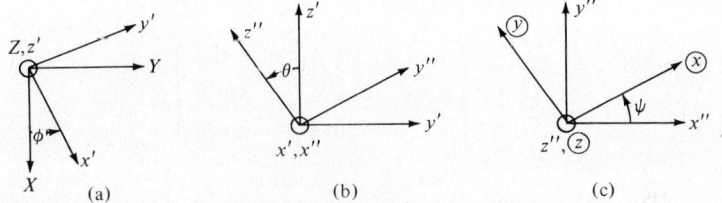

Figure 7.2

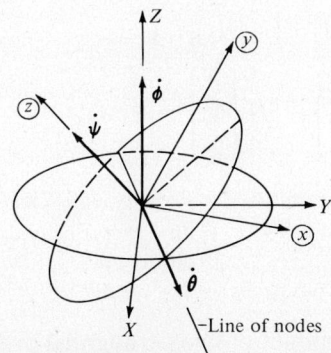

Figure 7.3

Sample Problem 7.7

Referring to Section 5.3C, find the kinetic energy of the spinning top shown in Figure 5.7.

Analysis: Since the top was assumed stationary at point O, it is convenient to use body-fixed principal axes referenced to this point to describe the kinetic energy. Here let $I_3 = I_{zz}$ and $I_1 = I_2 = I_{xx} = I_{yy}$ for a body of revolution.

The kinetic energy expression then is

$$T = \frac{I_3 \omega_z^2}{2} + \frac{I_1 \omega_x^2}{2} + \frac{I_2 \omega_y^2}{2}$$

$$= \tfrac{1}{2} I_3 (\dot{\psi} + \dot{\phi} \cos \theta)^2 + \tfrac{1}{2} I_1 [(\dot{\theta} \cos \psi + \dot{\phi} \sin \theta \sin \psi)^2$$
$$+ (-\dot{\theta} \sin \psi + \dot{\phi} \sin \theta \cos \psi)^2]$$

$$= \tfrac{1}{2} I_3 (\dot{\psi} + \dot{\phi} \cos \theta)^2 + \tfrac{1}{2} I_1 (\dot{\theta}^2 + \dot{\phi}^2 \sin \theta)$$

The total energy of the top is

$$E = \tfrac{1}{2} I_3 (\dot{\psi} + \dot{\phi} \cos \theta)^2 + \tfrac{1}{2} I_1 (\dot{\theta}^2 + \dot{\phi}^2 \sin^2 \theta) + mgl \cos \theta \qquad \blacksquare$$

The total direction cosine matrix for a body's rotation can be obtained through three successive individual transformations. Refer to Figure 7.2a to c.

$$[l_1] = \begin{bmatrix} \cos \phi & \sin \phi & 0 \\ -\sin \phi & \cos \phi & 0 \\ 0 & 0 & 1 \end{bmatrix} \qquad (7.37)$$

$$[l_2] = \begin{bmatrix} 1 & 0 & 0 \\ 0 & \cos \theta & \sin \theta \\ 0 & -\sin \theta & \cos \theta \end{bmatrix} \qquad (7.38)$$

$$[l_3] = \begin{bmatrix} \cos \psi & \sin \psi & 0 \\ -\sin \psi & \cos \psi & 0 \\ 0 & 0 & 1 \end{bmatrix} \qquad (7.39)$$

$$[l] = [l_3][l_2][l_1] \qquad (7.40)$$

$$\begin{Bmatrix} x \\ y \\ z \end{Bmatrix} = [l] \begin{Bmatrix} X \\ Y \\ Z \end{Bmatrix} \qquad (7.41)$$

$$[l] = \begin{bmatrix} \cos \psi \cos \phi - \cos \theta \sin \phi \sin \psi & \cos \psi \sin \phi + \cos \theta \cos \phi \sin \psi & \sin \psi \sin \theta \\ -\sin \psi \cos \phi - \cos \theta \sin \phi \cos \psi & -\sin \psi \sin \phi + \cos \theta \cos \phi \cos \psi & \cos \psi \sin \theta \\ \sin \theta \sin \phi & -\sin \theta \cos \phi & \cos \theta \end{bmatrix}$$

$$(7.42)$$

Remembering that the product of two orthogonal matrices is itself orthogonal leads to

$$[l]^{-1} = [l]^t \qquad (7.43)$$

Before leaving this section, one last point should be mentioned. From equations (7.31) to (7.33) one can write

$$\begin{bmatrix} \omega_x \\ \omega_y \\ \omega_z \end{bmatrix} = \begin{bmatrix} \sin\theta\sin\psi & \cos\psi & 0 \\ \sin\theta\cos\psi & -\sin\psi & 0 \\ \cos\theta & 0 & 1 \end{bmatrix} \begin{Bmatrix} \dot\phi \\ \dot\theta \\ \dot\psi \end{Bmatrix} \quad (7.44)$$

$$\begin{Bmatrix} \omega_x \\ \omega_y \\ \omega_z \end{Bmatrix} = [B] \begin{Bmatrix} \dot\phi \\ \dot\theta \\ \dot\psi \end{Bmatrix}$$

Because the coordinates associated with $\dot\phi$, $\dot\theta$, and $\dot\psi$ are *not orthogonal*, the transformation matrix $[B]$ is not orthogonal.

$$[B]^{-1} \neq [B]^t \quad (7.45)$$

Further, $(\dot\phi^2 + \dot\theta^2 + \dot\psi^2)^{1/2}$ does not equal to ω. The reader should be able to show by trigonometric means that

$$[B]^{-1} = \frac{1}{\sin\theta} \begin{bmatrix} \sin\psi & \cos\psi & 0 \\ \sin\theta\cos\psi & -\sin\theta\sin\psi & 0 \\ -\cos\theta\sin\psi & -\cos\theta\cos\psi & \sin\theta \end{bmatrix} \quad (7.46)$$

7.5 GENERAL FORCE-FREE MOTION OF A RIGID BODY: GEOMETRIC SOLUTION

In the last section, it was seen that to describe the absolute motion of a rotating body, Euler angles had to be introduced and the moment equation integrated. Closed-form solutions are not available to most three-dimensional rotation problems because of the inseparability of variables within the equations. In practice, moment equations are reduced first to quadrature; i.e., first integrals are obtained and then numerical methods are applied to solve the resulting equations on a digital computer. All present numerical methods for continuous time differential equations are applicable only to *first-order* state variable equations. Each general three-dimensional rotation problem is unique. There are, however, so-called classic problems that, because of some form of simplification in the dynamical equations, are solvable in closed form. The moment-free case for axially symmetric bodies is one such example. Since a very large class of dynamic problems involves force-free motion or motion where the only force acting is that due to gravity, this is not a trivial case.

There is *no resultant moment* about the center of mass. Euler's equation for principal axes located at the center of mass then reduces to

$$I_1 \dot\omega_1 = \omega_2 \omega_3 (I_2 - I_3) \quad (7.47)$$

$$I_2 \dot\omega_2 = \omega_1 \omega_3 (I_3 - I_1) \quad (7.48)$$

$$I_3 \dot\omega_3 = \omega_1 \omega_2 (I_1 - I_2) \quad (7.49)$$

222 ANALYTICAL MECHANICS

$\mathbf{H}_{c.m.}$ must be constant. Its magnitude and *direction* do not vary. Similarly, the system must be conservative:

$$T + V = \text{constant}$$

and the kinetic energy of rotation must be constant

$$\frac{I_1\omega_1^2}{2} + \frac{I_1\omega_2^2}{2} + \frac{I_3\omega_3^2}{2} = T_{rot} \tag{7.50}$$

Analytical solutions in terms of elliptic functions exist for equations (7.47) to (7.49), but they are quite lengthy and complex. To help visualize the motion, the classical method of Poinsot (1779–1859), introduced in 1834, will be employed instead. It is a geometrical model of the dynamical equations. Referring to equation (7.50), it can be rewritten as

$$I_1\omega_1^2 + I_2\omega_2^2 + I_3\omega_3^2 = 2T_{rot} \tag{7.51}$$

which is the same as

$$\mathbf{H}_{c.m.} \cdot \boldsymbol{\omega} = 2T_{rot} \tag{7.52}$$

Since both T_{rot} and $\mathbf{H}_{c.m.}$ are constants, the component of ω along the direction of $\mathbf{H}_{c.m.}$, which is fixed in space, must be constant. It *does not mean that ω is constant*. ω can change magnitude and direction; however, it must always touch a unique plane, determined by the initial conditions, that lies perpendicular to $\mathbf{H}_{c.m.}$. This plane is called the *invariable plane*. The *invariable line* is defined to be the line in the direction of $\mathbf{H}_{c.m.}$. Refer to Figure 7.4.

Further information can be obtained from Equation (7.51). This is an equation of an ellipsoid in terms of body-fixed axes where the tip of the vector ω develops the ellipsoid surface. This Poinsot ellipsoid is proportional to and concentric with the inertia ellipsoid described in Section 6.3, equation (6.80). The vector ω is related to the vector $\mathbf{a}$ by

$$\boldsymbol{\omega} = \mathbf{a}\omega\sqrt{I_{x'x'}} \tag{7.53}$$

To describe rotation, ω and $\mathbf{a}$ would lie in the same direction, that of the instantaneous axis of rotation along which the body's mass moment of inertia can be described by $I_{x'x'}$.

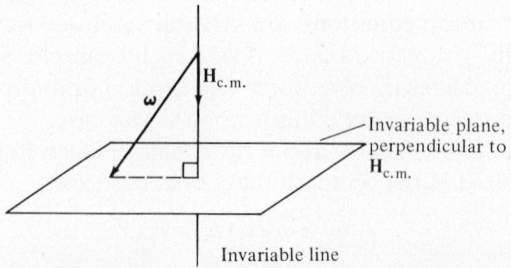

Figure 7.4

The gradient of the functional representation for the inertia ellipsoid, equation (6.80), is of interest.

$$F = I_1 a_1'^2 + I_2 a_2'^2 + I_3 a_3'^2 = 1 \tag{7.54}$$

$$\nabla F = \frac{\partial F}{\partial x} \mathbf{i}_1 + \frac{\partial F}{\partial y} \mathbf{i}_2 + \frac{\partial F}{\partial z} \mathbf{i}_3 \tag{7.55}$$

$$\nabla F = 2 I_1 a_1' \mathbf{i}_1 + 2 I_2 a_2' \mathbf{i}_2 + 2 I_3 a_3' \mathbf{i}_3 \tag{7.56}$$

which, because of equation (7.53), can be written as

$$\nabla F = \frac{2}{\omega \sqrt{I_{x'x'}}} (I_1 \omega_1 \mathbf{i}_1 + I_2 \omega_2 \mathbf{i}_2 + I_3 \omega_3 \mathbf{i}_3) \tag{7.57}$$

or
$$\nabla F = \frac{2}{\omega \sqrt{I_{x'x'}}} \mathbf{H}_{\text{c.m.}} \tag{7.58}$$

It is seen that the gradient of the inertia ellipsoid function acts parallel to the angular momentum vector. Since the Poinsot ellipsoid is proportional to the inertia ellipsoid, one concludes that the normal at the surface of the Poinsot ellipsoid at the tip of the vector $\boldsymbol{\omega}$ is parallel to the angular momentum vector. Further, *the tangent plane to the surface must be the aforementioned invariable plane* that lies perpendicular to the angular momentum vector, at a fixed distance from the origin of the Poinsot ellipsoid. Because the angular velocity vector touches the invariable plane at all times, the ellipsoid can be said to roll on the fixed invariable plane. The center of the ellipsoid remains at a constant height above the plane. *Since the Poinsot ellipsoid is fixed to the body, this construction in effect describes the motion of the real body in space.*

The curve traced out by the ellipsoid tangent point on the invariable plane is called the *herpolhode*, and the simultaneous curve traced out by the plane's contact point on the ellipsoid is called the *polhode*. See Figures 7.5 and 7.6.

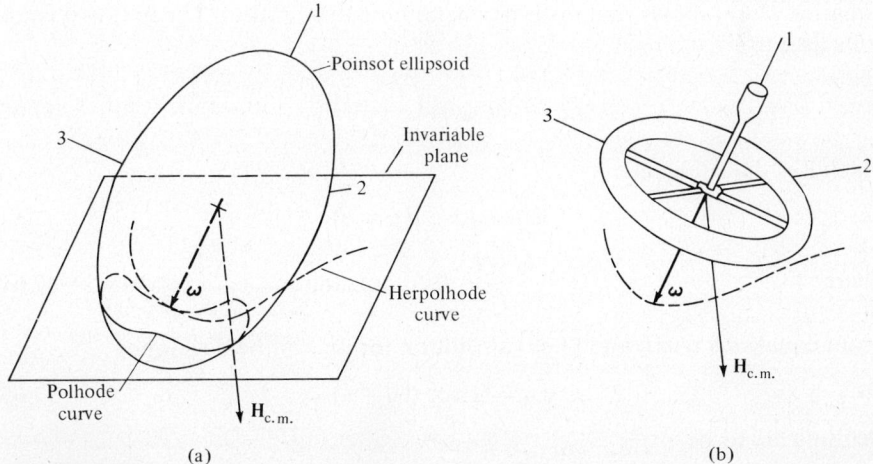

Figure 7.5

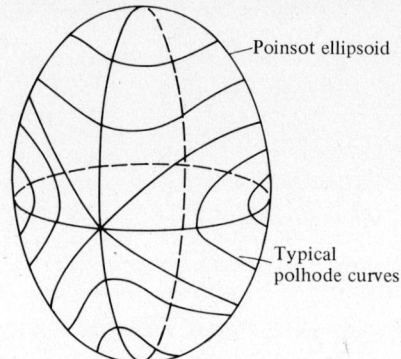

Figure 7.6

The surface traced out by the angular velocity vector in space is called the *space cone*. From the viewpoint of an observer fixed to the body, both $\mathbf{H}_{c.m.}$ and $\boldsymbol{\omega}$ turn, along with the invariable plane. The surface traced out by $\boldsymbol{\omega}$ *relative* to the body is called the *body cone*. The body cone must roll on the space cone.

The next section concerning the force-free motion of axially symmetric shapes should clear up some of the discussion presented here.

7.6 FORCE-FREE MOTION OF AXIALLY SYMMETRIC SHAPES

For axially symmetric shapes ($I_1 = I_2$) equations (7.47) to (7.49) simplify to

$$I_1 \dot{\omega}_1 = \omega_2 \omega_3 (I_1 - I_3) \tag{7.59}$$

$$I_1 \dot{\omega}_2 = \omega_1 \omega_3 (I_3 - I_1) \tag{7.60}$$

$$I_3 \dot{\omega}_3 = 0 \tag{7.61}$$

Equation (7.61) shows that ω_3 is a constant for this motion. The first two equations lead to

$$\ddot{\omega}_2 = -\omega_2 \omega_3^2 \left(\frac{I_3 - I_1}{I_1}\right)^2 \tag{7.62}$$

which has the solution

$$\omega_2 = A \sin(\Omega t + \alpha) \tag{7.63}$$

where

$$\Omega = \frac{I_3 - I_1}{I_1} \omega_3 = \text{constant} \tag{7.64}$$

From equations (7.60) and (7.63) a solution for ω_1 can be developed:

$$\omega_1 = A \cos(\Omega t + \alpha) \tag{7.65}$$

Defining $\boldsymbol{\omega}_T$ to be

$$\boldsymbol{\omega}_T = \omega_1 \mathbf{i}_1 + \omega_2 \mathbf{i}_2 \tag{7.66}$$

then
$$\omega_T = (\omega^2 - \omega_3^2)^{1/2} \tag{7.67}$$
and is constant in magnitude. Further,
$$\omega_2 = \omega_T \sin(\Omega t + \alpha) \tag{7.68}$$
$$\omega_1 = \omega_T \cos(\Omega t + \alpha) \tag{7.69}$$

The last two equations indicate that ω rotates relative to the body frame at a rate Ω. The total angular velocity then describes a right circular cone relative to the body. The angular velocity is said to precess about the axis of symmetry at a rate Ω. This cone is the *body cone* described in the previous section. Now refer to Figure 7.7. In this and in subsequent figures, the crosshatching on axis Z indicates that it is fixed in direction (only). It is apparent from this figure that

$$\tan \gamma = \frac{\omega_T}{\omega_3} = \text{constant} \tag{7.70}$$

$$\tan \theta = \frac{I_1 \omega_T}{I_3 \omega_3} = \text{constant} \tag{7.71}$$

$$\tan \theta = \frac{I_1}{I_3} \tan \gamma \tag{7.72}$$

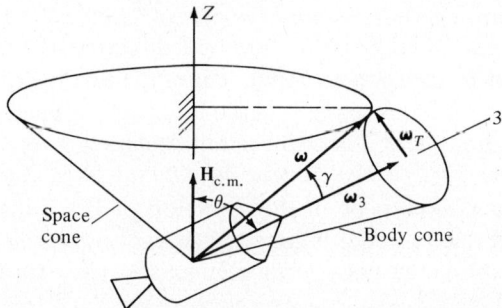

Figure 7.7

and that $\mathbf{H}_{c.m.}$, ω, and axis 3 must be in a plane. Since ω is constant in magnitude, and recalling the discussion of the invariable plane and the herpolhode curve, it is seen that the fixed space cone is also a right circular cone and that the circular herpolhode curve must have the equation

$$\omega_X^2 + \omega_Y^2 = \omega^2 - \omega_Z^2 = \text{constant radius} \tag{7.73}$$

The polhode curve is circular, being the curve of intersection between the body cone and the Poinsot ellipsoid. Since the plane containing ω, ω_T, and axis 3 rotates about the fixed direction defined by the *constant angular momentum*, the body cone is said to roll on the space cone. The relative turning of this plane as seen from the rotating body is Ω. The turning rate of this plane as seen by a *nonrotating* observer will be discussed shortly, and since the symmetry axis rotates with this plane, it, too, describes a cone in its absolute motion about the fixed direction. Before proceeding, one final comment should be made. Note that in equation (7.72) if $I_1 > I_3$, θ is greater than γ. If $I_3 > I_1$, the body cone has the space cone rolling inside its shape. The first case is called *direct precession*; the second case is defined as *retrograde precession*. See Figure 7.8.

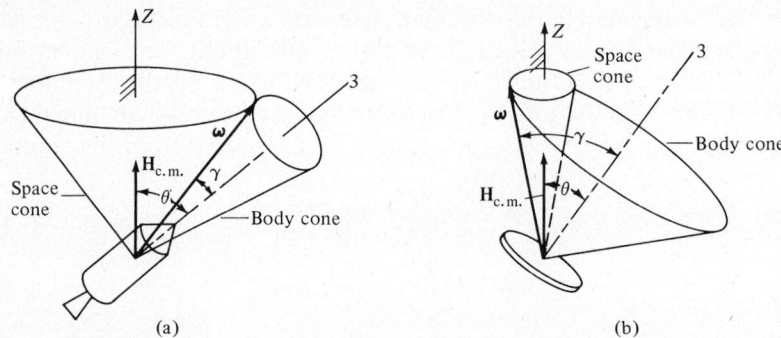

Figure 7.8

The discussion so far has not included a description of the *absolute motion* of the body. To accomplish that, the Euler angles of Section 7.4 will be introduced. With axis Z arbitrarily selected as the fixed direction defined by the vector $\mathbf{H}_{c.m.}$, it becomes the ϕ rotation axis that defines the turning of the line of nodes. (Recall that the line of nodes represents the intersection line between the transverse plane of the body and a fixed horizontal plane.) *ϕ represents the rate of turning of the symmetry axis as it "cones" around the fixed Z direction.*

By taking the Z axis to be in the direction of $\mathbf{H}_{c.m.}$, the second angle of turning θ about the line of nodes is constant (*steady precession*). The third angle of turning ψ acts about the body symmetry axis 3. Then from equation (7.44),

$$\omega_1 = \dot{\phi} \sin \theta \sin \psi \tag{7.74}$$

$$\omega_2 = \dot{\phi} \sin \theta \cos \psi \tag{7.75}$$

$$\omega_3 = \dot{\phi} \cos \theta + \dot{\psi} \tag{7.76}$$

and

$$\boldsymbol{\omega} = \dot{\boldsymbol{\phi}} + \dot{\boldsymbol{\psi}} \tag{7.77}$$

$$\dot{\boldsymbol{\omega}} = \dot{\boldsymbol{\phi}} \times \dot{\boldsymbol{\psi}} \tag{7.78}$$

From the previous results one can write

$$\omega_1^2 + \omega_2^2 = \dot{\phi}^2 \sin^2 \theta = \text{constant} \tag{7.79}$$

$$\omega_T = \dot{\phi} \sin \theta \tag{7.80}$$

$$\omega_3 = \dot{\phi} \cos \theta + \dot{\psi} = \text{constant} \tag{7.81}$$

$$H_3 = H \cos \theta = I_3 \omega_3 \tag{7.82}$$

$$H_T = H \sin \theta = I_1 \omega_T \tag{7.83}$$

or
$$H = \frac{I_3(\dot\phi \cos \theta + \dot\psi)}{\cos \theta} = \frac{I_1 \dot\phi \sin \theta}{\sin \theta} \tag{7.84}$$

Then
$$\dot\phi = \frac{I_3 \dot\psi}{(I_1 - I_3) \cos \theta} \tag{7.85}$$

is the turning rate of the symmetry axis as it cones about the Z direction. For $I_1 > I_3$, *direct precession*, the spin, and the precession rate $\dot\phi$ have the same sense. For $I_3 > I_1$, *retrograde precession*, the spin, and the precession rates have opposite senses. (See Figures 7.9 and 7.10.) Further, since $[\dot\omega]_{\rm rot} = \dot\omega$ for body-fixed axes,

$$[\dot\omega]_{\rm rot} = \dot\phi \times \dot\psi = \dot\omega \tag{7.86}$$

The observer in the rotating frame of the body sees the $\dot\phi$ vector turn at a rate

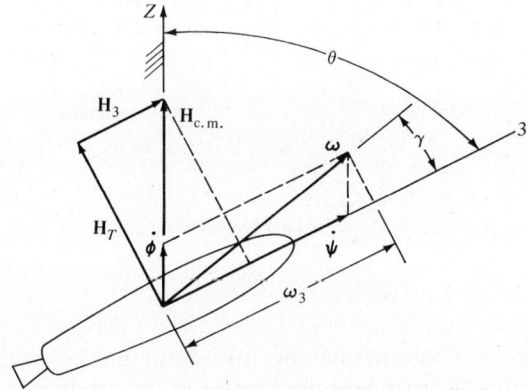

Figure 7.9

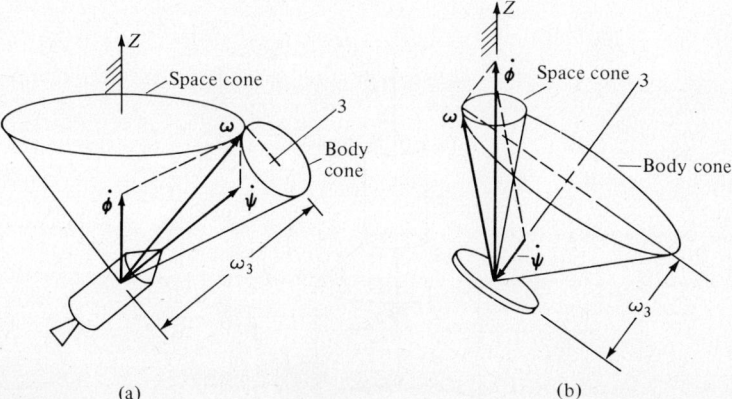

(a)   (b)

Figure 7.10

228 ANALYTICAL MECHANICS

$-\dot{\psi}$. Therefore, equation (7.86) would be more descriptive if rewritten as

$$[\dot{\omega}]_{\text{rot}} = -\dot{\psi} \times \dot{\phi} \tag{7.87}$$

It is apparent then that

$$-\dot{\psi} = \Omega = \frac{I_3 - I_1}{I_1}\omega_3 \tag{7.88}$$

It is also seen now that the plane containing $\mathbf{H}_{\text{c.m.}}$, $\boldsymbol{\omega}$, and axis 3 rotates relative to the body axes at a relative angular speed Ω that is equal in magnitude to the spin rate $\dot{\psi}$ of the body but opposite in sign. This can also be proved another way by taking the time derivative of equations (7.74) and (7.75), namely,

$$\dot{\omega}_1 = \dot{\psi}\dot{\phi}\sin\theta\cos\psi \tag{7.89}$$

$$\dot{\omega}_2 = -\dot{\psi}\dot{\phi}\sin\theta\sin\psi \tag{7.90}$$

The results correspond to a negative rotation of ω_T relative to the coordinate system (opposite to the actual body rotation) (see Figure 7.11). This by definition must be Ω, the turning rate of the ω vector relative to the body axes.

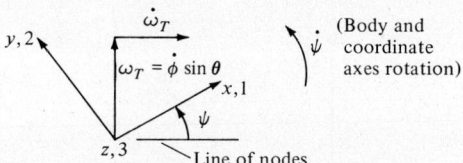

Figure 7.11

Sample Problem 7.8

The satellite of mass M shown has no initial angular velocity until a vernier steering rocket with a thrust F is ignited for 1 s. If the satellite has mass moments of inertia $I_1 = I_2 = M$ and $I_3 = \frac{25}{16}M$, find the spin rate $\dot{\psi}$, the subsequent precession rate $\dot{\phi}$, and the direction of the precession axis after the rocket is turned off.

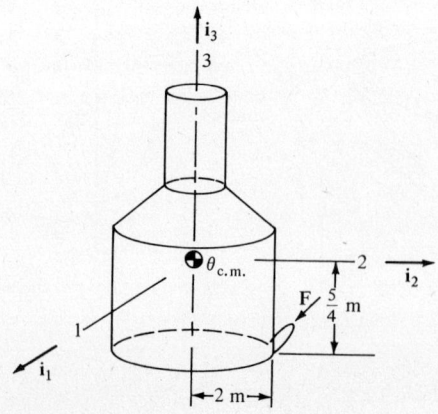

Analysis: The principle of angular impulse and momentum is applicable here.

$$\int_1^2 \mathbf{M}_{c.m.}\,dt = (\mathbf{H}_{c.m.})_2 - (\mathbf{H}_{c.m.})_1$$

$$\int_0^1 F(-\tfrac{5}{4}\mathbf{i}_2 - 2\mathbf{i}_3)\,dt = (\mathbf{H}_{c.m.})_2 - \mathbf{0}$$

$$(\mathbf{H}_{c.m.})_2 = -\tfrac{5}{4}F\mathbf{i}_2 - 2F\mathbf{i}_3$$

but
$$\mathbf{H}_{c.m.} = I_1\omega_1\mathbf{i}_1 + I_2\omega_2\mathbf{i}_2 + I_3\omega_3\mathbf{i}_3$$

where $I_1 = I_2 = M$ and $I_3 = \tfrac{25}{16}M$. Then

$$\omega_1 = 0 \qquad \omega_2 = \omega_T = -\frac{5}{4}\frac{F}{M} \qquad \omega_3 = -\frac{32}{25}\frac{F}{M}$$

After the vernier rocket is turned off, $\mathbf{H}_{c.m.}$ remains constant since the satellite is in the moment-free state. With $\boldsymbol{\omega}$ not acting along a principal axis, the satellite precesses and spins. Defining $\mathbf{H}_{c.m.}$ to be the fixed Z direction and, therefore, the precession axis, the angle θ can be found from equation (7.71).

$$\theta = \tan^{-1}\frac{I_1\omega_T}{I_3\omega_3} = \tan^{-1}\frac{-5F/4}{-2F} = \tan^{-1}\frac{5}{8}$$

$$\theta = 32° \qquad \text{(a constant)}$$

The angle γ is given by

$$\gamma = \tan^{-1}\frac{\omega_T}{\omega_3} = 44.3° \qquad \text{(constant)}$$

and from Figure (b), or from equation (7.80), it is seen that

$$\omega_T = \dot{\phi}\sin\theta = -\frac{5}{4}\frac{F}{M}$$

$$\dot{\phi} = -\frac{\tfrac{5}{4}F/M}{\sin 32°} = -2.36\frac{F}{M} \qquad \text{(a constant)}$$

One can solve for $\dot{\psi}$ from equation (7.76):

$$\omega_3 = \dot{\phi}\cos\theta + \dot{\psi} = -\frac{32}{25}\frac{F}{M}$$

$$\dot{\psi} = -\frac{32}{25}\frac{F}{M} + 2.36\frac{F}{M}\cos 32° = 0.72\frac{F}{M} \quad \text{(a constant)}$$

This is retrograde precession with $I_3 > I_1$ and $\dot{\psi}$ having opposite sense to $\dot{\phi}$. As a check, equation (7.85) states that

$$\dot{\phi} = \frac{I_3\dot{\psi}}{(I_1 - I_3)\cos\theta}$$

$$= \frac{\frac{25}{16}}{1 - \frac{25}{16}\cos 32°}\, 0.72 \left(\frac{F}{M}\right) = -2.36\frac{F}{M}$$

To summarize, after the vernier rocket is turned off, the satellite spins about its axis of symmetry in a positive sense at a constant rate $\dot{\psi} = 0.72 F/M$ while precessing at a faster rate, $-2.36 F/M$, in the opposite direction. The precession rate $\dot{\phi}$, the angle θ that the axis of symmetry forms with the fixed Z direction, and the angle γ that ω makes with the symmetry axis are all constant. The satellite's axis of symmetry describes a cone in its motion in space. ∎

7.7 THE MOTION OF A SYMMETRICAL TOP

This section investigates another classic problem for which a solution has been obtained: the motion of an axisymmetrical top moving about a fixed point. A symmetrical top with one point O on the symmetry axis fixed in space is shown in Figure 7.12. The only external moment acting about point O is caused by the gravity force. Since $\mathbf{H}_{\text{c.m.}}$ is not constant here, the fixed Z direction will now be conveniently selected as vertical.

The use of the Euler equations for body-fixed principal axes is in this case not the simplest way to approach this problem since ultimately Euler angles

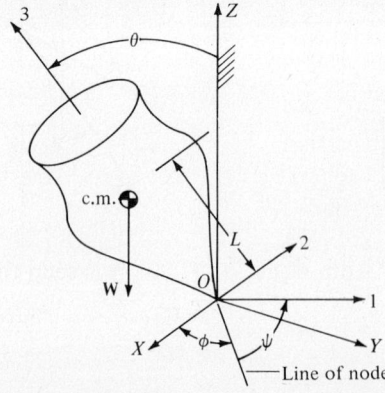

Figure 7.12

must be introduced. If the modified Euler equations are employed instead, having one coordinate axis the symmetry axis, the second axis, the line of nodes, and the third axis orthogonal to the other two, two of the three coordinate axes fall in the direction about which Euler angles are measured. The integrals to the moment equations and angular acceleration about these axes directly yield a description of the body's absolute motion. (In the next chapter, Lagrange's equations will be introduced. They provide a much better means to derive the results that follow.) Referring to Sections 7.3 and 7.4, one can write

$$H'_x = I_1 \omega'_x = I_1(\dot{\theta}) \tag{7.91}$$

$$H'_y = I_1 \omega'_y = I_1(\dot{\phi} \sin \theta) \tag{7.92}$$

$$H'_z = I_3 \omega'_z = I_3(\dot{\psi} + \dot{\phi} \cos \theta) \tag{7.93}$$

$$\Omega = \dot{\theta} \mathbf{i}' + \dot{\phi} \sin \theta \mathbf{j}' + \dot{\phi} \cos \theta \mathbf{k}' \tag{7.94}$$

$$\omega = \dot{\theta} \mathbf{i}' + \dot{\phi} \sin \theta \mathbf{j}' + (\dot{\phi} \cos \theta + \dot{\psi}) \mathbf{k}' \tag{7.95}$$

$$[\dot{\omega}]_{\text{rot}} = \dot{\omega} - \Omega \times \omega = \dot{\omega} - (\omega - \dot{\psi}\mathbf{k}) \times \omega$$
$$= \dot{\omega} + \dot{\psi}\mathbf{k}' \times \omega = \dot{\omega} + \dot{\psi} \times (\dot{\phi} + \dot{\theta} + \dot{\psi})$$
$$= [\ddot{\theta}\mathbf{i}' + \ddot{\phi} \sin \theta \mathbf{j}' + (\ddot{\phi} \cos \theta + \ddot{\psi})\mathbf{k}'$$
$$+ \dot{\phi} \times (\dot{\theta}) + (\dot{\phi} + \dot{\theta}) \times \dot{\psi}] + \dot{\psi} \times (\dot{\phi} + \dot{\theta} + \dot{\psi}) \tag{7.96}$$

$$[\dot{\omega}]_{\text{rot}} = \ddot{\theta}\mathbf{i}' + \ddot{\phi} \sin \theta \mathbf{j}' + (\ddot{\phi} \cos \theta + \ddot{\psi})\mathbf{k}'$$
$$+ \dot{\theta}\dot{\phi} \cos \theta \mathbf{j}' - \dot{\theta}\dot{\phi} \sin \theta \mathbf{k}' \tag{7.97}$$

This last equation could have been more simply derived by taking the time derivative of ω from equation (7.95) holding all the unit vectors *constant*. (This is the mathematical definition of the operator $[d/dt]_{\text{rot.}}$) Now

$$\mathbf{M} = [\dot{\mathbf{H}}]_{\text{rot}} + \Omega \times \mathbf{H}$$

$$M_{x'} = I_1 \ddot{\theta} + (I_3 - I_1)\dot{\phi}^2 \sin \theta \cos \theta + I_3 \dot{\psi}\dot{\phi} \sin \theta \tag{7.98}$$

$$M_{y'} = I_1(\ddot{\phi} \sin \theta + \dot{\theta}\dot{\phi} \cos \theta) + (I_1 - I_3)\dot{\theta}\dot{\phi} \cos \theta - I_3 \dot{\psi}\dot{\theta} \tag{7.99}$$

$$M_{z'} = I_3(\ddot{\phi} \cos \theta - \dot{\phi}\dot{\theta} \sin \theta + \ddot{\psi}) \tag{7.100}$$

Since $M_{z'} = 0$,

$$I_3(\ddot{\phi} \cos \theta - \dot{\phi}\dot{\theta} \sin \theta + \ddot{\psi}) = 0$$

$$\frac{d}{dt} I_3(\dot{\phi} \cos \theta + \dot{\psi}) = H_{z'} = \text{constant} \tag{7.101}$$

$$\dot{\phi} \cos \theta + \dot{\psi} = \omega'_z = \text{constant} \tag{7.102}$$

The body's total turning rate ω'_z about its axis of symmetry is seen to be constant. Equations (7.98) and (7.99) can be rewritten as

$$mgl \sin \theta = I_1(\ddot{\theta} - \dot{\phi}^2 \sin \theta \cos \theta) + I_3 \dot{\phi}\omega'_z \sin \theta$$
$$0 = I_1(\ddot{\phi} \sin \theta + 2\dot{\phi}\dot{\theta} \cos \theta) - I_3 \dot{\theta}\omega'_z \tag{7.103}$$

232 ANALYTICAL MECHANICS

To assist in the integration another first integral is needed to reduce the number of variables in the equations. Since no moments act about the Z axis, H_z is equal to a constant. Written in terms of ϕ this becomes

$$H_Z = \mathbf{H} \cdot \mathbf{K} = \text{constant}$$
$$= I_3(\dot{\psi} + \dot{\phi} \cos\theta)\cos\theta + I_1(\dot{\phi}\sin\theta)\sin\theta \tag{7.104}$$
$$H_Z = H_{z'}\cos\theta + I_1 \dot{\phi} \sin^2\theta = \text{constant} \tag{7.105}$$

or
$$\dot{\phi} = \frac{H_Z - H_{z'}\cos\theta}{I_1 \sin^2\theta} \tag{7.106}$$

The top's precession is, therefore, a function of the nutation angle θ. Now if equation (7.103) is integrated, a first integral for the Euler angle θ is obtained.

$$mgl \sin\theta = I_1 \ddot{\theta} - I_1 \frac{(H_Z - H_{z'}\cos\theta)^2}{I_1^2 \sin^4\theta} \sin\theta \cos\theta + I_3 \omega'_z \frac{H_Z - H_{z'}\cos\theta}{I_1 \sin^2\theta}\sin\theta$$

Multiplying by $\dot{\theta}$ yields

$$mgl \sin\theta\, \dot{\theta} = I_1 \ddot{\theta}\dot{\theta} - \frac{(H_Z - H_{z'}\cos\theta)^2}{I_1 \sin^3\theta}\cos\theta\,\dot{\theta} + \frac{H_Z(H_Z - H_{z'}\cos\theta)}{I_1 \sin^2\theta}\sin\theta\,\dot{\theta}$$

Integrating the above equation gives

$$-mgl \cos\theta = \frac{I_1 \dot{\theta}^2}{2} + \frac{1}{2}\frac{(H_Z - H_{z'}\cos\theta)^2}{I_1 \sin^2\theta} - E'$$

The constant of integration E' then is

$$E' = \frac{I_1 \dot{\theta}^2}{2} + \frac{1}{2}\frac{(H_Z - H_{z'}\cos\theta)^2}{I_1 \sin^2\theta} + mgl \cos\theta$$
$$= \frac{I_1 \dot{\theta}^2}{2} + \frac{I_1(\dot{\phi}^2 \sin^2\theta)}{2} + mgl \cos\theta \tag{7.107}$$

It is seen that the constant E' is related to the total energy by

$$E' = E - \frac{I_3 \omega_z^2}{2} \tag{7.108}$$

The three first integrals of the motion about the θ, ϕ, and ψ directions have been obtained. If θ could be found as a function of time, ϕ and ψ could also be determined through equations (7.102) and (7.106).
From equation (7.107),

$$\dot{\theta}^2 = \frac{2E'}{I_1} - \frac{2mgl \cos\theta}{I_1} - \frac{(H_Z - H_{z'}\cos\theta)^2}{I_1^2 \sin^2\theta} \tag{7.109}$$

or

$$t = \int_{\theta_0}^{\theta(t)} \frac{d\theta}{[2E'/I_1 - (2mgl \cos\theta)/I_1 - (H_Z - H_{z'}\cos\theta)^2/I_1^2 \sin^2\theta]^{1/2}} \tag{7.110}$$

Defining the following constants

$$\alpha = \frac{2E'}{I_1} \qquad \beta = \frac{2mgl}{I_1} \qquad a = \frac{H_{z'}}{I_1} \qquad b = \frac{H_z}{I_1} \qquad (7.111)$$

and letting $u = \cos\theta$, then

$$t = \int_{u_0}^{u(t)} \frac{du}{[(1-u^2)(\alpha - \beta u) - (b - au)^2]^{1/2}} \qquad (7.112)$$

Though solutions are readily available in the literature for this equation, they are lengthy and in terms of elliptic integrals. A description of the motion, however, can be obtained without resorting to closed-form analytical solutions. Rewriting equation (7.109) in terms of u leads to

$$\dot{u}^2 = (1 - u^2)(\alpha - \beta u) - (b - au)^2 \qquad (7.113)$$

or with $u = \cos\theta$,

$$\dot{u}^2 = f(u) = (-\dot\theta \sin\theta)^2 \qquad (7.114)$$

The function of u represented by equation (7.114) is a cubic polynomial. Though u is physically limited to values of ± 1, the function itself, equation (7.113), becomes positive for large values of positive u and negative for large negative u because of the term βu^3 that dominates. (β is always positive.) For $u = \pm 1$ ($\theta = 0°, 180°$), $\dot{u}^2$ is always negative or zero because of the term $-(b-au)^2$. Since the cubic has three roots, one root must lie in the region $u \geqslant 1$ (see Figure 7.13). The motion is limited to real θ; hence the only solutions attainable are for positive $\dot{u}^2$. This is the shaded region in the figure. One can see that there are two roots, u_1 and u_2, that limit the physical motion.

The limiting values are $\cos\theta_1$ and $\cos\theta_2$, which the symmetry axis of the top cannot exceed in its motion. The angle θ is the nutation angle. The symmetry axis "nods," or nutates, as it spins and precesses around the vertical within the limiting values θ_1 and θ_2, the turning points, respectively. The location of the roots and the possibility of a multiplicity provide a valuable description of the

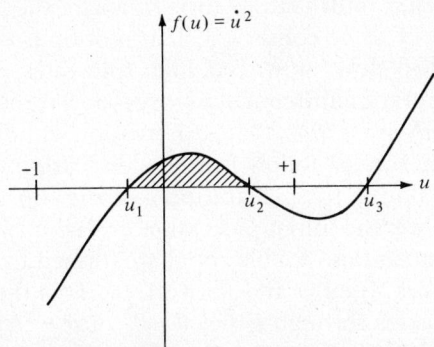

Figure 7.13

motion. One can imagine the top to be centered and enclosed in a transparent sphere. The symmetry axis' motion could be traced on the sphere's surface. It turns out that the possible path of the figure axis on the sphere falls into three categories depending upon when, and if, the precession rate becomes zero. Referring to equation (7.106), one can write this as

$$\dot{\phi} = \frac{b - au}{1 - u^2} \qquad (7.115)$$

When $\dot{\phi} = 0$, the value of $u = u_s = b/a$. If $u_s > u_2 > u_1$, the precession never stops, or $\dot{\phi}$ is always greater than zero between the limiting angles of θ_1 and θ_2. This is shown in Figure 7.14a. If $u_2 > u_s > u_1$, the precession rate must reverse itself between the limiting values of θ_1 and θ_2 (see Figure 7.14b). The figure axis forms loops as it precesses and nutates. Finally, if $u_s = u_2 > u_1$ or $u_s = u_1$, the top ceases to precess as it touches the limiting boundary circle. The figure axis locus exhibits cusps (see Figure 7.14c).

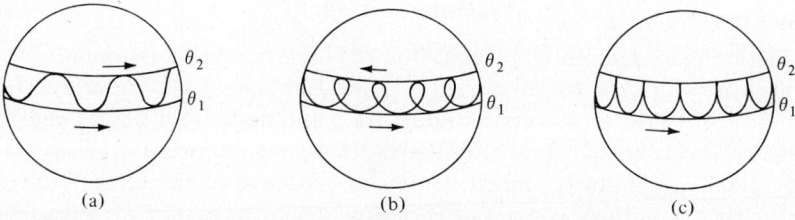

Figure 7.14

One should not lose sight of the fact that energy is conserved during the motion. This means that as $\dot{\theta}$ varies between the limits at θ_1 and θ_2, the precession rate and the potential energy change at the same time. The precession rate is, therefore, not constant during the motion. The case of *cuspidal motion*, where $\theta_s = \theta_2 < \theta_1$, is perhaps the most common to everyone's experience. When a top is spun and then released, it falls vertically first; it has *no initial precession* rate. As it falls, it gains kinetic energy in the form of nutational and precessional angular velocity. When it returns to its former height, the precession rate must again be zero for energy to be conserved. This motion is only observed at low spin rates when the top slows down. At high spin rates, θ_1 and θ_2 approach each other, and the nutational frequency increases. To the naked eye, the top seems to describe *steady precession*, that is, precession without nutation. This is a misperception though. It must follow from the discussion that the only way to initiate steady precession is to spin and precess the top simultaneously as it is released. What's more, the initial precession and spin rate and the potential energy must be chosen so that double roots are formed from equation (7.113). Cuspidal motion occurs when $\dot{\phi}$ and $\dot{u} = f(u) = 0$. This then also includes the case where the top passes through either $\theta = 0°$ or $\theta = 180°$ with a nutational velocity. Recall that $\dot{u}^2 = f(u)$ and $\dot{u} = \dot{\theta} \sin \theta$. Then $\theta = 0°$ or $180°$ can be a root of the equation $f(u) = 0$ even when $\dot{\theta}$ does not equal to zero.

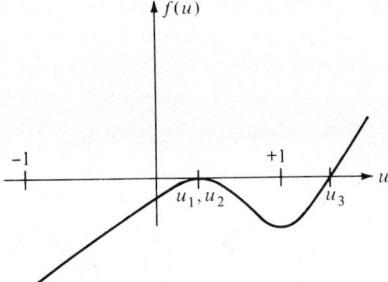

Figure 7.15

A more detailed look at steady precession will prove rewarding. In order for the function to have double roots, the curve of $f(u)$ must appear as shown in Figure 7.15.

Note that in this figure there is no possible nutational motion, $f(u)$ never becomes positive between the limits of ± 1. This means that both $f(u)$ and $f'(u)$ must equal zero at the double root, or

$$f(u) = (1-u^2)(\alpha - \beta u) - (b-au)^2 = 0$$
$$f'(u) = -2u(\alpha - \beta u) - \beta(1-u^2) + 2a(b-au) = 0$$

These two equations yield

$$\frac{(b-au)^2}{1-u^2} = -\frac{\beta(1-u^2) + 2a(b-au)}{2u} \tag{7.116}$$

Recalling that

$$b - au = \dot{\phi}(1-u^2) = \dot{\phi}\sin^2\theta$$

equation (7.116) becomes

$$\dot{\phi}^2 \sin^2\theta = \frac{-(2mgl/I_1)\sin^2\theta + (2H_z/I_1)\dot{\phi}\sin^2\theta}{2\cos\theta} \tag{7.117}$$

or

$$mgl = -\dot{\phi}^2 I_1 \cos\theta + \dot{\phi} I_3(\dot{\psi} + \dot{\phi}\cos\theta) \tag{7.118}$$

Then

$$\dot{\phi} = \frac{I_3 \dot{\psi}}{2(I_1 - I_3)\cos\theta}\left(1 \pm \sqrt{1 - \frac{4mgl(I_1 - I_3)\cos\theta}{I_3^2 \dot{\psi}^2}}\right) \tag{7.119}$$

Physically possible solutions are only allowable for

$$\dot{\psi}^2 \geqslant \frac{4mgl(I_1 - I_3)\cos\theta}{I_3^2} \tag{7.120}$$

Two roots are possible from equation (7.119) and are called the *fast* and *slow* precession rates. The slow precession rate usually prevails. For a top supported at its center the potential term $mgl\cos\theta$ is zero, and the top is moment-free. The steady precession rate (θ = constant) for the *moment-free top* of the last section is then

$$\dot{\phi} = \frac{I_3 \dot{\psi}}{(I_1 - I_3)\cos\theta} \qquad (7.121)$$

This is exactly equation (7.85). Similarly, equation (7.103) with $\ddot{\theta}$ set to zero for steady precession yields the same equation as equation (7.118). If $\theta = 90°$, then the result from equation (7.119) is indeterminate. One can resort instead to equation (7.103) or (7.118) to resolve the impasse. Namely, one obtains

$$mgl = I_3 \dot{\phi}\dot{\psi} \qquad (7.122)$$

or in its most common form familiar to all elementary physics students,

$$\dot{\phi} = \frac{mgl}{I_3 \dot{\psi}} \qquad (7.123)$$

For $\theta = 0°$, another interesting case develops. The top spins in a vertical position and is called a *sleeping top*. Figure 7.16 shows three possible curves with roots at $+1$ that can describe this motion. Two of the curves, (1) and (2), represent stable motion since nutation cannot build up between the double (or triple) roots at $+1$ and a third root. The third curve (3) represents unstable motion. Returning to equation (7.105) with $\theta = 0°$,

$$H_z = H'_z \qquad \text{or} \qquad a = b$$

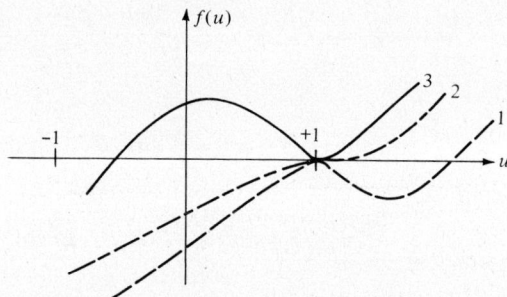

Figure 7.16

Similarly, from equation (7.107),

$$E' = mgl \qquad \text{or} \qquad \alpha = \beta$$

Then equation (7.113) becomes

$$\dot{u}^2 = (1 - u^2)\alpha(1 - u) - a^2(1 - u)^2$$

or
$$\dot{u}^2 = (1 - u)^2[\alpha(1 + u) - a^2] = 0 \qquad (7.124)$$

The three roots of this equation are $u = 1, 1, a^2/\alpha - 1$. Stable motion is possible only when $a^2/\alpha \geq 2$. This corresponds to a requirement set on the angular velocity ω'_z of

$$\omega'^2_z \geq \frac{4mglI_1}{I_3^2} \tag{7.125}$$

One can start a top spinning rapidly in the vertical position, and it will "sleep" until the friction at the support point slows it down. Below a certain spin rate the top begins to nutate, or becomes unstable.

Sample Problem 7.9

The right-circular cone shown precesses about the vertical at the same time that it nutates between the angles $\theta_2 = 90°$ and $\theta_1 = 120°$. If it is known that the precession rate at θ_2 is $\dot{\phi}_0$, find the precession and spin rates at $\theta_1 = 120°$. Find the spin rate at $\theta_2 = 90°$. The principal moment of inertia of the cone of mass M about the spin axis is given as C; the transverse moment of inertia about point O is given as A. The height of the cone is h.

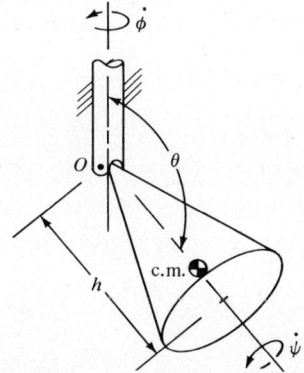

Analysis: For the top's motion, it is known that the energy and angular momentum terms about the vertical and body's symmetry axis are constant. Also, at the limiting angles θ_2 and θ_1, $\dot{\theta} = 0$. Then at $\theta_2 = 90°$,

$$H'_z = C(\dot{\phi}_0 \cos 90° + \dot{\psi}_0) = C\dot{\psi}_0$$

$$H_z = A\dot{\phi}_0$$

$$E' = \frac{A\dot{\phi}_0^2}{2}$$

At $\theta_1 = 120°$,

$$H'_z = C\left(-\frac{\dot{\phi}}{2} + \dot{\psi}\right)$$

$$H_z = -\frac{C}{2}\left(-\frac{\dot{\phi}}{2} + \dot{\psi}\right) + \frac{3}{4} A\dot{\phi}$$

$$E' = \frac{A\dot{\phi}^2}{2}\left(\frac{3}{4}\right) - Mg\left(\frac{3h}{4}\right)\left(\frac{1}{2}\right)$$

with $l = \tfrac{3}{4}h$. At 120°, then

$$\dot{\phi} = \left(\frac{4\dot{\phi}_0^2}{3} + \frac{Mgh}{A}\right)^{1/2}$$

$$\dot{\psi} = \left[\left(\frac{4\dot{\phi}_0^2}{3} + \frac{Mgh}{A}\right)^{1/2} \frac{C+3A}{4} - A\dot{\phi}_0\right]\frac{2}{C}$$

At 90°,

$$\dot{\psi}_0 = -\frac{1}{2}\left(\frac{4\dot{\phi}_0^2}{3} + \frac{Mgh}{A}\right)^{1/2} + \left[\left(\frac{4\dot{\phi}_0^2}{3} + \frac{Mgh}{A}\right)^{1/2}\frac{C+3A}{4} - A\dot{\phi}\right]\frac{2}{C}$$

With these quantities known, $\dot{\theta}$, $\dot{\phi}$, and $\dot{\psi}$ can be found for other θ values. ∎

Sample Problem 7.10
If the top of the last problem was known to steadily precess instead at an angle of $\theta = 60°$, find the top's slow precession rate. The spin rate is assumed to be $\dot{\psi}_0$. What would the value be at $\theta = 90°$?

Analysis: For steady precession one can use equation (7.119):

$$\dot{\phi} = \frac{C\dot{\psi}_0}{A - C}\left(1 - \sqrt{1 - \frac{3Mgh(A-C)}{2C^2\dot{\psi}_0^2}}\right)$$

For the special case of $\theta = 90°$, equation (7.123) can be used. It follows then that

$$\dot{\phi} = \frac{3Mgh}{4C\dot{\psi}_0}$$ ∎

7.8 GYROSCOPIC INSTRUMENTS

Because the motion of a precessing spinning wheel is so unique, it has inspired several generations of engineers to put gyroscopic effects to practical use. Perhaps its most important property is the spinning wheel's tendency to maintain a constant direction in space. When a moment is applied, it does not tip; instead it precesses. Engineers have tried to "tame" the spinning wheel to have it level passenger cabins on ocean-going ships. They have used this property to stabilize artillery shells and rocket payloads during flight. If the payloads are spun, wind disturbances will cause precession but not toppling. (This physical effect helps to explain why footballs remain stable when tossed and why bicycles are held upright and how they are turned.) Foucault (1819–1868) in France used a moment-free top, therefore having a fixed direction of angular momentum coinciding with its spin axis, to demonstrate the earth's rotation. The top *appears* to rotate relative to the earth at a rate of 15°/h (360°/24 h) when in actuality it is the earth's motion relative to the top that is viewed. It was he who joined the two greek words *gyros*, meaning turning, and *skopein*, meaning to view, to form the new word *gyroscope*.

At the turn of this century, with the simultaneous development of the airplane, a new application for the gyroscope developed: navigation. Its value in navigation as an instrument used to detect rotation has perhaps subsequently overshadowed all other areas of its technical importance. It is the intent of this section to qualitatively describe some of the most common gyroscopic instruments that are in commercial service today.

Gyroscopic navigational instruments can generally be categorized by the number of degrees of rotational freedom they have, excluding the spinning motion.

A single degree of freedom gyro, shown in Figure 7.17, has for all practical purposes its angular momentum pointed along its spin axis. This is achieved by spinning the rotor at very high speeds in "frictionless" bearings. The most commonly used rotor device for a gyro is a small electric motor having the rotor placed on the outside, around the stator, to provide a greater moment of inertia. It can turn at speeds of up to 24 000 rpm. It is supported in a frame called a *gimbal*. The gimbal is free to turn about the *gimbal axis*, also in frictionless bearings. In some applications where airjets are used to impinge on a rotor that has been designed in principle as a small gas turbine, speeds of 100 000 rpm are attained. In Figure 7.17, the single degree of freedom gyro is employed as a *rate gyro*. If, for example, the base surface is fixed to an aircraft turning about the z direction (yaw) at a rate $\dot{\phi}$, the moment on the gimbal will cause the angular momentum vector to rotate about the gimbal axis, the single degree of freedom. In doing so it will compress the provided torsional spring an amount θ. The moment from the spring will, in turn, cause the gyro to precess at the aircraft's yaw rate $\dot{\phi}$, or

$$\mathbf{M} = \dot{\mathbf{H}} = \dot{\boldsymbol{\phi}} \times \mathbf{H} \qquad M_x = k\theta = \dot{\phi} I_3 \dot{\psi}$$

$$\theta = \frac{I_3 \dot{\phi} \dot{\psi}}{k} \tag{7.126}$$

By installing a rate gyro in the plane's instrument panel, the pilot can directly see the y-axis motion; the rate gyro acts as an instrument needle, turning through the angle θ. By suitable calibration, the instrument dial can measure the yaw

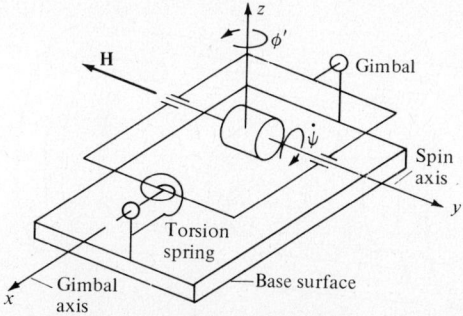

Figure 7.17

rate $\dot{\phi}$. However, modern applications use an E pickoff and a voltage reading device instead. An E pickoff is a transformer built in the shape of an E with excitation windings on the center arm. When the gimbal is at rest, it lies at the center leg of the pickoff. The voltages induced in the windings of the outside legs by the excitation windings of the center arm are equal, and no output signal results. When the gimbal turns, a flux unbalance results leading to a voltage signal output. The signal output is proportional to θ and can be read on a meter on the cockpit's instrument panel.

If the airplane *pitches* about the y axis, no moment is transferred to the gyro. The base surface just turns with the aircraft. If the plane *rolls*, or *banks* about the x axis, which is needed when turning at a given yaw rate, the base surface's motion will stretch the spring, which in turn will tend to precess the rotor about the vertical. The gimbal's support resists this motion with a torque that causes the rotor to unwind the spring. The net effect on the instrument of a roll maneuver is zero. The gimbal frame follows the base surface and remains parallel to it.

If the spinning rotor and gimbal were now enclosed by a case that was floated inside another case (see Figure 7.18), the gimbal axis and bearings could be eliminated and frictional effects subsequently reduced. Now, instead of a restoring spring force, there is a viscous damping force acting. Equation (7.126) becomes

$$M_x = D\dot{\theta} = \dot{\phi} I_3 \dot{\psi}$$

$$\dot{\theta} = \frac{\dot{\phi} I_3 \dot{\psi}}{D} \qquad (7.127)$$

or

$$\theta = \int \frac{\dot{\phi} I_3 \dot{\psi}\, dt}{D} + C \qquad (7.128)$$

The θ motion now measured is the integral of the precession rate. The gyro is called a *floated integrating gyro* and has many applications.

A 2 degrees of freedom gyro is shown in Figure 7.19. The spinning rotor is free to rotate in the inner gimbal. The inner gimbal axis can rotate relative to the outer gimbal, which in turn is free to rotate about the outer gimbal axis relative

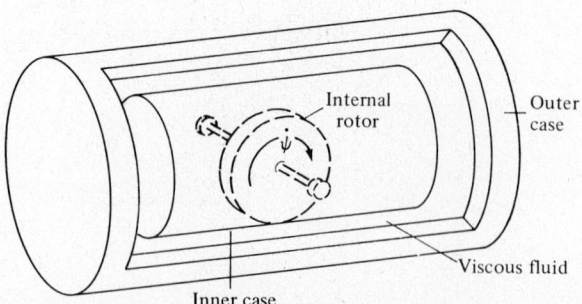

Figure 7.18

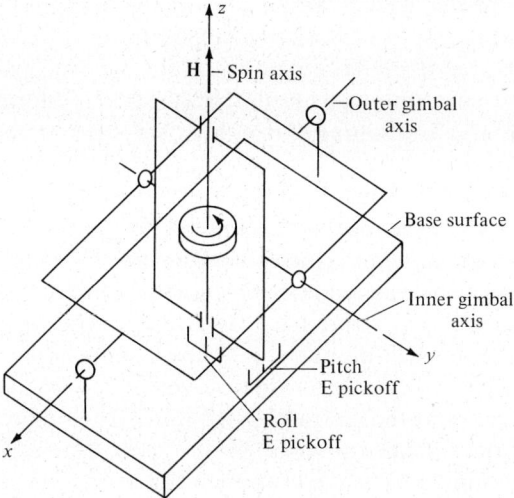

Figure 7.19

to the base surface. Two E pickoffs can detect rotation about the inner gimbal axis or the outer gimbal axis. If the base surface (the vehicle) turns about either the outer or the inner gimbal axis, the rotor does not precess. It remains pointed vertically upward. It is therefore called a *vertical gyro* and provides a stabilized reference line along the earth's vertical. The motion of the base relative to the inner gimbal axis is detected by the two E pickoffs and thus provides a measurement of vehicle roll or pitch. In the early days of blind flying, it was critical for the pilot's safety to have such a local horizon provided. Now such a stabilized reference line is used in the airplane's autopilot system.

If the spin axis is set horizontal, such as is shown in Figure 7.20, a *directional gyro* results. Here, the stabilized reference line could be pointed to true north.

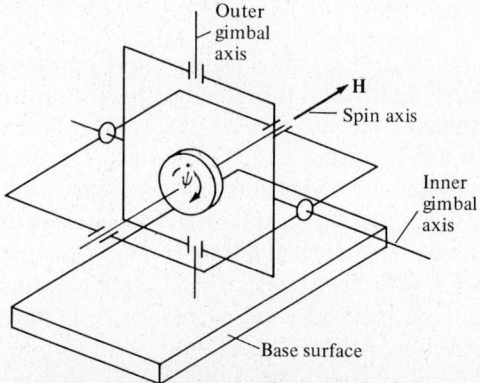

Figure 7.20

242 ANALYTICAL MECHANICS

The directional gyro would in effect become a form of compass if some provision was made to keep it in the local horizontal plane.

As attractive for navigational purposes as the vertical gyro may seem, it does have inherent errors, known as *drift*, that affect its performance and must be compensated for. The first source of drift is called *apparent drift* and is due to either the earth's rotation or the airplane's velocity. The gyro maintains a true vertical reference line, but the earth rotates underneath it causing it to move apparently with respect to the local horizontal plane.

In Figure 7.21 the angular velocity components of the earth at latitude angle λ and for a ship's heading angle β are shown. About the outer gimbal axis the gyro appears to rotate at a rate of $\Omega \cos \lambda \cos \beta$, whereas about the inner gimbal axis it appears to drift at a rate of $\Omega \cos \lambda \sin \beta$. Now if the airplane's motion relative to the earth is taken into account, it is seen that an additional apparent drift occurs about the aircraft's pitch axis. This apparent drift is equal to the airplane's forward speed divided by the radius of the earth. (The airplane's roll axis is assumed to be lined up with the vertical gyro's outer gimbal axis.)

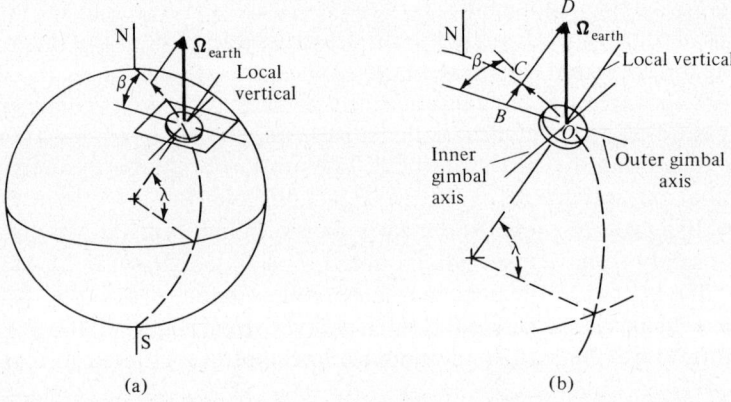

Figure 7.21

The second source of drift is called *random drift* and is caused by the presence of spurious disturbing torques inherent to the mechanical operation of the gyro itself. For example, the E pickoff exerts a reacting moment when measuring angular rotation that can be mistaken for actual vehicle motion. The same can be said for gimbal unbalance, or bearing friction. The random drift rate plus the apparent drift rate could lead to total drifting rates of up to 30°/h. This is obviously unacceptable. To correct for this, some sort of pendulous device is needed to act in concert with the gyro that would provide an alternative source of vertical reference. One method used to reduce drift, the employment of a servo system and pendulum, is schematically shown in Figure 7.22. Due to error, the gyro shown has turned about the outer gimbal axis relative to the local vertical. The pendulum on the outer gimbal representing the local vertical

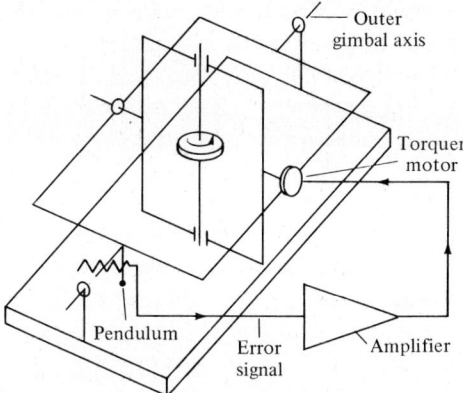

Figure 7.22

registers an error signal through the pickoff. The signal is then amplified and sent to a torquer located on the inner gimbal axis. The torque provided will cause the outer gimbal axis (the roll axis) to precess until the rotor's spin axis and the pendulum are in alignment once again. At that time the pickoff signal from the pendulum will go to zero. The drift has effectively been canceled. A more refined system would have a pendulous device for the pitch axis as well. An error signal for drift about the pitch axis (the inner gimbal axis) would be amplified and sent to a torquer motor on the outer gimbal axis. This torque would cause the spin axis to precess about the inner gimbal axis until the rotor axis and the inner gimbal pendulum were in alignment.

It should be pointed out that a system consisting of only a pendulum cannot be used as a navigational device. It is too sensitive to the motion of the vehicle. When the vehicle accelerates in straight flight, the pendulum deflects causing erroneous signals to be produced. The same is true when the vehicle turns. For an aircraft, the pendulum lines up with the bank angle. Many methods have been used to correct for these *sudden* false signals produced by pendulum surges, but the gyroscope is still needed ultimately to store information about the pendulum's slowly varying true position.

For voyages over large distances an entirely different method of establishing a local horizon is used. Three single degree of freedom gyros pointed along mutually perpendicular directions are first used to establish a *stable platform*, i.e., a platform kept in its original orientation. The three spin axes of the gyros in essence form a nonrotating coordinate axes reference. The apparent drift rates caused by the earth's rotation and the vehicle's motion are calculated. This information is used to control and rotate a second platform an appropriate amount relative to the reference orientation of the stable platform to keep the second platform always parallel to the local horizontal plane. This second platform provides a vertical reference. A simplified schematic of a stable platform is shown in Figure 7.23.

It is seen that a stable platform is unaffected by vehicle motion as no external

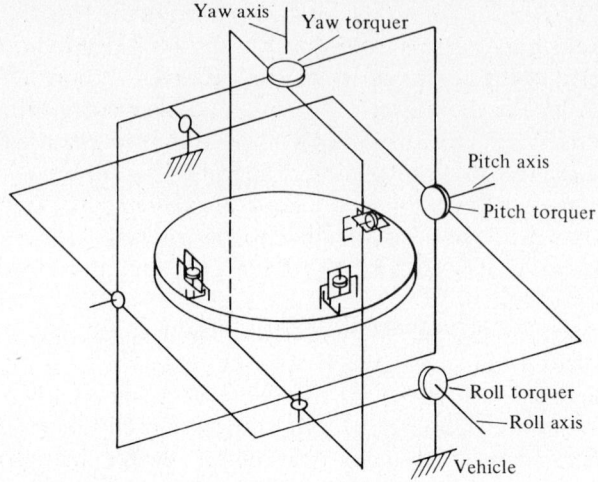

Figure 7.23

moments can reach the platform. However, friction, unbalance, and other disturbing internal torques will cause the platform to rotate. The gyros provided for each axis of rotation basically detect this motion through the E pickoffs and send amplified signals to the appropriate torquer axis to turn the platform back to its original reference position.

As mentioned, with a nonrotating reference frame established, the apparent drift caused by the earth's rotation and the vehicle's velocity can be calculated and corrected for, and the moving second platform can be kept locally horizontal. In principle it functions as follows. If the stable platform is initially established with its roll axis pointing due north, any northerly velocity causes an apparent drift about the pitch axis by an amount V_N/R. Similarly, any easterly velocity causes an apparent drift of $V_E/R \cos \lambda$. Using accelerometers on the rotating *horizontal table*, acceleration in these two directions are constantly monitored and their output signals integrated to produce velocity magnitudes. With the aid of an onboard computer, latitude angles and the necessary rotation rates needed to counter the apparent drift rate due to the vehicle's motion can be calculated. Proper voltage signals are then sent to torquers to affect table motion, keeping it continuously parallel to the local horizontal plane, while the stable platform remains unchanged in orientation.

The second source of drift is the earth's rotation of 15°/h. The horizontal table must be made to turn about an axis parallel to the earth's rotation axis at the same rate. If, for instance, the stable platform was oriented initially so that the roll axis was pointed parallel to the earth's rotation axis, for the entire time in flight, a constant rate of turning of 15°/h of the moving table about the stable platform's roll axis would affect cancellation of the drift. The moving table, initially established horizontal, would remain locally horizontal. The total rate of corrective turning needed about the stable platform's roll axis would be $\Omega + V_E/R \cos \lambda$. About the pitch axis it would be V_N/R.

By integrating the signals from the accelerometer again, information as to distance traveled to the north and to the east relative to a nonrotating earth can be obtained. A locally maintained horizontal accelerometer table is obviously mandatory here. With the addition of a clock to the system, the rotation of the earth's surface relative to the airplane can be traced, and the craft's location over the earth's surface can be established. The whole system consisting of the closed-loop servo-controlled stable platform, the motor-driven horizontal accelerometer table, the accelerometers, the computer, and the clock is called an *inertial navigation* system.

For journeys of several days, even this system will fail, and the stable platform will drift. Because of this, periodic adjustments by the crew using star sights as a reference are needed for the stable platform. Unmanned spacecraft must rely on other means to navigate. One such method takes advantage of the speed and storage capacity of digital computers. Before the journey, countless star maps are memorized by the computer. During the flight, a telescope mounted to the ship focuses its images on a grid network of infrared-detecting photovoltaic semiconductors. The electronic pattern that emerges is rapidly compared with all the star maps in the computer's memory to establish the one pattern that is in agreement. When that is accomplished, a fixed orientation or position in space can be ascertained for the ship.

Sample Problem 7.11

Shown in the figure is a directional gyro with a pendulous weight added to the rotor. If the gyro spin axis is initially pointed to the north in a local horizontal plane, derive the equations of motion for the gyro at any arbitrary latitude angle assuming small angular disturbances of the rotor from the equilibrium position. Show that the directional gyro will return to point north if small damping is present. A directional gyro employed in this manner is called a *gyrocompass*.

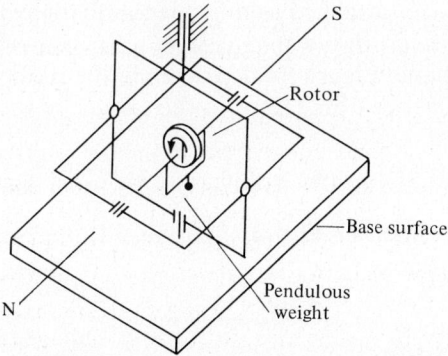

Analysis: A rotating frame of reference attached to the inner gimbal will be used for this analysis (see Section 7.3) as shown in Figures (a) and (b). Here

$$I_{xx} = I_{yy} = I_1 \qquad I_{zz} = I_3$$

246 ANALYTICAL MECHANICS

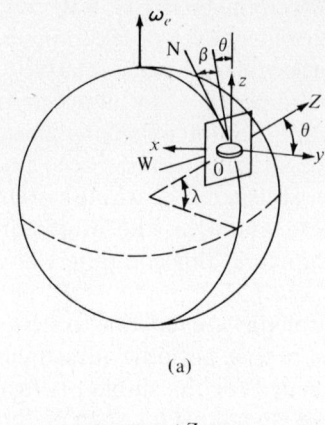

(a)

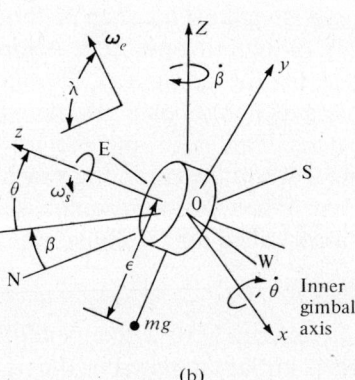

(b)

Assume point O is fixed or nonaccelerating and that the weight imbalance has a moment arm ε about the inner gimbal axis. Including the earth's rotation, the rotating coordinate axes attached to the inner gimbal have an angular velocity $\boldsymbol{\Omega} = \boldsymbol{\omega}_e + \dot{\boldsymbol{\theta}} + \dot{\boldsymbol{\beta}}$. Figure (a) shows the gyrocompass' axes orientation relative to the local meridian plane. Figure (b) depicts only the motion of the coordinate axes relative to a reference frame fixed to the surface of the earth, X, Y, Z. For small angles,

$$\boldsymbol{\Omega} = (\beta \omega_e \cos \lambda - \dot{\theta})\mathbf{i} + (\omega_e \sin \lambda - \theta \omega_e \cos \lambda - \dot{\beta})\mathbf{j} + (\theta \omega_e \sin \lambda + \omega_e \cos \lambda - \theta \dot{\beta})\mathbf{k}$$

The rotor's angular velocity is $\boldsymbol{\Omega} + \boldsymbol{\omega}_s$. Since $\omega_s \gg \Omega$, the angular momentum of the rotor can be written as

$$\mathbf{H}_O \simeq I_3 \omega_s \mathbf{k}$$

Since

$$\mathbf{M}_O = \dot{\mathbf{H}}_O = [\dot{\mathbf{H}}_O]_{\text{rot}} + \boldsymbol{\Omega} \times \mathbf{H}_O$$

Euler's modified equations become, with $\omega_s \gg \Omega$,

$$M_x = (\omega_e \sin \lambda - \theta \omega_e \cos \lambda - \dot{\beta})I_3 \omega_s = mg\varepsilon\theta$$

$$M_y = -(\beta\omega_e \cos\lambda - \dot\theta)I_3\omega_s = 0$$
$$M_z = 0$$

These equations reduce further to

$$\dot\theta - \beta\omega_e \cos\lambda = 0 \tag{a}$$

$$\dot\beta - \omega_e \sin\lambda + \theta\omega_e \cos\lambda + \frac{mg\varepsilon\theta}{I_3\omega_s} = 0 \tag{b}$$

Then
$$\ddot\theta + \left(\frac{mg\varepsilon + I_3\omega_s\omega_e \cos\lambda}{I_3\omega_s}\omega_e \cos\lambda\right)\theta = \omega_e^2 \cos\lambda \sin\lambda$$

which is a second-order linear differential equation having a solution of the form

$$\theta = A\cos(kt + \theta_0) + \frac{I_3\omega_s\omega_e \sin\lambda}{mg\varepsilon + I_3\omega_s\omega_e \cos\lambda} \tag{c}$$

where
$$k = \sqrt{\frac{I_3\omega_s(\omega_e \cos\lambda)^2 + mg\varepsilon\omega_e \cos\lambda}{I_3\omega_s}} \tag{d}$$

The solution for β, the heading angle, becomes, from equation (a),

$$\beta = -\frac{kA \sin(kt+\theta_0)}{\omega_e \cos\lambda} \tag{e}$$

Since both angular oscillations have the same frequency, the end of the spin axis will describe an ellipse about the northerly direction. With any sort of damping present, the motion will die out leaving $\beta = 0°$ (pointing north). In effect the pendulous weight causes the spin axis to precess back to the meridian line.

The final angle θ, with damping present, is

$$\theta_p = \frac{I_3\omega_s\omega_e \sin\lambda}{mg\varepsilon + I_3\omega_s\omega_e \cos\lambda} \tag{f}$$

The frequency of oscillation given by equation (d) is very small. The period of oscillation is an inverse function of the cosine of the latitude angle. For gyrocompass location near the north pole, the gyrocompass will not function since $\cos\lambda$ approaches zero. (The magnetic compass also fails near the magnetic north pole. The early explorers to the north pole were dependent upon clear weather to navigate, presenting yet another problem on their hazardous voyages.) For a ship's velocity v at a heading γ east of north from the local meridian, the angular velocity of the gimbal is changed by the amount

$$\Delta\omega_x \simeq \frac{v \cos\gamma}{R}$$

about the inner gimbal axis. Neglecting other changes produced by this motion, this velocity introduces a steady-state heading error, found from equation (a),

$$\beta_p = -\frac{v \cos\gamma}{R\omega_e \cos\lambda} \quad \text{(west of north)}$$

248   ANALYTICAL MECHANICS

A vehicle moving due north at a speed of 160 km/h at a latitude of 40° introduces a gyrocompass error of 7°. The use of the gyrocompass is, therefore, limited to slow-moving crafts such as boats. ∎

PROBLEMS

7.1 A homogeneous thin rod of mass m and total length $3b$ is bent as shown in the figure. What is the bending moment B on the rod at section A where the rod enters a chuck that drives it at constant angular speed ω. The moment caused by the static weight of the rod may be neglected.

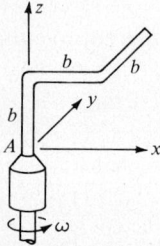

7.2 A homogeneous cylinder with mass m, radius r, and length h is split into two equal parts along its length. One of the parts is then displaced a distance h_1 axially and then welded back together with the other half. If the resulting shape is rotated about the axial direction with angular speed ω, find the bending moment at one end caused by rotation that would have to be resisted by the support bearings.

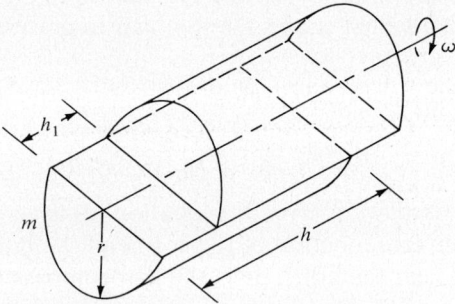

7.3 A thin right isosceles triangular plate of mass m and sides of length b rotates about the horizontal axis AB with constant angular speed ω. What are the bearing forces at A and B caused by the rotation?

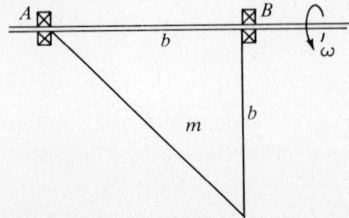

7.4 Repeat problem 7.3, but use Euler's equation for principal axes.

7.5 A thin rectangular plate having mass m and dimensions $a \times b$ rotates about a vertical axis with constant angular speed $\dot\psi$. At the same time, it is free to rotate about a pin located at point O without friction acting. Set up Euler's equation of motion for principal axes that pass through point O. For small θ, what are the conditions that must be met to have stable motion?

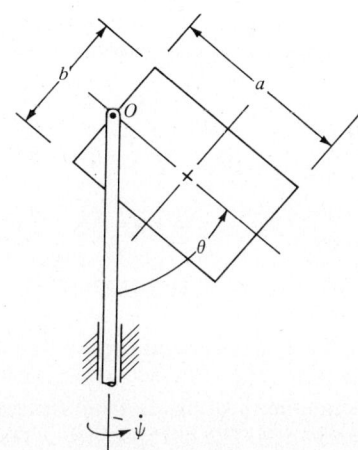

*7.6 If in problem 7.5 no moment acted along the vertical axis so that $\dot\psi$ were allowed to vary, write a computer program that would give θ as a function of time. At $t=0$, $\dot\psi = 100$ rad/s, $\theta = 1°$ and $\dot\theta = 0$. Take $m = 1$ kg, $a = 20$ cm, and $b = 10$ cm. Compare your results with an exact solution when θ becomes $\pi/2$. Use a communication interval of 0.01 s, and terminate the program after 0.25 s.

*7.7 Repeat problem 7.6 with $\dot\psi = 5$ rad/s at $t = 0$. Use a computer solution to find when $\dot\theta = 0$ again. Take 0.01 s as the communication interval and 0.3 s as the final time.

7.8 Two disks of mass m and radius r are welded to the rigid thin bar AB of length $2L$ and mass m_1. The system is known to roll without slipping in a straight line with a speed v. At the instant shown, the connecting bar is horizontal. Find, for this moment, the normal and friction forces acting at C and D.

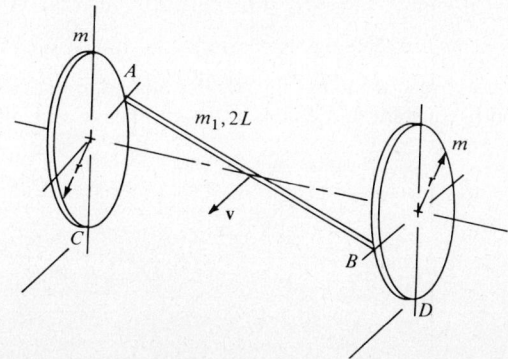

7.9 A vertical rod AC rotates with angular speed $\omega = ct$ (c is constant). Pinned at B is another thin rod of mass m and length L that can swing in a vertical plane that is rotating with rod AC. Initially, rod BD remains in contact with the smooth surface, but after time t_1, it rises. Find what t_1 equals. The system starts from rest, and all frictional effects can be neglected.

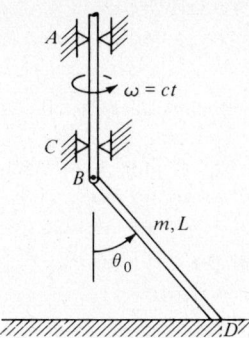

7.10 A gyroscopic testing machine constructed to give fatigue data is shown schematically below. A horizontal table rotates about a vertical axle with a speed of n rpm. On the table are two friction-free bearings at A and B placed symmetrically a distance L apart about the vertical axis. The test specimen is the long shaft supported by the bearings. One end of the shaft is driven by a motor at a constant angular speed ω rad/s; the other end of the shaft carries a thin flywheel of mass m whose polar moment of inertia is J. Calculate the flexural fiber stresses S on the specimen at A ($S = 4Mr/\pi r^4$).

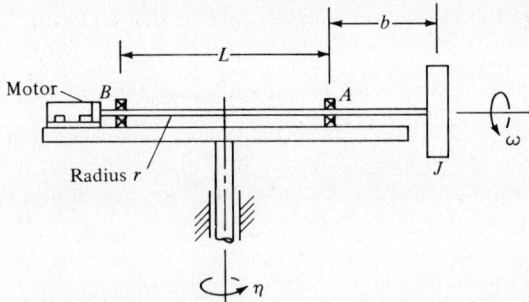

7.11 A ship's turbine is placed so that its axle lies along the boat's length. The turbines' angular speed is constant and equals ω. Due to heavy seas, the boat pitches. (Roll

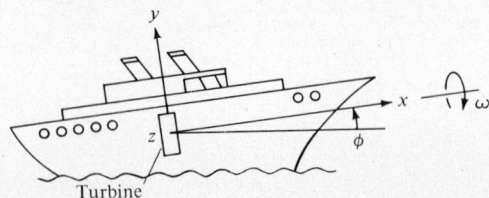

can be neglected.) This motion can be approximated by $\phi = \phi_0 \sin(2\pi/T)t$. What will the largest moment acting about the boat's yaw axis be if the turbine's polar moment of inertia is known to be J?

7.12 After takeoff, a landing wheel and its support structure are folded into the wing by rotation about an axle AB with a constant angular speed Ω. The wheel itself has mass m, radius R, and polar moment of inertia J. If the plane's takeoff speed is v and the landing wheel rolls without slipping, determine the moment acting on the wheel when it begins to lift.

7.13 A small sphere of mass m and radius R attached to a massless rod of length $2R$ is free to swing about a pin A in a vertical plane. The pin itself is free to rotate about a vertical axis resisted only by the torsional stiffness k_θ of member AB. Write Euler's equation of motion for the system, and determine the moment acting on the pin. Transform the equations of motion so that they refer to a rotating vertical x, y, z axes set. The total moment about the vertical is $k_\theta \theta$.

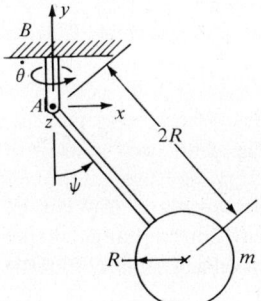

7.14 A thin rod of mass m_1 and length $2a$ is welded at its midpoint O to a light vertical axis so that it forms a fixed angle α. At the lower end of the rod is fastened a point mass m_2. The entire system rotates about the vertical at a constant angular speed Ω. Write Euler's equations of motion for the system. What is the bending moment acting at O?

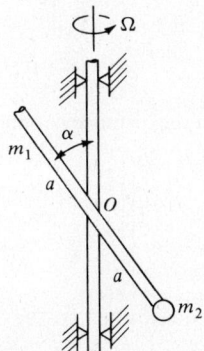

7.15 A circular vertical ring with radius R rotates about a fixed vertical axis with constant angular speed Ω. In the ring is a wheel whose axis of rotation forms a 60° angle with the vertical. Its axle passes through the center of the ring and is supported by frictionless bearings on the ring's circumference. The wheel has mass m, polar moment of inertia J, and transverse moment of inertia $J/2$. Its center of mass lies on the vertical axis. With what speed and direction must the wheel rotate so that the upper bearing at A carries no load?

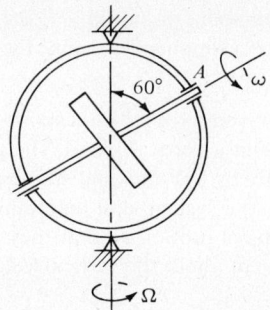

7.16 Repeat problem 7.10 using the modified Euler equation.

7.17 Repeat problem 7.15 using the modified Euler equation.

7.18 A rotationally symmetric wheel turns with constant angular speed ω about the horizontal shaft AB. The shaft itself is fixed to ring ACB, which is free to rotate about a vertical axis that passes through the center of mass of the wheel. The center of mass lies halfway between A and B. What are the dynamic bearing forces on the shaft if the ring turns with constant angular speed Ω? The wheel's polar moment of inertia is J, and its transverse moment of inertia is I. Use the modified Euler equations.

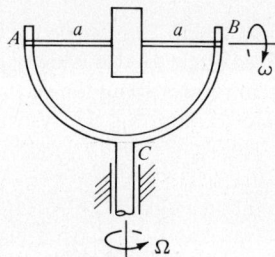

7.19 Repeat problem 7.9, but let the distance between the pin at B and the vertical rod be d.

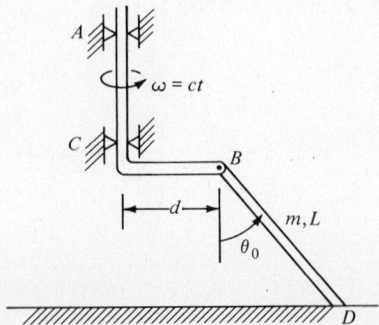

7.20 The right circular cone having altitude h, base radius R, and mass m rolls without slipping along the ground. If it turns about the vertical axis with a constant angular speed Ω, find the total moment acting on the cone from the ground about point O.

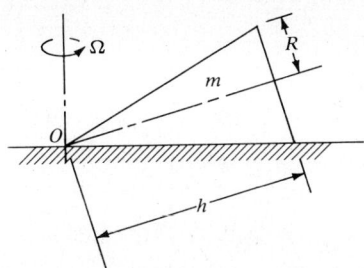

7.21 The thin disk shown of mass m and radius R rotates with constant angular spin $\dot{\psi}$ about a pin at A. The pin itself twists with the light shaft OA of length L at a constant rate $\dot{\theta}$ about a rotating horizontal axis. The axis is driven about the vertical direction at a constant precession rate of $\dot{\phi}$. All motion is friction-free. Find (a) the kinetic energy of the system for this instant and (b) the total moment acting on the pin at A.

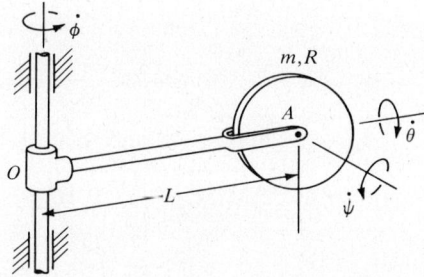

7.22 Repeat problem 7.21 using the modified Euler equations.

7.23 Repeat problem 7.21 for the instant when the spin axis makes an angle θ with the vertical about the line of nodes.

7.24 Repeat problem 7.23 using the modified Euler equations.

7.25 The uniform thin rod of mass m and length L swings in a vertical plane about pivot point A while the plane of oscillation itself rotates with angular speed $\dot{\phi}$. Find the rod's kinetic energy.

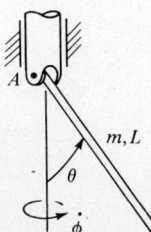

7.26 A circular vertical ring of radius R and mass m is free to rotate about a vertical axis. In the ring is a light axle that makes an angle of 60° with the vertical and spins relative to the ring at an angular speed $\dot{\psi}$. Fixed to the axle is a thin disk of mass m_1 and radius r whose center of gravity lies on the vertical axis of rotation. Find the kinetic energy of the system.

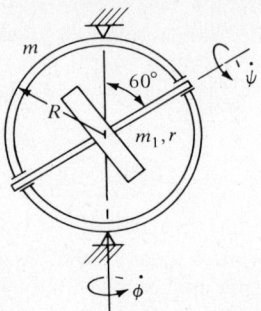

7.27 A thin rod of mass m and length L turns about a light rotating horizontal axle with an angular speed $\dot{\theta}$. This axle is imbedded in a vertical ring of mass m_1 and radius R that turns about the vertical at a rate $\dot{\phi}$. Find the kinetic energy of the system.

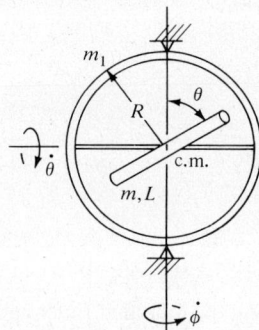

7.28 A thin ring of negligible mass rotates about a vertical axis at an angular speed $\dot{\phi}$.

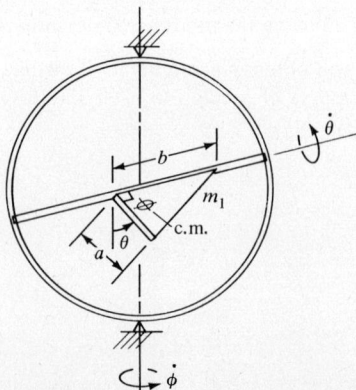

Fixed to a light horizontal axle along a diameter within the ring is a thin right triangular plate having mass m_1 and dimensions $a \times b$. The side of the plate fixed to the axle has dimensions b, and the plate's center of mass lies in the same plane as the vertical axis of rotation. If the axle turns relative to the ring at a rate $\dot{\theta}$, find the kinetic energy of the system.

7.29 The cylinder shown of mass m, length $3R$, and radius R describes force-free motion about its center of mass. If the spin rate is $\dot{\psi}_0$ and the angle θ is constant and equal to $60°$, find the precession rate. What is the angle γ? Does the cylinder have direct or retrograde precession?

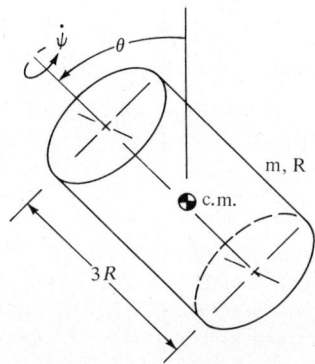

7.30 In problem 7.29, what does ω_T equal? What does H_3 equal? Draw the space and body cones. Show the body spin axis and the fixed Z direction. Label the angles γ and θ. Draw and label the angular rates $\dot{\psi}$, $\dot{\phi}$, ω, ω_T, and ω_3. Draw and label the vector components H_T and H_3 and their resultant H.

7.31 A satellite of mass m has approximately the shape of a thin ring with radius R. It spins about its symmetry axis, axis 3, with a constant rate $\dot{\psi}$. Suddenly it is hit by a meteorite that imparts an angular impulse of magnitude equal to 10% of the total initial angular momentum. This short-lived impulse acts in a positive sense about axis 1. After the impulse, the satellite begins to precess. Find the precession rate, the angles θ and γ, and ω_T, ω_3, H_T, and H for this motion. Does the satellite perform direct or retrograde precession? Draw the space and body cones.

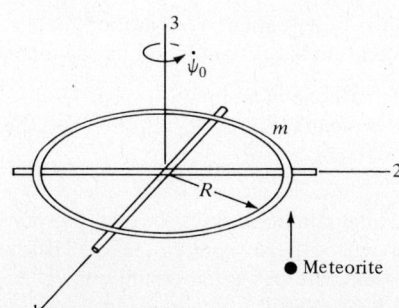

7.32 A gyroscope in torque-free cardanic suspension performs steady precession at an angle of 30°. If its precession rate is known to be $\dot{\phi}_0$ and if $I_3/I = \sqrt{3}$, find the spin rate present. Find ω_T and γ.

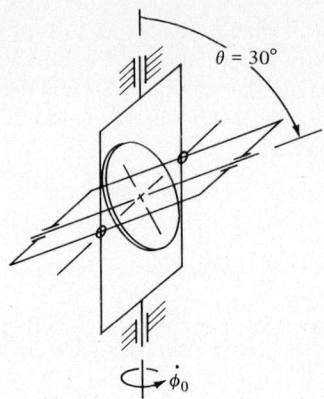

7.33 At one end of a thin light shaft OA of length $4R$ sits a disk of mass m_1 and radius R. The disk spins relative to the shaft about its own axis at an angular rate $\dot{\psi}$. The other end of the shaft is joined to a vertical axis by a frictionless pin at O so that the shaft may turn freely about the vertical axis at a precession rate $\dot{\phi}$ and may turn about the line of nodes at the rate $\dot{\theta}$. The angular rates $\dot{\phi}, \dot{\psi}$, and $\dot{\theta}$ are not constant. Determine the kinetic energy of the disk.

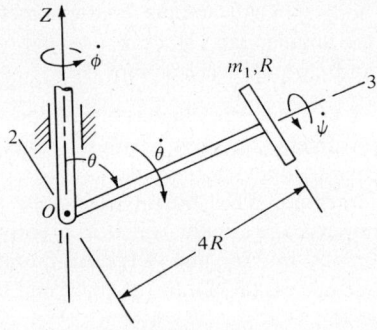

7.34 Derive the modified Euler equations of motion for the disk of problem 7.33. Show that angular momentum is constant about axis 3 and fixed axis Z.

7.35 Let the right circular cone of Sample Problem 7.9 describe steady precession at an angle of 120°. Find the minimum spin required and the period of slow precession if $\dot{\psi} = 100$ rad/s, $h = 10$ cm, and $R = 6$ cm.

7.36 In problem 7.34, what conditions must be met in order to have steady precession at an angle θ_0? What will be the slow precession rate? If θ_0 were to equal 90°, what would the steady precession rate be? What conditions on the total spin must be met for stable motion with $\theta_0 = 180°$?

*7.37 For problem 7.33, write a computer program that will give θ and ϕ as output as t varies. Take $m_1 = 5$ kg, and $R = 9.81$ cm. At $t = 0$, $\dot{\theta} = \dot{\phi} = 0$, $\dot{\psi} = 40\sqrt{65}$, and $\theta = 60°$. Show that the exact solution has $\dot{\theta} = 0$ when $\theta = 90°$. Use a communication interval of 0.01 s, and terminate the program at 0.5 s.

*7.38 Repeat problem 7.37, but take $\dot{\theta} = 0$, $\dot{\phi} = 2.0$ rad/s, $\dot{\psi} = 155.8128$ rad/s, and $\theta = 60°$ as initial conditions. R is changed here to 19.62 cm, and m_1 to 15 kg. Show that the exact solution has $\dot{\theta} = 0$ when $\theta = 90°$. Compare with computer results. Use a communication interval of 0.02 s, and terminate the program at 1.0 s.

*7.39 Repeat problem 7.38 with $\dot{\psi} = -352.8128$ rad/s. Show that the exact solution has $\dot{\theta} = 0$ when $\theta = 90°$ and $\dot{\phi} = 0$ when $\theta = 68.81°$. Compare with computer results. Use a 0.01 s communication interval, but terminate the program at 0.4 s.

*7.40 Refer to problem 7.38. At $t = 0$, the disk is known to perform steady precession with $\dot{\theta} = 0$, $\theta = 60°$, $\dot{\phi} = 2.48713$ rad/s, and $\dot{\psi} = 200$ rad/s. If a sudden angular impulse of 14.75 N·m·s is applied about axis 1 for a very brief moment of time at $t = 0$, write a computer program that will solve for subsequent values of θ with increasing time for one cycle of motion. Use a communication interval of 0.02 s, and terminate the program at 2.4 s.

ENERGY METHODS 8

8.1 INTRODUCTION

This chapter will explore the last fundamental developments in classical mechanics. It begins with contributions from Joseph Lagrange (1736–1813), in the early nineteenth century. By reformulating mechanics in terms of a scalar quantity, the *lagrangian*, and in terms of nonspecific generalized coordinates, Lagrange was one of the first theoreticians to develop a general law of nature, i.e., a law that is invariant under a coordinate transformation. This work was so concise that with just one equation, Lagrange successfully described all the past equations of motion of newtonian mechanics. Be they for rotation or translation, be they in cartesian or cylindrical coordinates, the appropriate equations can all be derived from *Lagrange's equation*. Lagrange's work holds forth the possibility of solving dynamical problems in terms of the most economical coordinates, that is, coordinates suitably selected so that they do not violate the physical constraints of the system. The subsequent number of equations of motion derived are then equal to the number of degrees of freedom the problem has; the constrained coordinates have been removed. Further, as will be seen, the constraint forces do not appear in the problem either, thus reducing the number of unknowns.

The functional makeup of the lagrangian itself—the lagrangian is the difference between the kinetic and potential energies of a system—also gives insight into which *generalized momenta* are conserved. Edward J. Routh (1831–1907) extended Lagrange's work by introducing a function similar to the lagrangian called the *routhian*. The routhian has all the velocity terms for coordinates with associated constant momentum replaced by functions of the generalized momentum terms themselves. Routh's equations of motion then become functions solely of the remaining independent generalized coordinates and their derivatives (velocities) and the constants representing the existing constant generalized momentum. In total, this reduces the complexity of the equations of motion even further.

Sir William Rowan Hamilton (1805–1865) also contributed immeasurably to the development of mechanics. *Hamilton's principle* presents an entirely new formulation to mechanics; it is an *integral principle*. Whereas the newtonian and lagrangian equations of motion are based on an instantaneous state description and are then integrated to obtain the path coordinates, Hamilton's principle predicts that the actual path taken by the system will be such that the *integral* of the lagrangian with respect to time between two specified time limits is an extremum. Using the *calculus of variations* and the work contributed in this field by Euler, it is possible to show that the integral becomes an extremum when Lagrange's equations are satisfied. Though Hamilton's principle in effect provides no new information, for all its elegance, it does mark a sharp conceptual departure from newtonian mechanics. Hamilton's work prepared the way for the eventual modification of the classical theory others would accomplish many years later. Schrödinger's (1887–1961) wave equation and the field of quantum mechanics trace their beginnings to Hamilton's efforts that had as a goal the capability of describing analogies between the laws of general optics and mechanics.

Hamilton also described a scalar function, similar to the lagrangian, that was dependent upon generalized momentum (instead of velocity) and generalized coordinates. The function is called the *hamiltonian*. The equations of motion subsequently derived from the hamiltonian, *Hamilton's equations*, are a first-order set. For m degrees of freedom, Lagrange's equations provide m second-order differential equations. Hamilton's equations result in $2m$ first-order differential equations for the $2m$ independent generalized momenta and generalized coordinates that describe phase space.

For all problems in classical mechanics the results provided by the Lagrange or Hamilton methods are essentially the same. The equations of motion obtained from the hamiltonian, however, are perhaps more directly suitable for digital computer calculations, being in first-order form to begin with. It is also possible to transform the hamiltonian, dependent upon both momenta and coordinates, by means of *canonical transformations* to a new set of $2m$ coordinates in phase space; this may make problem solution more tractable. As an example, one can transform the hamiltonian to a new set of coordinates that are all constants. These constants are the initial conditions to the problem, and the time development of the solution may be found by reversing the transformation. The emphasis then lies in finding the proper transformation that works rather than solving the equations of motion.

8.2 INDEPENDENT GENERALIZED COORDINATES AND CONSTRAINT EQUATIONS

Before developing Lagrange's equations of motion, it is important to review the concept of degrees of freedom (see Section 5.1). The number of degrees of freedom equals the minimum number of independent coordinates needed to completely describe the position of system. For a set of coordinates that are not entirely independent, it is equal to the number of coordinates used to describe the system minus the number of constraint equations present governing the system's motion. It is a fixed quantity independent of the coordinate set selected to describe the motion. *Generalized coordinates* are unspecified coordinates. By deriving equations of motion in terms of a general set of coordinates, the results found will be valid for any coordinate system that is ultimately specified.

The derivations that follow depend upon the coordinates being indepen-

260 ANALYTICAL MECHANICS

dent, and it is best at this time to speak of *independent generalized coordinates*. The transformation equations between the old coordinates and the new independent generalized coordinates q_j can be represented as follows:

$$\mathbf{r}_1 = \mathbf{r}_1(q_1, q_2, q_3, \ldots, q_m, t)$$
$$\mathbf{r}_2 = \mathbf{r}_2(q_1, q_2, q_3, \ldots, q_m, t)$$
$$\mathbf{r}_n = \mathbf{r}_n(q_1, q_2, q_3, \ldots, q_m, t) \tag{8.1}$$

where n particles and m degrees of freedom are present. For example, the plane motion of the simple pendulum in Figure 8.1 has 1 degree of freedom. If we specify the angle θ to be the independent generalized coordinate q_1, then the transformation equation, equation (8.1), becomes

$$\mathbf{r}_1 = l \sin \theta \mathbf{i} + l \cos \theta \mathbf{j}$$

where
$$x = l \sin \theta$$
$$y = l \cos \theta$$

If now the simple pendulum's pivot point at A was constrained to move in a vertical direction as shown in Figure 8.2, the system would have 2 degrees of freedom. The length and angular measures $q_1 = y$ and $q_2 = \theta$, respectively, could be selected as the independent generalized coordinate set. The transformation equation, equation (8.1), would then become

$$\mathbf{r}_1 = l \sin \theta \mathbf{i} + (y + l \cos \theta)\mathbf{j}$$

In equation (8.1), the number of constraint equations present, k, must equal

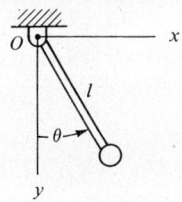

Figure 8.1

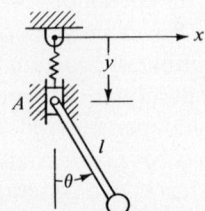

Figure 8.2

$3n - m$. Constraint equations can be classified in two main categories: holonomic and nonholonomic. *Holonomic* constraint equations can be written, in terms of n generalized coordinates, in the form

$$f_s(q_1, q_2, \ldots, q_n, t) = 0 \quad (s = 1, 2, \ldots, k) \tag{8.2}$$

Constraint equations of this type may be used to solve for k coordinates in terms of m independent generalized coordinates. This is usually not done. Instead, independent generalized coordinates are selected by inspection from the beginning of the problem. For example, Figure 8.3 shows a barbell constrained to move in a plane. For the coordinate set x_1, x_2, y_1, y_2, there is one holonomic constraint present, namely,

$$\sqrt{(x_2 - x_1)^2 + (y_2 - y_1)^2} = L$$

The number of coordinates (4) minus the number of constraint equations (1) equals the number of degrees of freedom (3). Given this set, it is possible to reduce the number of coordinates needed by employing the constraint equation. Another approach, which is superior, is to select x_1, y_1, and θ to be the independent generalized coordinates (see Figure 8.4). If this is done at the beginning of the problem solution, no further algebraic reductions are necessary or possible since no constraint equations exist in this set. Constraint equations that cannot be used to eliminate the dependent coordinates are called *nonholonomic*. They are of the form

$$\sum_{i=1}^{n} a_{si} \, dq_i + a_{st} \, dt = 0 \quad (s = 1, 2, \ldots, k) \tag{8.3}$$

and are *nonintegrable*. For a wheel having radius r rolling without slipping in a straight line, the constraint equation relating the velocity v of the center of the

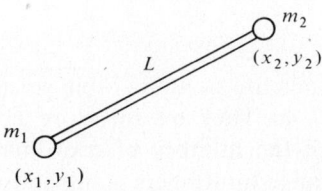

Figure 8.3

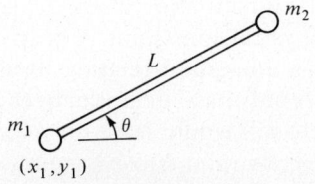

Figure 8.4

262 ANALYTICAL MECHANICS

wheel to its angular velocity $\dot\theta$ can be written as

$$v = \dot x = r\dot\theta$$

or
$$dx - r\, d\theta = 0$$

which is integrable and can be used to replace x by θ, or vice versa. The constraint is holonomic. If the wheel rolled along an unknown curved path instead, the two constraint equations for rolling would be given by (see Figure 8.5)

$$\dot x = v \sin\phi = r\dot\theta \sin\phi$$
$$\dot y = v \cos\phi = r\dot\theta \cos\phi$$

where ϕ is the angle the tangent to the wheel's path makes with the y axis. In differential form the constraint equations can be written as

$$dx - r \sin\phi\, d\theta = 0 \qquad dy - r \cos\phi\, d\theta = 0$$

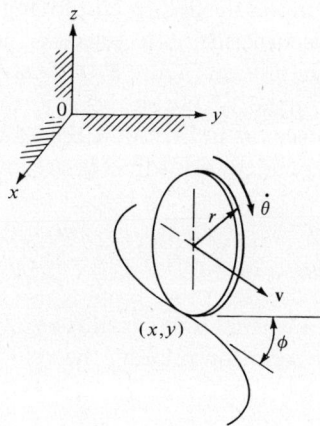

Figure 8.5

The two constraint equations are nonintegrable since they cannot be derived from any function $F(x, y, \theta, \phi)$. They are, therefore, nonholonomic and cannot be used directly to reduce the number of coordinates in a given problem. When the method of Lagrange multipliers is introduced, a means will become available whereby nonholonomic constraints can be actively employed.

Constraint equations that are not functions of time are called *scleronomic*. Constraint equations that are functions of time are called *rheonomic* and indicate the presence of a moving physical constraint. A bead constrained to slide upon a fixed rod in a plane has a constraint relating its x and y position, namely, $y = x \tan\alpha$ (see Figure 8.6). It only has 1 degree of freedom. If the rod turned with constant angular speed ω, it still would have only 1 degree of freedom since α is still constrained; but the constraint equations now are rheonomic:

$$y = x \tan \omega t \qquad \alpha = \omega t$$

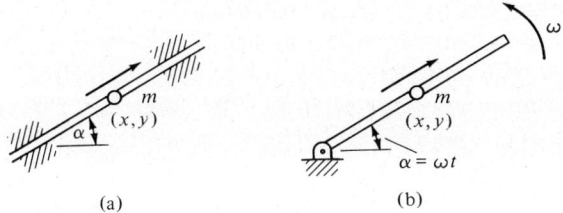

Figure 8.6

A unilateral constraint of the form

$$f(q_1, q_2, \ldots, q_m, t) \leq 0$$

is piecewise holonomic. For a particle moving inside a sphere, for instance, the equation of constraint is

$$x^2 + y^2 + z^2 - r^2 \leq 0$$

Inside the sphere it will have 3 degrees of freedom; on the surface of the sphere, just 2. The coefficient of restitution for impact provides the means of coupling the two regimes of motion.

8.3 VIRTUAL WORK AND GENERALIZED FORCES

A *virtual displacement* $\delta \mathbf{r}_i$ of the system is an assumed infinitesimal change of system coordinates occurring while time is held constant. It is called *virtual* rather than *real* since no actual displacement can take place without the passage of time.

For a system in static equilibrium the *total* force $\mathbf{F}_i^{(T)}$ on each particle must be zero.

$$\mathbf{F}_i^{(T)} = \mathbf{0} \tag{8.4}$$

The *virtual work* brought about by a virtual displacement of a system of n particles in static equilibrium must also be zero.

$$\delta W = \sum_{i=1}^{n} \mathbf{F}_i^{(T)} \cdot \delta \mathbf{r}_i = 0 \tag{8.5}$$

If the forces are separated into those that are applied forces $\mathbf{F}_i$ and those that are constraint forces $\mathbf{C}_i$, then

$$\delta W = \sum_{i=1}^{n} \mathbf{F}_i \cdot \delta \mathbf{r}_i + \sum_{i=1}^{n} \mathbf{C}_i \cdot \delta \mathbf{r}_i = 0 \tag{8.6}$$

If now the work done by the constraint forces is *assumed* to be zero for virtual displacements that are taken consistent with the physical constraints they impose, the virtual work done by the applied forces will be zero.

$$\delta W = \sum_{i=1}^{n} \mathbf{F}_i \cdot \delta \mathbf{r}_i = 0 \tag{8.7}$$

264 ANALYTICAL MECHANICS

Equation (8.7) is called the *principle of virtual work for applied forces*. Great simplification in many statics problems is afforded by this principle since all constraint forces are eliminated from consideration in the problem solution.

The restriction in equation (8.7) that the constraint forces do no virtual work on a system whose virtual displacements are consistent with the constraints is easily met by simple constraint forces. A particle constrained to move on a frictionless rod, for example, has its allowed virtual displacement tangent to the rod and perpendicular to the constraining normal force. The constraint force does no real or virtual work. If the rod were to rotate instead at a constant angular speed, the virtual displacement of the particle would again be tangent to the rod. The constraining normal force does do *real* work with the passage of time since the particle's *real* displacement does in part act perpendicular to the rod. For a virtual displacement though, with time held constant, the rod is imagined fixed momentarily. The particle is still free to slide tangent to the rod, but the constraining normal force does no virtual work. See Figure 8.7.

For a rigid body, the internal constraint forces also do no work since they must act perpendicular to the allowable displacements of the body. The same result is obtained for the frictional constraint force that is present to prevent sliding when an object rolls.

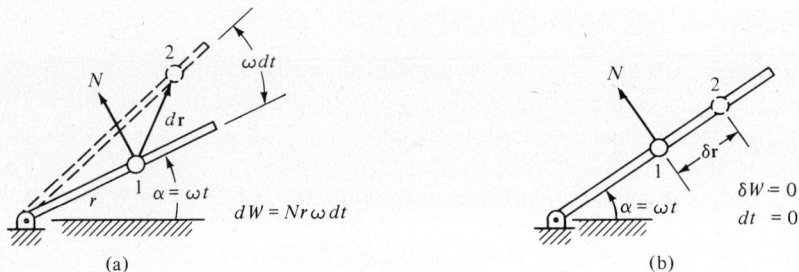

Figure 8.7

The *generalized force* Q_j is defined with the aid of the virtual work expression and the transformation equations, equation (8.1).

$$\mathbf{r}_i = \mathbf{r}_i(q_1, q_2, \ldots, q_m, t) \tag{8.1}$$

From equation (8.1), a velocity expression can be derived in terms of the m generalized independent coordinates,

$$\mathbf{v}_i = \sum_{j=1}^{m} \frac{\partial \mathbf{r}_i}{\partial q_j} \dot{q}_j + \frac{\partial \mathbf{r}_i}{\partial t} \tag{8.8}$$

and the displacement is given by

$$d\mathbf{r}_i = \sum_{j=1}^{m} \frac{\partial \mathbf{r}_i}{\partial q_j} dq_j + \frac{\partial \mathbf{r}_i}{\partial t} dt \tag{8.9}$$

Further, the virtual displacement can be written in terms of virtual displacements of the generalized coordinates.

$$\delta \mathbf{r}_i = \sum_{j=1}^{m} \frac{\partial \mathbf{r}_i}{\partial q_j} \delta q_j \qquad (8.10)$$

Since the virtual work of the applied forces is defined by

$$\delta W = \sum_{i=1}^{n} \mathbf{F}_i \cdot \delta \mathbf{r}_i$$

the expression can be rewritten as

$$\delta W = \sum_{i=1}^{n} \sum_{j=1}^{m} \mathbf{F}_i \cdot \frac{\partial \mathbf{r}_i}{\partial q_j} \delta q_j \qquad (8.11)$$

or

$$\delta W = \sum_{j=1}^{m} Q_j \, \delta q_j \qquad (8.12)$$

where the generalized force Q_j is defined to be

$$Q_j \equiv \sum_{i=1}^{n} \mathbf{F}_i \cdot \frac{\partial \mathbf{r}_i}{\partial q_j} \qquad (8.13)$$

Sample Problem 8.1

The simple pendulum shown has its pivot point connected to a vertical spring so that it has 2 degrees of freedom. Using the generalized coordinates $q_1 = y$ and $q_2 = \theta$, determine the generalized forces on the system, Q_y and Q_θ, brought about by the external force F.

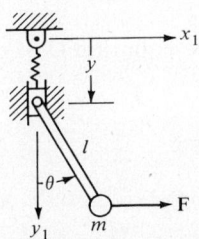

Analysis: The problem will be solved in two different ways. First the transformation equations will be obtained in terms of y and θ, namely,

$$x_1 = l \sin \theta \qquad y_1 = y + l \cos \theta$$

or
$$\mathbf{r} = (l \sin \theta)\mathbf{i} + (y + l \cos \theta)\mathbf{j}$$

Then
$$\frac{\partial \mathbf{r}}{\partial q_1} = \frac{\partial \mathbf{r}}{\partial y} = \mathbf{j}$$

$$\frac{\partial \mathbf{r}}{\partial q_2} = \frac{\partial \mathbf{r}}{\partial \theta} = l \cos \theta \, \mathbf{i} - l \sin \theta \, \mathbf{j}$$

266 ANALYTICAL MECHANICS

With $\mathbf{F} = F\mathbf{i}$ equation (8.13) can now be used:

$$Q_1 = Q_y = \mathbf{F} \cdot \frac{\partial \mathbf{r}}{\partial q_1} = \mathbf{F} \cdot \frac{\partial \mathbf{r}}{\partial y} = F\mathbf{i} \cdot \mathbf{j} = 0$$

$$Q_2 = Q_\theta = \mathbf{F} \cdot \frac{\partial \mathbf{r}}{\partial q_2} = \mathbf{F} \cdot \frac{\partial \mathbf{r}}{\partial \theta}$$

$$Q_\theta = Fl \cos \theta$$

Note that the generalized force for an angular coordinate is a moment.

The second way of arriving at these results is to find the total virtual work of the force F brought about by the independent virtual displacement δy and $\delta \theta$. For a virtual initial displacement δy, holding θ constant, the force F does no work; hence, Q_y must equal to zero. For a virtual displacement $\delta \theta$, holding y constant, the displacement of the particle in the direction of the force is

$$l \cos \theta \, \delta \theta$$

and the work expression must be

$$\delta W = Fl \cos \theta \, \delta \theta = Q_\theta \, \delta \theta$$

therefore

$$Q_\theta = Fl \cos \theta \qquad \blacksquare$$

In Sample Problem 8.1, the conciseness of using generalized coordinates and forces is clearly demonstrated. When the generalized coordinate is a distance, the generalized force associated with it will be a force. When the generalized coordinate is chosen as an angle, the associated generalized force will be a moment.

For *conservative forces* [see equation (3.31)],

$$\mathbf{F}_i = \nabla_i V$$

where $V = V(\mathbf{r}_1, \mathbf{r}_2, \ldots, \mathbf{r}_n)$. Then

$$Q_j = \sum_{i=1}^{n} \mathbf{F}_i \cdot \frac{\partial \mathbf{r}_i}{\partial q_j} = - \sum_{i=1}^{n} \nabla_i V \cdot \frac{\partial \mathbf{r}_i}{\partial q_j}$$

or

$$Q_j = - \frac{\partial V}{\partial q_j} \qquad (8.14)$$

where for conservative forces the potential function cannot depend upon time explicitly.

The principle of virtual work for conservative systems now becomes

$$\delta W = -\delta V = - \sum_{j=1}^{n} \frac{\partial V}{\partial q_j} \delta q_j = 0 \qquad (8.15)$$

Since the coordinates are independent, each term of equation (8.15) must equal zero.

ENERGY METHODS 267

$$\frac{\partial V}{\partial q_1} = \frac{\partial V}{\partial q_2} = \cdots = \frac{\partial V}{\partial q_m} = 0 \quad (8.16)$$

Equation (8.16) defines the possible states of static equilibrium. Which state of static equilibrium is actually *stable* can be determined by expanding the potential function in terms of a Taylor series about that questioned equilibrium point.

$$V = V_0 + \sum_{j=1}^{m} \left(\frac{\partial V}{\partial q_j}\right)_0 \delta q_j + \frac{1}{2} \sum_{j=1}^{m} \sum_{i=1}^{m} \left(\frac{\partial^2 V}{\partial q_j \partial q_i}\right)_0 \delta q_j \delta q_i + \cdots \quad (8.17)$$

Since by equation (8.15) the first summation on the right vanishes, the change in potential energy due to the virtual variation of coordinates is

$$\Delta V = \frac{1}{2} \sum_{j=1}^{m} \sum_{i=1}^{m} \left(\frac{\partial^2 V}{\partial q_j \partial q_i}\right)_0 \delta q_j \delta q_i + \cdots \quad (8.18)$$

A system is said to be *stable* if for all possible virtual displacements from an equilibrium point,

$$\Delta V > 0 \quad (8.19)$$

It is said to be a *neutral* position of equilibrium if for all possible virtual displacements,

$$\Delta V = 0 \quad (8.20)$$

Finally, if any virtual displacement condition exists where

$$\Delta V < 0 \quad (8.21)$$

the system is *unstable* about that equilibrium point.

Sample Problem 8.2
Find the possible equilibrium points, and discuss the stability of the spring-mass system shown. The spring is nonlinear and has a force displacement relationship $F_x = -x + 2x^2$. Ignore friction.

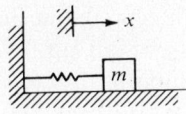

Analysis: Since the spring force is $F_x = -x + 2x^2$, it can be shown to be a conservative force. Further since

$$\frac{dV}{dx} = -F_x$$

$$V = \frac{x^2}{2} - \frac{2x^3}{3}$$

For one independent generalized coordinate, $q_1 = x$,

$$\frac{\partial V}{\partial q_1} = \frac{\partial V}{\partial x} = 0$$

defines the equilibrium states, or

$$x - 2x^2 = 0$$

and the two states are at

$$x_1 = 0 \qquad x_2 = \tfrac{1}{2}$$

Now

$$\frac{\partial^2 V}{\partial q_1^2} = \frac{\partial^2 V}{\partial x^2} = 1 - 4x$$

About $x_1 = 0$,

$$\Delta V = \tfrac{1}{2}(1 - 4x_1)\,\delta x^2 = \frac{\delta x^2}{2} > 0$$

and the system is stable for all δx. About $x_2 = \tfrac{1}{2}$,

$$\Delta V = \tfrac{1}{2}(1 - 4x_2)\,\delta x^2 = -\frac{\delta x^2}{2} < 0$$

and the system is unstable for all δx. A plot of V versus x will show that stability occurs for an equilibrium position where the *potential energy is minimum*. The equilibrium positions are located at points of *zero slope*.

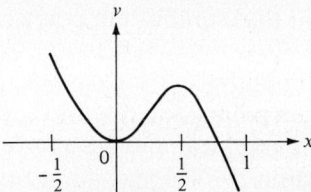

The instability at $x = \tfrac{1}{2}$ can be interpreted in the following manner. If the body is at rest at $x = \tfrac{1}{2}$ and is given a small positive displacement, the spring will continue to stretch to its failure point. If it is moved slightly to the left, it will move to point $x = 0$ and proceed to oscillate about it maintaining its initial total energy. With any sort of damping present, energy will be taken out of the system, and the mass will stop at $x = 0$. ∎

8.4 LAGRANGE'S EQUATIONS

The groundwork has now been laid for the derivation of Lagrange's equations. It was seen in the last section that the principle of virtual work provided a means of simplifying static problems by eliminating the constraint forces from problem consideration. It is planned now to extend this principle to accelerating particles.

D'Alembert (1717–1783) suggested a means whereby any dynamic system can be described by the equations of statics if reverse effective forces (inertia forces) equal to $-m_1 \mathbf{a}_i$ are introduced. This follows directly from Newton's second law

$$\mathbf{F}_i^{(T)} = m_i \mathbf{a}_i$$

Adding reverse effective forces produces a quasi-static equilibrium state,

$$\mathbf{F}_i^{(T)} - m_i \mathbf{a}_i = \mathbf{0} \qquad (8.22)$$

Proceeding as before, since equation (8.22) holds,

$$\sum_{i=1}^{n} (\mathbf{F}_i^{(T)} - m_i \mathbf{a}_i) \cdot \delta \mathbf{r}_i = 0 \qquad (8.23)$$

In terms of applied and constraint forces, this becomes

$$\sum_{i=1}^{n} \mathbf{F}_i \cdot \delta \mathbf{r}_i + \sum_{i=1}^{n} \mathbf{C}_i \cdot \delta \mathbf{r}_i - \sum_{i=1}^{n} m_i \mathbf{a}_i \cdot \delta \mathbf{r}_i = 0 \qquad (8.24)$$

Since the work of the constraint forces caused by a virtual displacement is assumed to be zero,

$$\sum_{i=1}^{n} \mathbf{F}_i \cdot \delta \mathbf{r}_i - \sum_{i=1}^{n} m_i \mathbf{a}_i \cdot \delta \mathbf{r}_i = 0 \qquad (8.25)$$

Equation (8.25), with the constraint forces absent, is known as *D'Alembert's principle*.

Equation (8.25) can now be written in terms of independent *generalized coordinates* and generalized forces. Substituting equations (8.10) and (8.12) leads to

$$\sum_{j=1}^{m} Q_j \, \delta q_j = \sum_{j=1}^{m} \left(\sum_{i=1}^{n} m_i \mathbf{a}_i \cdot \frac{\partial \mathbf{r}_i}{\partial q_j} \right) \delta q_j \qquad (8.26)$$

The quantity in parenthesis on the right-hand side of equation (8.26) can be simplified further. The kinetic energy of the system is given by

$$T = \sum_{i=1}^{n} \tfrac{1}{2} m_i \mathbf{v}_i \cdot \mathbf{v}_i \qquad (8.27)$$

The partial derivative of T with respect to $\dot{q}_j$ leads to

$$\frac{\partial T}{\partial \dot{q}_j} = \sum_{i=1}^{n} m_i \mathbf{v}_i \cdot \frac{\partial \mathbf{v}_i}{\partial \dot{q}_j} \qquad (8.28)$$

From equation (8.8),

$$\frac{\partial \mathbf{v}_i}{\partial \dot{q}_j} = \frac{\partial \mathbf{r}_i}{\partial q_j} \qquad (8.29)$$

Substituting this result back into equation (8.28) yields

$$\frac{\partial T}{\partial \dot{q}_j} = \sum_{i=1}^{n} m_i \mathbf{v}_i \cdot \frac{\partial \mathbf{r}_j}{\partial q_j} \qquad (8.30)$$

270 ANALYTICAL MECHANICS

Further

$$\frac{d}{dt}\left(\frac{\partial T}{\partial \dot{q}_j}\right) = \sum_{i=1}^{n}\left[m_i\mathbf{a}_i\cdot\frac{\partial \mathbf{r}_i}{\partial q_i} + m_i\mathbf{v}_i\cdot\frac{d}{dt}\left(\frac{\partial \mathbf{r}_i}{\partial q_j}\right)\right] \qquad (8.31)$$

The last term

$$\frac{d}{dt}\left(\frac{\partial \mathbf{r}_i}{\partial q_j}\right) = \sum_{k=1}^{m}\frac{\partial^2 \mathbf{r}_i}{\partial q_j \partial q_k}\dot{q}_k + \frac{\partial^2 \mathbf{r}_i}{\partial q_j \partial t} \qquad (8.32)$$

with the aid of equation (8.8) can be seen to be

$$\frac{d}{dt}\left(\frac{\partial \mathbf{r}_i}{\partial q_j}\right) = \frac{\partial \mathbf{v}_i}{\partial q_j} \qquad (8.33)$$

Equation (8.31) then becomes

$$\frac{d}{dt}\left(\frac{dT}{\partial \dot{q}_j}\right) = \sum_{i=1}^{n}\left(m_i\mathbf{a}_i\cdot\frac{\partial \mathbf{r}_i}{\partial q_j} + m_i\mathbf{v}_i\cdot\frac{\partial \mathbf{v}_i}{\partial q_j}\right) \qquad (8.34)$$

Since

$$\frac{\partial T}{\partial q_j} = \sum_{i=1}^{n} m_i\mathbf{v}_i\cdot\frac{\partial \mathbf{v}_i}{\partial q_j} \qquad (8.35)$$

then

$$\frac{d}{dt}\left(\frac{\partial T}{\partial \dot{q}_j}\right) - \frac{\partial T}{\partial q_j} = \sum_{i=1}^{n} m_i\mathbf{a}_i\cdot\frac{\partial \mathbf{r}_i}{\partial q_j} \qquad (8.36)$$

Substituting this identity into equation (8.26) leads to the result

$$\sum_{j=1}^{m} Q_j\,\delta q_j = \sum_{j=1}^{m}\left[\frac{d}{dt}\left(\frac{\partial T}{\partial \dot{q}_j}\right) - \frac{\partial T}{\partial q_j}\right]\delta q_j \qquad (8.37)$$

Since all the generalized coordinates were selected as *independent*, each coefficient of the individual varied coordinate must equal zero, independent of the others. That means that m equations of motion result having the form

$$\frac{d}{dt}\left(\frac{\partial T}{\partial \dot{q}_j}\right) - \frac{\partial T}{\partial q_j} = Q_j \qquad (8.38)$$

For *conservative systems* V is a function of position only, and

$$Q_j = -\frac{\partial V}{\partial q_j}$$

Defining the *lagrangian* as

$$L = T - V \qquad (8.39)$$

and with

$$\frac{\partial V}{\partial \dot{q}_j} = 0 \qquad (8.40)$$

equation (8.38) can be rewritten as

$$\frac{d}{dt}\left(\frac{\partial L}{\partial \dot{q}_j}\right) - \frac{\partial L}{\partial q_j} = 0 \qquad (j = 1, 2, \ldots, m) \tag{8.41}$$

The m equations of motion resulting from equation (8.41) are known as *Lagrange's equations*.

If the system has forces Q'_j that are not represented or derivable from potential functions, then one can separate the generalized force into two parts:

$$Q_j = -\frac{\partial V}{\partial q_j} + Q'_j \tag{8.42}$$

Equation (8.38) becomes

$$\frac{d}{dt}\left(\frac{\partial L}{\partial \dot{q}_j}\right) - \frac{\partial L}{\partial q_j} = Q'_j \tag{8.43}$$

For example, Q'_j may represent nonconservative friction forces or time-dependent driving forces. For *viscous damping*, Rayleigh (1842–1919) suggested a function

$$D = \frac{1}{2} \sum_{j=1}^{m} \sum_{k=1}^{m} C_{jk} \dot{q}_j \dot{q}_k \tag{8.44}$$

called the *Rayleigh dissipation function*, whose partial derivatives would be equal to the friction forces acting and could be added directly to equation (8.41),

$$\frac{d}{dt}\left(\frac{\partial L}{\partial \dot{q}_j}\right) - \frac{\partial L}{\partial q_j} + \frac{\partial D}{\partial \dot{q}_j} = 0 \tag{8.45}$$

Before proceeding to some sample problems, it is best to list the advantages of the results that have been derived. Lagrange's equation, equation (8.41), represents an entirely new formulation of classical mechanics. It is scalar in nature being a function of simply defined kinetic and potential energy terms. It is independent of the coordinate system used, maintaining the same form "generally." If the generalized coordinate used is an angle, moment equations result (even moment equations referenced to arbitrary points); if the coordinate used is a distance, force equations are produced. Since constraint forces were eliminated during the derivation, Lagrange's equations of motion are simpler to use than Newton's equations because fewer unknowns are present. With fewer unknowns to solve for, there are fewer equations of motion to deal with, and therefore, problem algebra is reduced.

Sample Problem 8.3
The two masses m_1 and m_2 shown are free to slide without friction on the horizontal surface. The spring constants are known to be k_1 and k_2. If x_1 and x_2 are measured from the equilibrium positions of the springs, derive Lagrange's equations of motion for the system.

272 ANALYTICAL MECHANICS

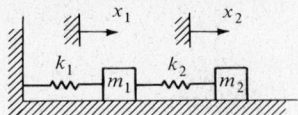

Analysis: The kinetic energy for this 2 degrees of freedom system can be expressed in terms of the generalized coordinates x_1 and x_2 as

$$T = \frac{m_1 \dot{x}_1^2}{2} + \frac{m_2 \dot{x}_2^2}{2}$$

and

$$V = \frac{k_1 x_1^2}{2} + \frac{k_2(x_2 - x_1)^2}{2}$$

Then with

$$L = T - V$$
$$= \frac{m_1 \dot{x}_1^2}{2} + \frac{m_2 \dot{x}_2^2}{2} - \frac{k_1 x_1^2}{2} - \frac{k_2(x_2 - x_1)^2}{2}$$

The Lagrange equation for the x_1 coordinate is given by

$$\frac{d}{dt}\left(\frac{\partial L}{\partial \dot{x}_1}\right) - \frac{\partial L}{\partial x_1} = 0$$

or

$$\frac{d}{dt}(m_1 \dot{x}_1) - [-k_1 x_1 + k_2(x_2 - x_1)] = 0$$

$$m_1 \ddot{x}_1 + k_1 x_1 - k_2(x_2 - x_1) = 0$$

Newton's equation of motion for mass m_1 leads to identical results.

$$F_x = m_1 \ddot{x}_1 \qquad k_2(x_2 - x_1) - k_1 x_1 = m_1 \ddot{x}_1$$

For coordinate x_2,

$$\frac{d}{dt}\left(\frac{\partial L}{\partial \dot{x}_2}\right) - \frac{\partial L}{\partial x_2} = 0$$

or

$$m_2 \ddot{x}_2 + k_2(x_2 - x_1) = 0$$

Suppose now a force $F = A \cos \omega t$ were acting on mass m_1. Find the generalized forces $Q_{x,1}$ and $Q_{x,2}$ as a function of F. Write the new equations of motion for the system.

$$\delta W = Q_{x,1} \delta x_1 + Q_{x,2} \delta x_2 = F \delta x_1 + 0$$

where δx_1 and δx_2 are varied independently. Then $Q_{x,1} = A \cos \omega t$, and $Q_{x,2} = 0$. Now

$$\frac{d}{dt}\left(\frac{\partial L}{\partial \dot{x}_1}\right) - \frac{\partial L}{\partial x_1} = Q_{x,1}$$

or
$$m_1\ddot{x}_1 + k_1 x_1 - k_2(x_2 - x_1) = A \cos \omega t$$

The equation for x_2 remains the same. ∎

Sample Problem 8.4
Derive Lagrange's equation of motion for Sample Problem 8.1. The unstretched length of the spring is l_0. The pendulum has mass m and length l.

Analysis: In terms of the two independent generalized coordinates y and θ, the kinetic energy can be derived as follows:

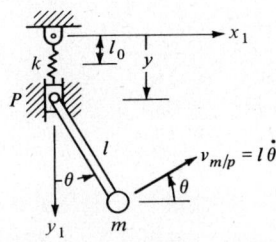

$$\mathbf{v}_m = \mathbf{v}_p + \mathbf{v}_{m/p}$$
$$\mathbf{v}_m = \dot{y}\mathbf{j} + l\dot{\theta}(\cos\theta\mathbf{i} - \sin\theta\mathbf{j})$$
$$\mathbf{v}_m \cdot \mathbf{v}_m = v_m^2 = l^2\dot{\theta}^2 + \dot{y}^2 - 2\dot{y}l\dot{\theta}\sin\theta$$
$$T = \frac{mv_m^2}{2} = \frac{m}{2}(l^2\dot{\theta}^2 + \dot{y}^2 - 2\dot{y}l\dot{\theta}\sin\theta)$$

The potential energy V is
$$V = -mg(y + l\cos\theta) + \frac{k}{2}(y - l_0)^2$$

where l_0 is the unstretched length of the spring.

$$L = \frac{m}{2}(l^2\dot{\theta}^2 + \dot{y}^2 - 2\dot{y}l\dot{\theta}\sin\theta) + mg(y + l\cos\theta) - \frac{k}{2}(y - l_0)^2$$

The Lagrange equation of motion for the y coordinate is
$$\frac{d}{dt}\left(\frac{\partial L}{\partial \dot{y}}\right) - \frac{\partial L}{\partial y} = 0$$

$$\frac{d}{dt}[m(\dot{y} - l\dot{\theta}\sin\theta)] + k(y - l_0) - mg = 0$$

$$m(\ddot{y} - l\ddot{\theta}\sin\theta - l\dot{\theta}^2\cos\theta) + k(y - l_0) - mg = 0 \qquad (a)$$

Newton's equation of motion leads to the same results.

$$F_y = ma_y$$
$$-k(y-l_0) + mg = ma_y = m[\mathbf{a}_p + (\mathbf{a}_{m/p})_n + (\mathbf{a}_{m/p})_t] \cdot \mathbf{j}$$
$$-k(y-l_0) + mg = m(\ddot{y} - l\ddot{\theta}\sin\theta - l\dot{\theta}^2\cos\theta)$$

For the θ coordinate

$$\frac{d}{dt}\left(\frac{\partial L}{\partial \dot{\theta}}\right) - \frac{\partial L}{\partial \theta} = 0$$

$$\frac{d}{dt}m(l^2\dot{\theta} - \dot{y}l\sin\theta) + mgl\sin\theta + m\dot{y}l\dot{\theta}\cos\theta = 0$$

Note that the last term for $\partial L/\partial\theta$ comes from T, the kinetic energy expression. Then

$$m(l^2\ddot{\theta} - \ddot{y}l\sin\theta - \dot{y}l\dot{\theta}\cos\theta) + mgl\sin\theta + m\dot{y}l\dot{\theta}\cos\theta = 0$$

This reduces to

$$ml^2\ddot{\theta} - m\ddot{y}l\sin\theta + mgl\sin\theta = 0 \qquad (b)$$

which is a moment equation. Compare this result with that derived from equation (6.98) for an arbitrary reference point, namely,

$$\mathbf{M}_P = \dot{\mathbf{H}}_P + (\mathbf{R}_{c.m.} - \mathbf{r}_P) \times m\mathbf{a}_P \qquad (6.98)$$

$$mgl\sin\theta\mathbf{k} = -ml^2\ddot{\theta}\mathbf{k} + ml(\cos\theta\mathbf{j} + \sin\theta\mathbf{i}) \times \ddot{y}\mathbf{j}$$

or
$$mgl\sin\theta = -ml^2\ddot{\theta} + m\ddot{y}l\sin\theta$$

Suppose now that θ was constrained to move such that $\theta = \omega t$. There would be only 1 degree of freedom left resulting in just one Lagrange's equation. In terms of the independent coordinate y, T and V become

$$T = \frac{m}{2}(l^2\omega^2 + \dot{y}^2 - 2\dot{y}l\omega\sin\omega t)$$

$$V = -mg(y + l\cos\omega t) + \frac{k}{2}(y - l_0)^2$$

$$\frac{d}{dt}\left(\frac{\partial L}{\partial \dot{y}}\right) - \frac{\partial L}{\partial y} = 0$$

$$\frac{d}{dt}[m(\dot{y} - l\omega\sin\omega t)] + k(y - l_0) - mg = 0$$

$$m(\ddot{y} - l\omega^2\cos\omega t) + k(y - l_0) - mg = 0 \qquad (c)$$

This result is derivable from equation (a) by setting $\ddot{\theta}=0$, $\dot{\theta}=\omega$, and $\theta=\omega t$. However, equation (b) cannot be used by setting $\ddot{\theta}=0$ since there is an unknown external constraint force acting, i.e., a moment here, that is the cause of the constant angular speed. It is clearly seen that independent generalized coordinates must be used to establish the equation of motion. This limitation was established when deriving equation (8.38) from (8.37). It is a wise policy to describe the kinetic and potential energies in terms of independent coordinates as soon as possible in a problem solution. ∎

Sample Problem 8.5

The wheel shown of mass m_1, radius R, and mass moment of inertia $I_{c.m.}$ rolls without slipping on the horizontal floor. Attached to the center of the wheel is a spring having spring constant k that is unstretched when θ equals zero. Wrapped around the wheel is a wire also attached to the mass m_2. The wire does not slip on the wheel. (a) Develop Lagrange's equations for the system. (b) Find the tension in the wire.

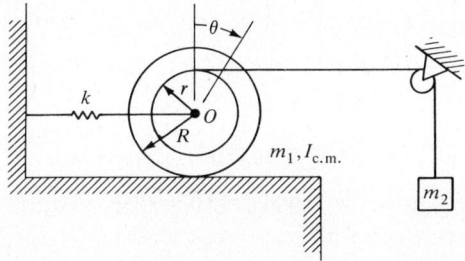

Analysis: There is only 1 degree of freedom present in this problem. θ is selected as the independent generalized coordinate and then the rolling constraint is written in term of $\dot{\theta}$.

$$v_O = \dot{x} = R\dot{\theta}$$

Similarly, the speed of mass m_2 is given by

$$v_2 = (r+R)\dot{\theta}$$

The kinetic energy becomes

$$T = \frac{I_{c.m.}\dot{\theta}^2}{2} + \frac{m_1(R\dot{\theta})^2}{2} + m_2\frac{[(r+R)\dot{\theta}]^2}{2}$$

The potential energy for the system is

$$V = \frac{k}{2}(R\theta)^2 - m_2 g(R+r)\theta$$

Lagrange's equation for the θ coordinate is

$$\frac{d}{dt}\left(\frac{\partial L}{\partial \dot{\theta}}\right) - \frac{\partial L}{\partial \theta} = 0$$

and leads to the answer

$$(I_{c.m.} + m_1 R^2)\ddot{\theta} + m_2(R+r)^2\ddot{\theta} + kR^2\theta - m_2 g(R+r) = 0$$

276 ANALYTICAL MECHANICS

This can be rewritten as

$$(I_{c.m.} + m_1 R^2)\ddot{\theta} = -kR^2\theta + m_2 g(R+r) - m_2(R+r)^2\ddot{\theta}$$

or, with subscript i.c. denoting the instantaneous center,

$$I_{i.c.}\ddot{\theta} = M_{i.c.} - (R+r)m_2 a_2$$

The tension in the wire cannot be found from Lagrange's equation directly. (See the discussion of Lagrange multipliers, Section 8.5.) Instead, returning to Newton's equation of motion for mass m_2, one can write

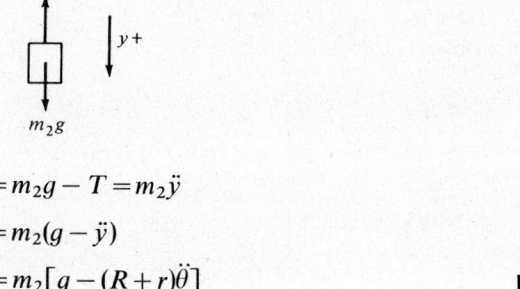

$$F_Y = m_2 g - T = m_2 \ddot{y}$$
$$T = m_2(g - \ddot{y})$$
$$T = m_2[g - (R+r)\ddot{\theta}]\quad\blacksquare$$

Sample Problem 8.6
The thin homogeneous rod of mass m and length l is free to turn in the gimbal. The massless gimbal in turn is free to rotate about the horizontal. Derive Lagrange's equations for the rod.

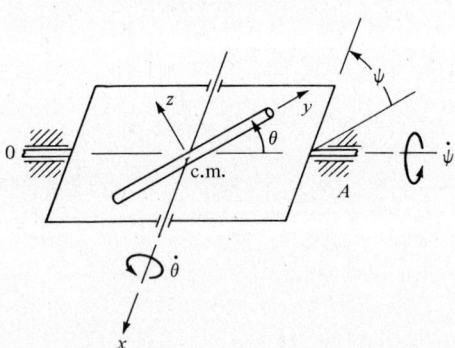

Analysis: In terms of principal moments of inertia, the kinetic energy of a body is given by

$$T = \frac{mv_{c.m.}^2}{2} + \frac{\hat{I}_{xx}\omega_x^2}{2} + \frac{\hat{I}_{yy}\omega_y^2}{2} + \frac{\hat{I}_{zz}\omega_z^2}{2}$$

Here, for the rod and the principal axis shown, $\hat{I}_{yy} \simeq 0$, and $\hat{I}_{xx} = \hat{I}_{zz} = ml^2/12$. Further

and
$$\mathbf{v}_{c.m.} = \mathbf{0}$$
$$\omega = \dot{\theta} + \dot{\psi} = \dot{\theta}\mathbf{i} + \dot{\psi}(\cos\theta\mathbf{j} - \sin\theta\mathbf{k})$$

Then
$$T = \frac{ml^2}{24}(\dot{\theta}^2) + \frac{ml^2}{24}(-\dot{\psi}\sin\theta)^2$$

and
$$V = 0 \qquad L = T$$

For the θ coordinate,
$$\frac{d}{dt}\left(\frac{\partial L}{\partial \dot{\theta}}\right) - \frac{\partial L}{\partial \theta} = 0$$

or
$$\frac{ml^2}{12}\ddot{\theta} - \frac{ml^2}{12}\dot{\psi}^2 \sin\theta \cos\theta = 0 \tag{a}$$

(Compare this result with results from Euler's equations.) For the ψ direction,
$$\frac{d}{dt}\left(\frac{\partial L}{\partial \dot{\psi}}\right) - \frac{\partial L}{\partial \psi} = 0$$

or since $\partial L/\partial \psi = 0$,
$$\frac{\partial L}{\partial \dot{\psi}} = \text{constant} = \left(\frac{ml^2}{12}\sin^2\theta\right)\dot{\psi}$$
$$= I_{\psi\psi}\dot{\psi} = H_\psi \tag{b}$$

The angular momentum about the ψ direction is seen to be constant. Rewriting equation (a) in terms of (b) yields
$$\frac{ml^2}{12}\ddot{\theta} - \frac{12 H_\psi^2 \cos\theta}{ml^2 \sin^3\theta} = 0 \tag{c}$$

■

Before proceeding to the next sample problem a discussion of the results of Sample Problem 8.6 will prove informative.

The partial derivative of the lagrangian with respect to the coordinate velocity is called the *conjugate*, or *canonical*, or *generalized momentum* p_j.

$$\frac{\partial L}{\partial \dot{q}_j} = \bar{p}_j \tag{8.46}$$

It is seen that for a conservative system, if the lagrangian does not contain a given coordinate but still has the coordinate velocity represented (degree of freedom), the *canonical momentum* for that coordinate is constant. Such a coordinate is called a *cyclic coordinate*. A formal proof shows

$$\frac{d}{dt}\left(\frac{\partial L}{\partial \dot{q}_j}\right) - \frac{\partial L}{\partial q_j} = 0$$

$$\frac{d}{dt}(p_j) = \frac{\partial L}{\partial q_j}$$

and for a cyclic coordinate $\partial L/\partial q_j = 0$; therefore, $p_j = $ constant.

Sample Problem 8.7
Referring to Sample Problem 8.6, assume now that $\dot{\psi}$ = constant. Write a Fortran program to numerically solve the new differential equation of motion. Use a Runge-Kutta fourth-order numerical integration method. For this problem take $m = 1$ kg, $x = 10$ cm, and $\dot{\psi} = 10$ rpm. At $t = 0$, $\dot{\theta}_0 = 0$ and $\theta_0 = 30°$.

Analysis: Equation (a) of Sample Problem 8.6 does not change. Now $\dot{\psi}$ is constant, and therefore equations (b) and (c) are no longer valid. Equation (a) rewritten here is

$$\frac{ml^2}{12}\ddot{\theta} - \frac{ml^2}{12}\dot{\psi}^2 \sin\theta \cos\theta = 0$$

The Runge-Kutta program used to solve this problem is shown in Figure 8.8. See Section 2.8. ■

Sample Problem 8.8
Write the lagrangian for the case of the spinning heavy top. Identify the cyclic coordinates present and the corresponding conserved momentums.

Analysis: With one point fixed

$$T = \frac{I_{xx}\omega_x^2}{2} + \frac{I_{yy}\omega_y^2}{2} + \frac{I_{zz}\omega_z^2}{2}$$

where the mass moments of inertia are referenced to the fixed point.

With the z axis taken as the spin axis and with $I_{xx} = I_{yy}$ for a body of revolution, this becomes

$$T = \frac{I_{xx}}{2}(\omega_x^2 + \omega_y^2) + \frac{I_{zz}}{2}\omega_z^2$$

In terms of Euler angles

$$\omega_x = \dot{\theta}\cos\psi + \dot{\phi}\sin\theta\sin\psi$$
$$\omega_y = -\dot{\theta}\sin\psi + \dot{\phi}\sin\theta\cos\psi$$
$$\omega_z = \dot{\psi} + \dot{\phi}\cos\theta$$

$$T = \frac{I_{xx}}{2}(\dot{\theta}^2 + \dot{\phi}^2\sin^2\theta) + \frac{I_{zz}}{2}(\dot{\psi} + \dot{\phi}\cos\theta)^2$$

$$V = mgl\cos\theta$$

$$L = \frac{I_{xx}}{2}(\dot{\theta}^2 + \dot{\phi}^2\sin^2\theta) + \frac{I_{zz}}{2}(\dot{\psi} + \dot{\phi}\cos\theta)^2 - mgl\cos\theta$$

The cyclic coordinates are ϕ and ψ; therefore,

$$\frac{\partial L}{\partial \dot{\phi}} = p_\phi = I_{xx}\dot{\phi}\sin^2\theta + I_{zz}(\dot{\psi} + \dot{\phi}\cos\theta)\cos\theta = \text{constant}$$

```
00100 PROGRAM ROD(INPUT,OUTPUT)
00110 REAL K1,K2,K3,K4
00120 DDTH(X)=PSID*PSID*SIN(X)*COS(X)
00130 DATA PSID,TH,TFIN,N,DTH/10.,30.,2.,25,0./
00140 T=0.
00150 TH=TH/57.296
00160 PRINT 9,
00170 9 FORMAT(9X,"T",16X,"TH",15X,"DTH")
00180 PRINT 10,T,TH,DTH
00190 10 FORMAT(3X,F7.3,2(10X,F8.4))
00200 DT=TFIN/FLOAT(N)
00210 PSID=PSID*6.283/60.
00220 1 K1=DT*DDTH(TH)
00230 K2=DT*DDTH(TH+DT*DTH/2.)
00240 K3=DT*DDTH(TH+DT*DTH/2.+DT*K1/4.)
00250 K4=DT*DDTH(TH+DT*DTH+DT*K2/2.)
00260 TH=TH+DT*DTH+DT*(K1+K2+K3)/6.
00270 DTH=DTH+(K1+2.*(K2+K3)+K4)/6.
00280 T=T+DT
00290 IF(T.GT.TFIN) GO TO 2
00300 PRINT 10,T,TH,DTH
00310 GO TO 1
00320 2 STOP
00330 END
```

(a)

T	TH	DTH
0.000	.5236	0.0000
.080	.5251	.0380
.160	.5297	.0761
.240	.5373	.1145
.320	.5480	.1533
.400	.5619	.1926
.480	.5789	.2325
.560	.5991	.2730
.640	.6225	.3142
⋮	⋮	⋮
1.520	1.1061	.7760
1.600	1.1696	.8094
1.680	1.2355	.8389
1.760	1.3037	.8637
1.840	1.3736	.8834
1.920	1.4448	.8973
2.000	1.5170	.9051

(b)

Figure 8.8

280 ANALYTICAL MECHANICS

$$\frac{\partial L}{\partial \dot\psi} = p_\psi = I_{zz}(\dot\psi + \dot\phi \cos\theta) = \text{constant}$$

Note that

$$p_\psi = H_z \quad \text{and} \quad p_\phi = H_Z$$

For the θ coordinate,

$$\frac{d}{dt}\left(\frac{\partial L}{\partial \dot\theta}\right) - \frac{\partial L}{\partial \theta} = 0$$

or
$$I_{xx}(\ddot\theta - \dot\phi^2 \sin\theta \cos\theta) + I_{zz}(\dot\psi + \dot\phi \cos\theta)\dot\phi \sin\theta - mgl \sin\theta = 0$$

Compare these results with those obtained previously in Section 7.7. ∎

Sample Problem 8.9

Write a Fortran program that uses Euler's method to numerically integrate the equations of motion found in Sample Problem 8.8. At $t=0$, $\theta_0 = \pi/2$, $\dot\theta_0 = 0$, and $\dot\psi_0 = 2000$ rpm. It is also observed that the top passes through the angle $\theta = \pi$. $I_{zz} = 1.25 \times 10^{-5}$ kg·m² $I_{xx} = 23.1 \times 10^{-5}$ kg·m², and $mgl = 0.07$ N·m.

Analysis: To integrate the equations, $\dot\phi_0$ is needed. It is known that p_ϕ and p_ψ are constant. Therefore, with the given conditions at $t = 0$,

$$p_\phi = I_{xx}\dot\phi_0 \qquad p_\psi = I_{zz}\dot\psi_0$$

When $\theta = \pi$,

$$p_\phi = -p_\psi \quad \text{or} \quad \dot\phi_0 = \frac{-I_{zz}\dot\psi_0}{I_{xx}}$$

Further

$$p_\psi = I_{zz}(\dot\psi + \dot\phi \cos\theta) = I_{zz}\dot\psi_0$$

$$p_\phi = -I_{zz}\dot\psi_0$$

$$\dot\phi = \frac{-I_{zz}\dot\psi_0}{I_{xx}} \frac{1 + \cos\theta}{\sin^2\theta} \tag{a}$$

Equation (a) does not hold at $\theta = \pi$; instead one can write

$$\dot\phi = \frac{-I_{zz}\dot\psi_0}{I_{xx}} \frac{1}{1 - \cos\theta} = \frac{\dot\phi_0}{2} \tag{a'}$$

The equation for $\ddot\theta$ is

$$\ddot\theta = \left(\dot\phi \cos\theta - \frac{I_{zz}\dot\psi_0}{I_{xx}}\right)\dot\phi \sin\theta + \frac{mgl \sin\theta}{I_{xx}} \tag{b}$$

```
00100 PROGRAM TOP(INPUT,OUTPUT)
00110 REAL IXX,IZZ,MO
00120 DIMENSION Z(4)
00130 FDPH(TH)=-A/(1.-COS(TH))
00140 FDDTH(DPH,TH)=(DPH*COS(TH)-A)*DPH*SIN(TH)+MO*SIN(TH)/IXX
00150 DATA MO,IXX,IZZ,DPSI,TH,DTH,DT,TFIN/.07,23.1E-5,1.25E-5,
00160 12000.,,90.,,0.,,005,10./
00170 B=DPSI*0.1047
00180 A=IZZ*B/IXX
00190 TH=TH/57.29578
00200 PRINT 20,
00210 20 FORMAT(9X,"T",15X,"TH",14X,"DTH",14X,"DPH")
00220 T=0.
00230 Z(1)=TH
00240 Z(2)=DTH
00250 5 Z(4)=FDPH(Z(1))
00260 Z(3)=FDDTH(Z(4),Z(1))
00270 PRINT 21,T,Z(1),Z(2),Z(4)
00280 21 FORMAT(3X,F7.3,3(3X,E14.7))
00290 T=T+DT
00300 IF(Z(2).EQ.0.) GO TO 7
00310 DO 3 I=1,2
00320 Z(I)=Z(1)+Z(I+1)*DT
00330 3 CONTINUE
00340 IF(Z(1).GT.3.1415.OR.T.GT.TFIN) GO TO 11
00350 GO TO 5
00360 7 Z(2)=FDDTH(Z(4),Z(1))*DT/2.
00370 Z(1)=Z(1)+Z(2)*DT
00380 GO TO 5
00390 11 STOP
00400 END                    (a)
```

T	TH	DTH	DPH
0.000	.1570796E+01	0.	-.1133117E+02
.005	.1576189E+01	.1078564E+01	-.1127039E+02
.010	.1581382E+01	.3228793E+01	-.1121026E+02
.015	.1597726E+01	.5372170E+01	-.1103406E+02
.020	.1624587E+01	.7495304E+01	-.1073304E+02
.025	.1662063E+01	.9585567E+01	-.1038470E+02
⋮	⋮	⋮	⋮
.080	.2688638E+01	.2716586E+02	.5966417E+01
.085	.2824467E+01	.2790682E+02	-.5810451E+01
.090	.2964001E+01	.2843194E+02	-.5710492E+01
.095	.3106161E+01	.2872841E+02	-.5667363E+01

(b)

Figure 8.9

8.5 LAGRANGE MULTIPLIERS

The question of how to treat nonholonomic constraints will now be dealt with. The method described is equally effective for holonomic systems when knowledge of the acting constraint forces is desired. A nonholonomic constraint is defined by

$$\sum_{i=1}^{n} a_{si}\, dq_i + a_{st}\, dt = 0 \qquad (s = 1, 2, \ldots, k) \tag{8.3}$$

For virtual displacements only, this can be rewritten as

$$\sum_{i=1}^{n} a_{si}\, \delta q_i = 0 \tag{8.47}$$

All the n generalized coordinates are not independent because of the presence of the k constraint equations, each having the form given by equation (8.3). Since they are not independent, equation (8.37) cannot be reduced further to yield Lagrange's equation. Equation (8.37) rewritten is

$$\sum_{i=1}^{n} Q_i\, \delta q_i = \sum_{i=1}^{n} \left[\frac{d}{dt}\left(\frac{\partial T}{\partial \dot q_i}\right) - \frac{\partial T}{\partial q_i} \right] \delta q_i \tag{8.37}$$

Introducing the factor λ_s, the *Lagrange multiplier*, one can write

$$\lambda_s \sum_{i=1}^{n} a_{si}\, \delta q_i = 0 \tag{8.48}$$

and so

$$\sum_{s=1}^{k} \sum_{i=1}^{n} \lambda_s a_{si}\, \delta q_i = 0 \tag{8.49}$$

Subtracting this zero quantity from equation (8.37) yields

$$\sum_{i=1}^{n} Q_i\, \delta q_i = \sum_{i=1}^{n} \left[\frac{d}{dt}\left(\frac{\partial T}{\partial \dot q_i}\right) - \frac{\partial T}{\partial q_i} - \sum_{s=1}^{k} \lambda_s a_{si} \right] \delta q_i \tag{8.50}$$

For a conservative system equation (8.50) can be changed to

$$\sum_{i=1}^{n} \left[\frac{d}{dt}\left(\frac{\partial L}{\partial \dot q_i}\right) - \frac{\partial L}{\partial q_i} \right] \delta q_i = \sum_{i=1}^{n} \left(\sum_{s=1}^{k} \lambda_s a_{si} \right) \delta q_i \tag{8.51}$$

The k values of λ_s are chosen now so that for k coordinates,

$$\frac{d}{dt}\left(\frac{\partial L}{\partial \dot q_i}\right) - \frac{\partial L}{\partial q_i} = \sum_{s=1}^{k} \lambda_s a_{si} \qquad (i = 1, 2, \ldots, k) \tag{8.52}$$

These k relations between the coordinates can be used to reduce equation (8.51) to

$$\sum_{i=k+1}^{n}\left[\frac{d}{dt}\left(\frac{\partial L}{\partial \dot{q}_i}\right)-\frac{\partial L}{\partial q_i}\right]\delta q_i = \sum_{i=k+1}^{n}\left(\sum_{s=1}^{k}\lambda_s a_{si}\right)\delta q_i \qquad (8.53)$$

where now all the coordinates present are independent. Since they are independent,

$$\frac{d}{dt}\left(\frac{\partial L}{\partial \dot{q}_i}\right)-\frac{\partial L}{\partial q_i} = \left(\sum_{s=1}^{k}\lambda_s a_{si}\right) \qquad (i=k+1,\ldots,n) \qquad (8.54)$$

Then by virtue of equations (8.52) and (8.54), with properly selected λ_s values,

$$\frac{d}{dt}\left(\frac{\partial L}{\partial \dot{q}_i}\right)-\frac{\partial L}{\partial q_i} = \sum_{s=1}^{k}\lambda_s a_{si} \qquad (i=1,2,\ldots,n) \qquad (8.55)$$

These are n equations for the $n+k$ unknowns. (There are still k unknown values of λ_s.) By referring now to equation (8.3) rewritten as

$$\sum_{i=1}^{n} a_{si}\dot{q}_i + a_{st} = 0 \qquad (s=1,2,\ldots,k) \qquad (8.56)$$

the additional k equations to completely define the problem are available. The quantity

$$\sum_{s=1}^{k}\lambda_s a_{si} = C_i \qquad (8.57)$$

is called the *generalized constraint force*. Equation (8.55) becomes then

$$\frac{d}{dt}\left(\frac{\partial L}{\partial \dot{q}_i}\right)-\frac{\partial L}{\partial q_i} = C_i \qquad (i=1,2,\ldots,n) \qquad (8.58)$$

The resulting equation can be interpreted as describing an n-degrees-of-freedom problem having generalized constraint forces C_i steering the motion so that the geometric constraints of the system are met. This suggests an approach whereby Lagrange multipliers can be used to solve for unknown constraint forces. First the constrained coordinate is "freed," then Lagrange's equations are written for all the free coordinates using equation (8.55). Finally by applying equations (8.56) and (8.57), the constraint forces acting can be solved for.

Sample Problem 8.10
Referring to Sample Problem 8.5, use the method of Lagrange multipliers to solve for the constraint force acting in the wire.

Analysis: The constrained coordinate of interest here is y. The equation of constraint is

$$\dot{y} - (R+r)\dot{\theta} = 0$$

Since this is only one constraint equation, s in equation (8.56) has only one value here, unity.

$$a_{s1} = a_{11} = 1 \qquad q_1 = y$$
$$a_{12} = -(R+r) \qquad q_2 = \theta$$
$$a_{1t} = 0$$

With the y coordinate freed, L becomes

$$L = (I_{\text{c.m.}} + m_1 R^2)\frac{\dot{\theta}^2}{2} + \frac{m_2 \dot{y}^2}{2} - \frac{k(R\theta)^2}{2} + m_2 g y$$

$$\frac{d}{dt}\left(\frac{\partial L}{\partial \dot{y}}\right) - \frac{\partial L}{\partial y} = \lambda_1 a_{11} = \lambda_1$$

or
$$m_2 \ddot{y} - m_2 g = \lambda_1 = C_1 \tag{a}$$

Similarly

$$\frac{d}{dt}\left(\frac{\partial L}{\partial \dot{\theta}}\right) - \frac{\partial L}{\partial \theta} = \lambda_1 a_{12} = -\lambda_1 (R+r)$$

or
$$(I_{\text{c.m.}} + m_1 R^2)\ddot{\theta} + kR^2\theta = -\lambda_1(R+r) = C_2 \tag{b}$$

Finally, applying the constraint equation

$$\dot{y} - (R+r)\dot{\theta} = 0 \tag{c}$$

rewritten as

$$\ddot{y} = (R+r)\ddot{\theta}$$

Equations (a) and (b) become

$$m_2(R+r)\ddot{\theta} - m_2 g = C_1 \tag{d}$$
$$(I_{\text{c.m.}} + m_1 R^2)\ddot{\theta} + kR^2\theta = -C_1(R+r) = C_2 \tag{e}$$

C_1 clearly is equal in magnitude to the tension acting in the wire. The sign discrepancy comes from the selection of y positive downward. A positive C_1 must then also act downward, but this is in the opposite direction of the acting tension force.

Note that by eliminating C_1 from equations (d) and (e), the results obtained in Sample Problem 8.5 are duplicated.

The reader can now solve for the friction force acting by freeing the coordinate x and using a different Lagrange multiplier. ∎

8.6 THE ROUTHIAN

It was seen in Section 8.4 that for a conservative system the generalized momenta for cyclic coordinates were constant. Routh introduced a means of eliminating the associated coordinate velocities of the cyclic coordinates from the lagrangian. He suggested that constants representing the generalized momenta be substituted for the coordinate velocities in a new function called the routhian.

ENERGY METHODS 285

The subsequent equations of motion derived from the routhian are fewer in number and have the constant momentum terms already introduced into the equations. No subsequent substitutions are necessary. The routhian is defined by

$$R(q_1,\ldots,q_s,\dot{q}_1,\ldots,\dot{q}_s,p_{s+1},\ldots,p_m,t)=\sum_{j=s+1}^{m}p_j\dot{q}_j-L \qquad (8.59)$$

with $m-s$ cyclic coordinates present. Routh's equations for the s remaining coordinates are

$$\frac{d}{dt}\left(\frac{\partial R}{\partial \dot{q}_j}\right)-\frac{\partial R}{\partial q_j}=0 \qquad (j=1,2,\ldots,s) \qquad (8.60)$$

This can be seen from equation (8.59) since for $j=1,2,\ldots,s$,

$$\frac{\partial R}{\partial \dot{q}_j}=-\frac{\partial L}{\partial \dot{q}_j} \qquad \frac{\partial R}{\partial q_j}=-\frac{\partial L}{\partial q_j}$$

Sample Problem 8.11
For Sample Problem 8.6, write the routhian function and use it to derive the equations of motion.

Analysis:

$$L=\frac{ml^2}{24}\dot{\theta}^2+\frac{ml^2}{24}\dot{\psi}^2\sin^2\theta$$

The ψ coordinate is cyclic; hence,

$$p_\psi=\frac{ml^2}{12}\dot{\psi}\sin^2\theta=\text{constant} \qquad (a)$$

$$R=p_\psi\dot{\psi}-L$$

where from equation (a)

$$\dot{\psi}=\frac{12p_\psi}{ml^2\sin^2\theta}$$

so

$$R=\frac{12p_\psi^2}{ml^2\sin^2\theta}-\frac{ml^2\dot{\theta}^2}{24}-\frac{6p_\psi^2}{ml^2\sin^2\theta}$$

$$=-\frac{ml^2\dot{\theta}^2}{24}+\frac{6p_\psi^2}{ml^2\sin^2\theta} \qquad (b)$$

The variable $\dot{\psi}$ does not appear in the routhian, equation (b). Then

$$\frac{d}{dt}\left(\frac{\partial R}{\partial \dot{\theta}}\right)-\frac{\partial R}{\partial \theta}=0$$

286 ANALYTICAL MECHANICS

becomes

$$-\frac{ml^2\ddot{\theta}}{12} + \frac{12p_\psi^2 \cos\theta}{ml^2 \sin^3\theta} = 0$$

$$\frac{ml^2\ddot{\theta}}{12} - \frac{12p_\psi^2 \cos\theta}{ml^2 \sin^3\theta} = 0 \qquad (c)$$

Compare this result with equation (c) of Sample Problem 8.6.

The reader might note what *cannot* be done. It is incorrect to eliminate $\dot{\psi}$ in the lagrangian using the generalized momentum and then develop Lagrange's equation. One needs the routhian. To see this, take

$$\dot{\psi} = \frac{12p_\psi}{ml^2 \sin^2\theta}$$

substitute this result *directly* into the lagrangian

$$L = \frac{ml^2\dot{\theta}^2}{24} + \frac{6p_\psi^2}{ml^2 \sin^2\theta}$$

Then Lagrange's equation becomes

$$\frac{d}{dt}\left(\frac{\partial L}{\partial \dot{\theta}}\right) - \frac{\partial L}{\partial \theta} = 0$$

$$\frac{ml^2\ddot{\theta}}{12} + \frac{12p_\psi^2 \cos\theta}{ml^2 \sin^3\theta} = 0 \qquad (d)$$

Note the *incorrect* sign in equation (d). ∎

Sample Problem 8.12: The Two-Body Problem

Consider the motion of two spherical masses m_1 and m_2 free to move under the influence of their own mutual gravitational attraction. Develop an expression for the lagrangian and routhian functions for the system in terms of the relative positive vector from mass m_1 to mass m_2. Integrate the equations of motion for r in terms of θ to develop the orbit equation. What type of curves develop? Find an expression for the orbital period.

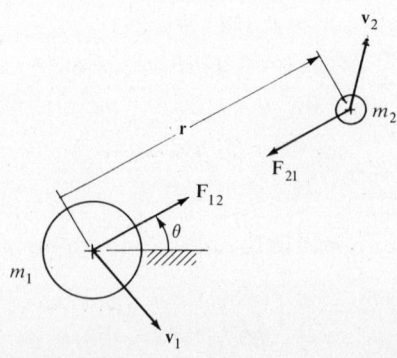

ENERGY METHODS 287

Analysis: The mass center of the system must move at constant speed since there are no external forces assumed acting. With the center of mass as reference point,

$$\mathbf{r}_1 = \mathbf{R}_{c.m.} + \boldsymbol{\rho}_1 \qquad (1)$$

$$\mathbf{r}_2 = \mathbf{R}_{c.m.} + \boldsymbol{\rho}_2 \qquad (2)$$

where
$$\mathbf{R}_{c.m.} = \frac{\mathbf{r}_1 m_1 + \mathbf{r}_2 m_2}{m_1 + m_2} \qquad (3)$$

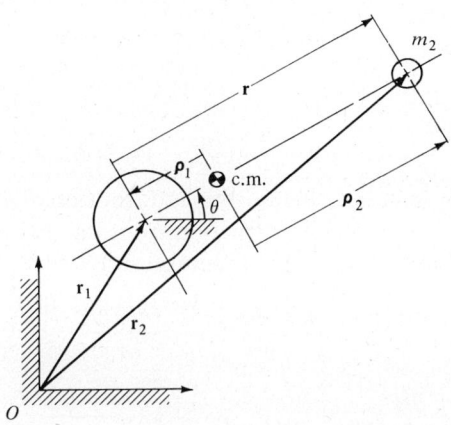

Also
$$0 = m_1 \boldsymbol{\rho}_1 + m_2 \boldsymbol{\rho}_2 \qquad (4)$$

Since
$$\mathbf{r} = \boldsymbol{\rho}_2 - \boldsymbol{\rho}_1 \qquad (5)$$

then
$$\boldsymbol{\rho}_1 = -\frac{m_2 \mathbf{r}}{m_1 + m_2} \qquad (6)$$

$$\boldsymbol{\rho}_2 = \frac{m_1 \mathbf{r}}{m_1 + m_2} \qquad (7)$$

From equations (1) and (2)

$$\frac{m_1 v_1^2}{2} = \frac{m_1}{2} \mathbf{v}_1 \cdot \mathbf{v}_1 = \frac{m_1}{2} (v_{c.m.}^2 + \dot{\boldsymbol{\rho}}_1 \cdot \dot{\boldsymbol{\rho}}_1 + 2 \mathbf{v}_{c.m.} \cdot \dot{\boldsymbol{\rho}}_1) \qquad (8)$$

$$\frac{m_2 v_2^2}{2} = \frac{m_2}{2} (v_{c.m.}^2 + \dot{\boldsymbol{\rho}}_2 \cdot \dot{\boldsymbol{\rho}}_2 + 2 \mathbf{v}_{c.m.} \cdot \dot{\boldsymbol{\rho}}_2) \qquad (9)$$

$$T = \frac{m_1 + m_2}{2} v_{c.m.}^2 + \frac{m_1 m_2^2}{2(m_1 + m_2)^2} \dot{\mathbf{r}} \cdot \dot{\mathbf{r}} + \frac{m_2 m_1^2}{2(m_1 + m_2)^2} \dot{\mathbf{r}} \cdot \dot{\mathbf{r}} \qquad (10)$$

$$= \frac{m_1 + m_2}{2} v_{c.m.}^2 + \frac{m_1 m_2}{2(m_1 + m_2)} (\dot{r}^2 + r^2 \dot{\theta}^2) \qquad (11)$$

Further, defining the *reduced mass* as

$$m_2' = \frac{m_1 m_2}{m_1 + m_2} \tag{12}$$

and the *increased mass* as

$$m_1' = m_1 + m_2 \tag{13}$$

equation (11) can be written as

$$T = \frac{m_1'}{2} v_{c.m.}^2 + \frac{m_2'}{2} (\dot{r}^2 + r^2 \dot{\theta}^2) \tag{14}$$

Note that if $m_1 \gg m_2$,

$$m_2' \sim m_2 \quad \text{and} \quad m_1' \sim m_1 \tag{15}$$

which would hold, for instance, in describing the motion of a satellite around the earth. (The effects of gravitational attraction from the other bodies in the solar system are neglected in the two-body assumption.) With

$$V = -\frac{G m_1 m_2}{r} = -\frac{G m_1' m_2'}{r} \tag{16}$$

then

$$L = \frac{m_1'}{2} v_{c.m.}^2 + \frac{m_2'}{2} (\dot{r}^2 + r^2 \dot{\theta}^2) + \frac{G m_1' m_2'}{r} \tag{17}$$

From this it can be seen that $v_{c.m.}$ is constant and can be ignored in the lagrangian. Rewriting equation (17), L becomes

$$L = \frac{m_2'}{2} (\dot{r}^2 + r^2 \dot{\theta}^2) + \frac{G m_1' m_2'}{r} \tag{18}$$

It is also observed that the θ coordinate is cyclic,

$$\frac{\partial L}{\partial \dot{\theta}} = m_2' r^2 \dot{\theta} = p_\theta = \text{constant} \tag{19}$$

Equation (19) states that the angular momentum of the system is constant. (This is a *central-force* problem; the motion must be planar.) In a slightly different form

$$\frac{p_\theta}{2 m_2'} = \frac{r^2 \dot{\theta}}{2} = \text{constant} = \dot{A} \quad \text{(areal velocity)} \tag{20}$$

The astronomer Johann Kepler (1571–1630) described this result before Newton in his famous three *laws of planetary motion*. The second law states that the radius vector from the sun to a planet sweeps out equal areas in equal units of time.

The routhian for this problem is

$$R = p_\theta \dot{\theta} - L \tag{21}$$

or
$$R = \frac{p_\theta^2}{m_2' r^2} - \frac{m_2' \dot{r}^2}{2} - \frac{p_\theta^2}{2 m_2' r^2} - \frac{G m_1' m_2'}{r} \tag{22}$$

$$= \frac{p_\theta^2}{2 m_2' r^2} - \frac{m_2' \dot{r}^2}{2} - \frac{G m_1' m_2'}{r} \tag{23}$$

The equation of motion for the r coordinate is

$$\frac{d}{dt}\left(\frac{\partial R}{\partial \dot{r}}\right) - \frac{\partial R}{\partial r} = 0 \tag{24}$$

$$-m_2' \ddot{r} + \frac{p_\theta^2}{m_2' r^3} - \frac{G m_1' m_2'}{r^2} = 0 \tag{25}$$

or
$$m_2' \ddot{r} - \frac{p_\theta^2}{m_2' r^3} = -\frac{G m_1' m_2'}{r^2} \tag{26}$$

Replacing r by $1/u$,

$$\dot{r} = \frac{dr}{dt} = \dot{\theta}\frac{d}{d\theta}\left(\frac{1}{u}\right) = -\frac{\dot{\theta}}{u^2}\frac{du}{d\theta} = -\frac{p_\theta}{m_2'}\frac{du}{d\theta} \tag{27}$$

$$\ddot{r} = \frac{d^2 r}{dt^2} = -\left(\frac{p_\theta u}{m_2'}\right)^2 \frac{d^2 u}{d\theta^2} \tag{28}$$

Equation (26) can be rewritten now as a function of θ:

$$-\frac{p_\theta^2 u^2}{m_2'}\frac{d^2 u}{d\theta^2} - \frac{p_\theta^2}{m_2'} u^3 = -G m_1' m_2' u^2 \tag{29}$$

$$\frac{d^2 u}{d\theta^2} + u = G m_1' \left(\frac{m_2'}{p_\theta}\right)^2 \tag{30}$$

Defining a new term

$$h = \frac{p_\theta}{m_2'} \tag{31}$$

then
$$\frac{d^2 u}{d\theta^2} + u = \frac{G m_1'}{h^2} \tag{32}$$

The solution to this equation with θ as the independent variable is

$$\frac{1}{r} = u = C \cos(\theta + \theta_0) + \frac{G m_1'}{h^2} \tag{33}$$

$$r = \frac{h^2 / G m_2'}{1 + (C h^2 / G m_1') \cos(\theta + \theta_0)} \tag{34}$$

Recognizing this equation to be that for a *conic section* with eccentricity ε, one can rewrite this as

$$r = \frac{h^2 / G m_1'}{1 + \varepsilon \cos(\theta + \theta_0)} \tag{35}$$

where
$$\varepsilon = \frac{Ch^3}{Gm'_1} \tag{36}$$

It is usually assumed that $\theta_0 = 0$ at the point of closest approach of body 2 relative to body 1. Then

$$r = \frac{h^2/Gm'_1}{1 + \varepsilon \cos \theta} \tag{37}$$

Equation (37) is a restatement of Kepler's first law in which he stated that the planets trace elliptic orbits about the sun. The closest distance to the sun, when θ is zero, is called the *perihelion*, and the farthest distance is called the *aphelion*, for $\theta = 180°$. Note that these two points are the turning points in the orbit where $\dot{r} = 0$.

Equation (37) represents a conic section with the attracting body located at the focus. For an ellipse with two foci, a negative eccentricity locates the closest distance of approach at $\theta = 180°$; hence, the attracting body must be at the other focus.

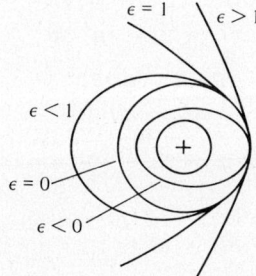

For $\varepsilon > 1$, the path described is a *hyperbola* having asymptotes at $\theta = \pm \cos^{-1}(-1/\varepsilon)$.
For $\varepsilon = 1$, the path is a *parabola* with an asymptote at $\theta = \pm 180°$.
For $\varepsilon < 1$, the path is an *ellipse*.
For $\varepsilon = 0$, the path is a *circular*.

The total energy (excluding the velocity of the center of mass for the two bodies) of the system is constant and is equal to

$$E = T + V = \frac{m'_2 v^2}{2} - \frac{Gm'_1 m'_2}{r} \tag{38}$$

The energy per unit mass e is defined to be

$$e = \frac{E}{m'_2} \tag{39}$$

It can be shown that

$$e = \mp \frac{Gm'_1}{2a} \tag{40}$$

for the motion in which a is the semimajor axis of an elliptic or hyperbolic path. The energy e equals zero for a parabolic path. For an elliptic or circular orbit, the total energy is negative. For a hyperbolic orbit, it is positive. The *vis viva* integral can be written for the velocity as

$$v^2 = Gm'_1 \left(\frac{2}{r} \mp \frac{1}{a}\right) \tag{41}$$

For a *circular orbit* of radius r_c,

$$F_n = \frac{Gm'_1 m'_2}{r_c^2} = m'_2 a_n = \frac{m'_2 v_c^2}{r_c} \tag{42}$$

or

$$v_c = \sqrt{\frac{Gm'_1}{r_c}} \tag{43}$$

Since at the surface of m'_1 with an assumed radius R and a local acceleration of gravity g the gravitational attractive force is

$$F = \frac{Gm'_1 m'_2}{R^2} = m'_2 g \tag{44}$$

one can write

$$Gm'_1 = gR^2 \tag{45}$$

or

$$v_c = \sqrt{\frac{gR^2}{r_c}} \tag{46}$$

For the *escape velocity* necessary to leave the gravity field of mass m'_1 one can obtain

$$v_{esc} = \sqrt{\frac{2gR^2}{r_p}} \tag{47}$$

where r_p is the distance of closest approach to m'_1. The escape velocity for mass m'_2 results in a parabolic trajectory.

The period T for an elliptic orbit is found from

$$\text{Area of the ellipse} = \dot{A}T = \pi ab \tag{48}$$

where b is the semiminor axis of the ellipse. For an ellipse one can show that

$$b = a\sqrt{1 - \varepsilon^2} \tag{49}$$

so

$$T = \frac{2\pi a^2 \sqrt{1 - \varepsilon^2}}{h} \tag{50}$$

Further, with $2a = r_p + r_a$,

$$2a = \frac{h^2}{Gm'_1} \left(\frac{1}{1 + \varepsilon} + \frac{1}{1 - \varepsilon}\right)$$

or

$$h = \sqrt{aGm'_1(1 - \varepsilon^2)} \tag{51}$$

Then
$$T = 2\pi \sqrt{\frac{a^3}{Gm'_1}} \tag{52}$$

Equation (52) is the last of Kepler's laws. It states that the squares of the periods of the planets are proportional to the cubes of the semimajor axes of their orbits in their motion about the sun.

$$\left(\frac{T_i}{T_j}\right)^2 = \left(\frac{a_i}{a_j}\right)^3 \tag{53}$$

If one integrates θ as a function of t, one can show that the time to travel from $\theta_0 = 0$ to θ in an elliptic orbit is

$$t = \frac{a^{3/2}}{\sqrt{Gm'_1}} \left[2 \tan^{-1} \sqrt{\frac{1-\varepsilon}{1+\varepsilon}} \left(\tan \frac{\theta}{2} \right) - \frac{\varepsilon \sqrt{1-\varepsilon^2} \sin \theta}{1 + \varepsilon \cos \theta} \right] \tag{54}$$

For a hyperbolic orbit,

$$t = \frac{a^{3/2}}{\sqrt{Gm'_1}} \left[\frac{\varepsilon \sqrt{\varepsilon^2 - 1} \sin \theta}{1 + \varepsilon \cos \theta} - \ln \frac{\sqrt{\varepsilon+1} + \sqrt{\varepsilon-1} \tan(\theta/2)}{\sqrt{\varepsilon+1} - \sqrt{\varepsilon-1} \tan(\theta/2)} \right] \tag{55}$$

and for a parabolic orbit,

$$t = r_p^{3/2} \sqrt{\frac{2}{Gm'_1}} \left(\tan \frac{\theta}{2} + \frac{1}{3} \tan^3 \frac{\theta}{2} \right) \tag{56}$$

For more information, the reader is referred to the many fine reference texts on the subject of orbital mechanics that are readily available. ∎

Sample Problem 8.13
Two earth satellites are injected into orbit at the same place and at the same time. Satellite A is injected into a 24-h equatorial circular orbit (staying over the same location on earth). Satellite B has the same speed as A initially but has its velocity pointed outward at a 45° angle. It too travels in an equatorial plane. Write a computer program that will numerically solve for the motion of both satellites as a function of time. Reference measurements to a fixed direction in space passing through the orbit's perigee point.

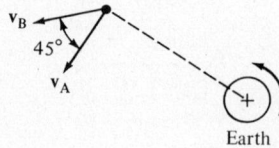
Earth

Analysis: The motion can be depicted as shown in the figure.

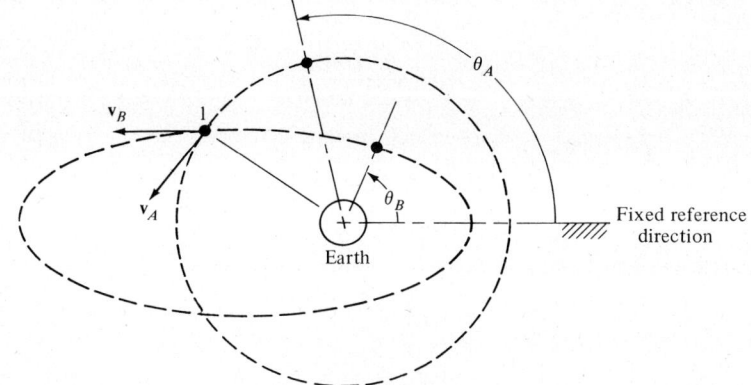

From Sample Problem 8.12, equation (46),

$$v_c = \sqrt{\frac{gR^2}{r_c}} \tag{a}$$

For a 24-h orbit,

$$v_c = r_c 7.3 \times 10^{-5} \tag{b}$$

Combining equations (a) and (b) and letting $g = 9.82$ m/s^2,

$$r_c = 42\,136 \text{ km} \qquad v_c = 3076 \text{ m/s}$$

From equations (38) to (41),

$$e = \frac{v_c^2}{2} - \frac{Gm_1'}{r} = \frac{1}{2}\frac{gR^2}{r_c} - \frac{gR^2}{r_c} = -\frac{1}{2}\frac{gR^2}{r_c} \tag{c}$$

but

$$e = \frac{-Gm_1'}{2a} = -\frac{1}{2}\frac{gR^2}{r_c} \tag{d}$$

Therefore,

$$a = r_c = 42\,136 \text{ km}$$

(it has the same energy as the satellite in a circular orbit). From equation (51),

$$h = \sqrt{aGm_1'(1 - \varepsilon^2)} \tag{e}$$

$$\varepsilon^2 = 1 - \frac{h^2}{aGm_1'} \qquad \varepsilon = \left(1 - \frac{h^2}{aGm_1'}\right)^{1/2}$$

where

$$h = r^2\dot\theta = rv \sin \gamma$$
$$= r_c v_c \sin 45°$$
$$= 42\,136(3076)(\sin 45°) \times 10^3 = 0.916 \times 10^{11}$$

$$\varepsilon = \left(1 - \frac{(0.916 \times 10^{11})^2}{42\,136(9.82)(6371^2) \times 10^9}\right)^{1/2} = 0.707$$

294 ANALYTICAL MECHANICS

From equation (37),

$$r = \frac{h^2/Gm_1'}{1 + \varepsilon \cos \theta} \tag{f}$$

$$\cos \theta = \left(\frac{h^2/Gm_1'}{r} - 1\right)\frac{1}{\varepsilon} = \left(\frac{(0.916 \times 10^{11})^2}{9.82(6371^2)(42\,136) \times 10^9} - 1\right)\frac{1}{0.707}$$

$$\theta = 135°$$

The equations of motion are

$$\ddot{r} = \frac{h^2}{r^3} - \frac{Gm_1'}{r^2} \tag{g}$$

and

$$\ddot{\theta} = -\frac{2\dot{r}\dot{\theta}}{r} \tag{h}$$

The initial conditions are at $t = 0$, $\theta_A = \theta_B = 135°$, and $r_0 = 42\,136$ km. Further

$$\dot{r}_0 = 3076 \cos 45° = 2175 \text{ m/s}$$

$$\dot{\theta}_0 = \frac{v_c \sin 45°}{r_c} = 0.52 \times 10^{-4} \text{ rad/s}$$

These values will now be nondimensionalized:

$$\hat{r} = \frac{r}{R}$$

$$(\hat{r})_0 = \frac{42\,136}{6371} = 6.614$$

$$\hat{v} = \frac{v}{\sqrt{gR}}$$

$$\hat{v}_c = \frac{3076}{\sqrt{9.82(6371) \times 10^3}} = 0.3889$$

$$\hat{h} = \frac{h}{\sqrt{gR^3}} = \frac{0.916 \times 10^{11}}{\sqrt{9.82(6371 \times 10^3)^3}} = 1.819$$

$$(\hat{\dot{\theta}}) = \frac{\dot{\theta}R}{\sqrt{gR}} = 0.52 \times 10^{-4}\sqrt{\frac{6371 \times 10^3}{9.82}} = 0.0419$$

$$(\dot{v}_r)_0 = \frac{2175}{\sqrt{9.82(6371 \times 10^3)}} = 0.2750$$

$$(\hat{\omega}_e) = 7.3 \times 10^{-5}\sqrt{\frac{6371 \times 10^3}{9.82}} = 0.0580$$

$$\hat{T} = T\sqrt{\frac{g}{R}} = 24 \times 3600\sqrt{\frac{9.82}{6371 \times 10^3}} = 107.27$$

$$D\hat{T} = 0.96275 \quad (DT \simeq 13 \text{ min})$$

Also, the force term in equation (g)

$$\frac{F}{m_2} = \frac{-gR^2}{r^2}$$

becomes in nondimensionalized form

$$\frac{\hat{F}}{m_2} = -\frac{1}{(\hat{r})^2}$$

An integration of these equations of motion using a Runge-Kutta method follows (Figure 8.10). See Section 2.8. ∎

8.7 HAMILTON'S PRINCIPLE

There is yet another way, a very elegant way, to develop the governing equations of motion for conservative systems. Employing the *m independent generalized coordinates* to specify a unique state of the system in an *m*-dimensional configuration space, Hamilton proposed that *the actual path taken by the system point in configuration space is such that the integral I of the lagrangian with respect to time between two fixed time limits,*

$$I = \int_{t_1}^{t_2} L\, dt \qquad (8.61)$$

is an extremum. That the integral is an extremum means that the *variation of I* must equal zero.

$$\delta I = 0 = \delta \int_{t_1}^{t_2} L\, dt \qquad (8.62)$$

Since the upper and lower time limits are fixed, an extremum of the integral is found when the variation of L, the integrand, is an extremum for every instant of time along the path in configuration space.

$$\delta I = 0 \qquad (8.63)$$

$$= \delta \int_{t_1}^{t_2} L(q_1, q_2, \ldots, q_m, \dot{q}_1, \dot{q}_2, \ldots, \dot{q}_m, t)\, dt \qquad (8.64)$$

The *calculus of variations* provides a means of evaluating this integral. The function L is varied about an assumed solution by varying each q_j and $\dot{q}_j$ along the solution path while holding time constant. No variations are permitted at the endpoints t_1 and t_2. This is accomplished by expressing the variation of the coordinates about the solution path in terms of a small parameter α

$$\begin{aligned} q_j &= q_j^* + \alpha n_j(t) \\ \dot{q}_j &= \dot{q}_j^* + \alpha \dot{n}_j(t) \end{aligned} \qquad (8.65)$$

In equation (8.65), q_j^* and $\dot{q}_j^*$ represent the coordinate and the coordinate velocity, respectively, along the solution path (when $\alpha = 0$). The function $n_j(t)$ is any

```
00100 PROGRAM ORBIT(INPUT,OUTPUT)
00110 REAL K1,K2,K3,K4,M1,M2,M3,M4
00120 DIMENSION Z(4)
00130 DDR(R)=(A/R-1.)/(R*R)
00140 DDTH(DTH,DR,R)=-2.*DR*DTH/R
00150 DATA H,R,DR,DTH,THD,OM/1.819,6.614,.2750,.0419,135.,.058/
00160 T=0.
00170 DT=0.96875
00180 T=0.
00190 THA=THD
00200 A=H*H
00210 TH=THD/57.29578
00220 THAR=TH
00230 TT=T
00240 Z(1)=R
00250 Z(2)=TH
00260 Z(3)=DR
00270 Z(4)=DTH
00280 PRINT 20,
00290 20 FORMAT(7X,'T',5X,'THAD',6X,'THD',12X,'R',11X,'DR',10X,'DTH')
00300 2 PRINT 21,TT,THA,THD,Z(1),Z(3),Z(4)
00310 21 FORMAT(3X,F5.2,2(2X,F7.3),3(2X,E11.4))
00320 K1=DT*DDR(Z(1))
00330 K2=DT*DDR(Z(1)+DT*Z(3)/2.)
00340 K3=DT*DDR(Z(1)+DT*Z(3)/2.+DT*K1/4.)
00350 K4=DT*DDR(Z(1)+DT*Z(3)+DT*K2/2.)
00360 M1=DT*DDTH(Z(4),Z(3),Z(1))
00370 M2=DT*DDTH(Z(4)+M1/2.,Z(3)+K1/2.,Z(1)+DT*Z(3)/2.)
00380 M3=DT*DDTH(Z(4)+M2/2.,Z(3)+K2/2.,Z(1)+DT*Z(3)/2.+DT*K1/4.)
00390 M4=DT*DDTH(Z(4)+M3,Z(3)+K3,Z(1)+DT*Z(3)+DT*K2/2.)
00400 Z(1)=Z(1)+DT*Z(3)+DT*(K1+K2+K3)/6.
00410 Z(2)=Z(2)+DT*Z(4)+DT*(M1+M2+M3)/6.
00420 Z(3)=Z(3)+(K1+2.*(K2+K3)+K4)/6.
00430 Z(4)=Z(4)+(M1+2.*(M2+M3)+M4)/6.
00440 THAR=THAR+OM*DT
00450 T=T+DT
00460 TT=T*.2237
00470 THA=THAR*57.29578
00480 THD=Z(2)*57.29578
00490 IF(THAR.GT.(6.2832+TH).OR.TT.GT.24) GO TO 5
00500 GO TO 2
00510 5 PRINT 21,TT,THA,THD,Z(1),Z(3),Z(4)
00520 STOP
00530C WITHOUT ROUND-OFF ERROR THE SECOND SIGHTING SHOULD BE IN 24 HOURS
00540 END
```

(a)

Figure 8.10

T	THAD	THD	R	DR	HTH
0.00	135.000	135.000	.6614E+01	.2750E+00	.4190E-01
.22	138.219	137.237	.6875E+01	.2642E+00	.3878E-01
.43	141.439	139.313	.7126E+01	.2537E+00	.3610E-01
.65	144.658	141.250	.7367E+01	.2437E+00	.3377E-01
.87	147.877	143.068	.7598E+01	.2341E+00	.3175E-01
1.08	151.097	144.779	.7820E+01	.2248E+00	.2997E-01
1.30	154.316	146.398	.8034E+01	.2158E+00	.2840E-01
1.52	157.535	147.935	.8239E+01	.2071E+00	.2700E-01
1.73	160.754	149.399	.8435E+01	.1987E+00	.2576E-01
1.95	163.974	150.797	.8624E+01	.1905E+00	.2465E-01
2.17	167.193	152.137	.8804E+01	.1826E+00	.2365E-01
2.38	170.412	153.424	.8978E+01	.1749E+00	.2274E-01
2.60	173.632	154.663	.9143E+01	.1674E+00	.2192E-01
2.82	176.851	155.859	.9302E+01	.1601E+00	.2118E-01
3.03	180.070	157.016	.9454E+01	.1530E+00	.2051E-01
3.25	183.290	158.137	.9598E+01	.1460E+00	.1989E-01
3.47	186.509	159.226	.9737E+01	.1392E+00	.1933E-01
3.68	189.728	160.284	.9868E+01	.1325E+00	.1882E-01
⋮	⋮	⋮	⋮	⋮	⋮
8.23	257.334	178.750	.1128E+02	.1092E-01	.1439E-01
8.45	260.553	179.548	.1129E+02	.5552E-02	.1437E-01
8.67	263.772	180.345	.1130E+02	.1824E-03	.1437E-01
8.89	266.992	181.143	.1129E+02	.5187E-02	.1437E-01
9.10	270.211	181.941	.1129E+02	.1056E-01	.1439E-01
9.32	273.430	182.741	.1127E+02	.1594E-01	.1442E-01
9.54	276.649	183.543	.1125E+02	.2133E-01	.1447E-01
⋮	⋮	⋮	⋮	⋮	⋮
21.24	450.492	424.534	.2490E+01	.3439E+00	.2958E+00
21.45	453.711	438.947	.2842E+01	.3779E+00	.2270E+00
21.67	456.931	450.093	.3214E+01	.3882E+00	.1774E+00
21.89	460.150	458.910	.3590E+01	.3862E+00	.1422E+00
22.10	463.369	466.064	.3960E+01	.3779E+00	.1169E+00
22.32	466.589	472.007	.4321E+01	.3666E+00	.9818E-01
22.54	469.808	477.047	.4670E+01	.3540E+00	.8405E-01
22.75	473.027	481.396	.5007E+01	.3409E+00	.7313E-01
22.97	476.247	485.207	.5331E+01	.3279E+00	.6451E-01
23.19	479.466	488.589	.5642E+01	.3151E+00	.5759E-01
23.40	482.685	491.623	.5941E+01	.3027E+00	.5193E-01
23.62	485.904	494.372	.6229E+01	.2908E+00	.4725E-01
23.84	489.124	496.882	.6505E+01	.2793E+00	.4332E-01
24.05	492.343	499.192	.6770E+01	.2683E+00	.4000E-01

(b)

Figure 8.10 (continued)

298 ANALYTICAL MECHANICS

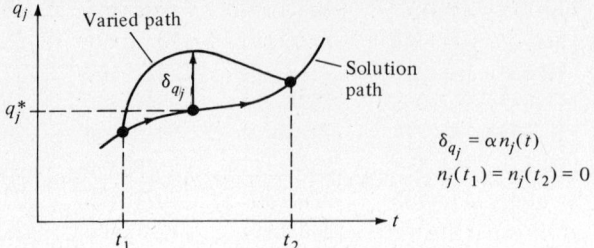

Figure 8.11

function of time continuous through the second derivative that vanishes at the endpoints t_1 and t_2.

$$n_j(t_1) = n_j(t_2) = 0 \tag{8.66}$$

See Figure 8.11. If the relations of equation (8.65) are substituted into the integrand, the integral now becomes a function solely of α for specified $n_j(t)$. The extremum, or *stationary value*, of the integral occurs under the condition

$$\left.\frac{dI}{d\alpha}\right|_{\alpha=0} = \int_{t_1}^{t_2} \left(\sum_{j=1}^{m} \frac{\partial L}{\partial q_j} \frac{\partial q_j}{\partial \alpha}\bigg|_{\alpha=0} + \sum_{j=1}^{m} \frac{\partial L}{\partial \dot{q}_j} \frac{\partial \dot{q}_j}{\partial \alpha}\bigg|_{\alpha=0} \right) dt = 0 \tag{8.67}$$

where derivatives are permitted under the integral sign.

$$\int_{t_1}^{t_2} \left[\sum_{j=1}^{m} \frac{\partial L}{\partial q_j} n_j(t) + \sum_{j=1}^{m} \frac{\partial L}{\partial \dot{q}_j} \dot{n}_j(t) \right] dt = 0 \tag{8.68}$$

The second term in the brackets can be integrated by parts leading to

$$\int_{t_1}^{t_2} \sum_{j=1}^{m} \frac{\partial L}{\partial q_j} n_j(t)\, dt + \sum_{j=1}^{m} \frac{\partial L}{\partial \dot{q}_j} n_j(t)\bigg|_{t_1}^{t_2} - \int_{t_1}^{t_2} \sum_{j=1}^{m} \frac{d}{dt}\left(\frac{\partial L}{\partial \dot{q}_j}\right) n_j(t)\, dt = 0 \tag{8.69}$$

and with equation (8.66) the results become

$$\int_{t_1}^{t_2} \sum_{j=1}^{m} \left[\frac{\partial L}{\partial q_j} - \frac{d}{dt}\left(\frac{\partial L}{\partial \dot{q}_j}\right) \right] n_j(t)\, dt = 0 \tag{8.70}$$

Since $n_j(t)$ was *arbitrary*, and since the coordinates q_j were *independent*, it must follow that

$$\frac{\partial L}{\partial q_j} - \frac{d}{dt}\left(\frac{\partial L}{\partial \dot{q}_j}\right) = 0 \quad (j = 1, 2, \ldots, m) \tag{8.71}$$

The last equations are the Euler-Lagrange equations and represent the necessary and sufficient conditions that must be met by the integral at each time interval between t_1 and t_2 so that the integral I is an extremum. Hamilton's principle in application then leads to Lagrange's equations. The integral represented by equation (8.61) is never evaluated. Instead, Hamilton's principle establishes a condition on the integral that must be met for the entire motion from t_1 to t_2.

Equation (8.62), a mathematical statement of Hamilton's principle, is valid only for conservative and holonomic systems. A more general expression of *Hamilton's principle* that is valid even for *nonconservative* systems is

$$\int_{t_1}^{t_2} \left(\delta T + \sum_{j=1}^{m} Q_j \, \delta q_j \right) dt = 0 \tag{8.72}$$

where T denotes the kinetic energy of the system, and $\sum_{j=1}^{m} Q_j \delta q_j$ represents the work done on the system by the external forces (any acting constraint forces are assumed not to perform work) during a virtual displacement $(\delta q_1, \delta q_2, \ldots, \delta q_m)$. The displacements are arbitrary, but they must be zero at times t_1 and t_2 and must be consistent with any constraints present. To verify the relations given by equation (8.72), first describe the variation of T.

$$\int_{t_1}^{t_2} \left[\sum_{j=1}^{m} \frac{\partial T}{\partial \dot{q}_j} \delta \dot{q}_j + \sum_{j=1}^{m} \left(\frac{\partial T}{\partial q_j} + Q_j \right) \delta q_j \right] dt = 0 \tag{8.73}$$

Substituting the result of equation (8.38) into equation (8.73) leads to

$$\int_{t_1}^{t_2} \left[\sum_{j=1}^{m} \frac{\partial T}{\partial \dot{q}_j} \delta \dot{q}_j + \frac{d}{dt}\left(\frac{\partial T}{\partial \dot{q}_j}\right) \delta q_j \right] dt = 0 \tag{8.74}$$

This can be rewritten as

$$\int_{t_1}^{t_2} \frac{d}{dt}\left(\sum_{j=1}^{m} \frac{\partial T}{\partial \dot{q}_j} \delta q_j \right) dt = 0$$

or

$$\sum_{j=1}^{m} \frac{\partial T}{\partial \dot{q}_j} \delta q_j \bigg|_{t_1}^{t_2} = 0 \tag{8.75}$$

which is correct since the coordinate variations are defined to be zero at the endpoints of the interval.

Note that the extended form of Hamilton's principle, equation (8.72), is no longer a statement about the variation δI of an integral. Only for holonomic systems will the variation of the integral be the same as the integral of the variations. The principles represented by equations (8.63) and (8.72) have two different perspectives. Equation (8.63), which concerns itself only with conservative holonomic systems, analyzes the entire solution path at once. It asserts that the actual path taken by the system point leads to a stationary value of the integral. The necessary and sufficient conditions for the integral to be stationary "happen to be" the Lagrange equations of motion. On the other hand, equation (8.72) takes an instantaneous lagrangian (or newtónian) view once more. It establishes conditions on the motion (equations of motion) at every instant of time. If these conditions are met, the integral, represented by equation (8.72), must be zero. Hamilton's principle for conservative holonomic systems is therefore of greater interest since it introduces an entirely new formulation of mechanics. For the sake of completeness, *nonholonomic* systems of the type

$$\sum_{i=1}^{n} a_{si} \, dq_i + a_{st} \, dt = 0 \qquad (s = 1, 2, \ldots, k)$$

will now be discussed.

300 ANALYTICAL MECHANICS

For virtual displacements only, this can be rewritten as

$$\sum_{i=1}^{n} a_{si}\,\delta q_i = 0 \qquad (8.47)$$

Employing Lagrange multipliers λ_s, one can write

$$\lambda_s \sum_{i=1}^{n} a_{si}\,\delta q_i = 0 \qquad (8.48)$$

$$\sum_{s=1}^{k}\sum_{i=1}^{n} \lambda_s a_{si}\,\delta q_i = 0 \qquad (8.49)$$

If the system is conservative, and if equation (8.49) is added to the integrand of equation (8.70), the results obtained are

$$\int_{t_1}^{t_2}\left[\sum_{i=1}^{n}\frac{\partial L}{\partial q_i} - \frac{d}{dt}\left(\frac{\partial L}{\partial \dot q_i}\right) + \sum_{s=1}^{k}\lambda_s a_{si}\right]\delta q_i\,dt = 0 \qquad (8.76)$$

If the k values of λ_s are selected such that

$$\frac{\partial L}{\partial q_i} - \frac{d}{dt}\left(\frac{\partial L}{\partial \dot q_i}\right) + \sum_{s=1}^{k}\lambda_s a_{si} = 0 \qquad (i = 1, 2, \ldots, k) \qquad (8.77)$$

then for the $i = k+1$ to n coordinates that are independent in equation (8.76),

$$\frac{\partial L}{\partial q_i} - \frac{d}{dt}\left(\frac{\partial L}{\partial \dot q_i}\right) + \sum_{s=1}^{k}\lambda_s a_{si} = 0 \qquad (i = k+1, \ldots, n) \qquad (8.78)$$

and the results already derived in Section 8.5 repeat themselves here; namely,

$$\frac{\partial L}{\partial q_i} - \frac{d}{dt}\left(\frac{\partial L}{\partial \dot q_i}\right) + \sum_{s=1}^{k}\lambda_s a_{si} = 0 \qquad (i = 1, 2, \ldots, n) \qquad (8.79)$$

with

$$\sum_{i=1}^{n} a_{si}\,dq_i + a_{st}\,dt = 0 \qquad (s = 1, 2, \ldots, k) \qquad (8.3)$$

Sample Problem 8.14

The Bernoulli brothers, Jakob (1654–1705) and Johann (1667–1748), are credited with introducing the calculus of variations. Because of a rivalry between the two, they constantly challenged each other with mathematical problems of greater and greater complexity. One of these problems became a classic; it is called the *brachistrochrone* problem. The question posed is, What is the equation of the curve joining two points along which a particle may slide without friction, starting from rest, to reach the other end in the least amount of time?

ENERGY METHODS

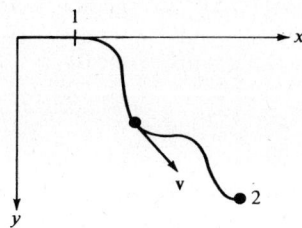

Analysis: The time required to travel from point 1 to point 2 is

$$t = \int_1^2 \frac{ds}{v}$$

Because energy is conserved,

$$T + V = E = \text{constant}$$

or

$$\frac{mv^2}{2} = mgy \qquad v = \sqrt{2gy}$$

Then

$$t = \int_1^2 \frac{ds}{\sqrt{2gy}} = \int_1^2 \sqrt{\frac{1+x'^2}{2gy}}\, dy$$

with $x' = dx/dy$. The minimum time can be found by setting the first variation of the integral equal to zero.

$$\delta t = 0 = \delta \int_1^2 \sqrt{\frac{1+x'^2}{2gy}}\, dy$$

$$0 = \int_1^2 \left[\frac{\partial}{\partial x} \left(\sqrt{\frac{1+x'^2}{2gy}} \right) - \frac{d}{dy}\left(\frac{\partial}{\partial x'} \sqrt{\frac{1+x'^2}{2gy}} \right) \right] dy$$

Then

$$\frac{\partial}{\partial x} \sqrt{\frac{1+x'^2}{2gy}} = \frac{d}{dy}\left(\frac{\partial}{\partial x'} \sqrt{\frac{1+x'^2}{2gy}} \right)$$

and since

$$\frac{\partial}{\partial x} \sqrt{\frac{1+x'^2}{2gy}} = 0$$

$$\frac{\partial}{\partial x'} \sqrt{\frac{1+x'^2}{2gy}} = c \qquad \text{(a constant)}$$

Taking the partial derivative of the square root's argument as required,

$$\frac{x'}{\sqrt{2gy(1+x'^2)}} = c$$

one finally obtains

$$\frac{dx}{dy} = \sqrt{\frac{2gc^2 y}{1 - 2gc^2 y}} \qquad \text{(a)}$$

Let now
$$y = a(1 - \cos\theta) \tag{b}$$
where
$$a = \frac{1}{4gc^2}$$
Then from equation (a) with $dy = a \sin\theta \, d\theta$,
$$x = \int a(1 - \cos\theta) \, d\theta$$
or
$$x = a(\theta - \sin\theta) + c_1 \tag{c}$$
Equations (b) and (c) are parametric equations describing a cycloid. This is the curve traced out by a particle on the circumference of a circle of radius a rolling on a flat plane.

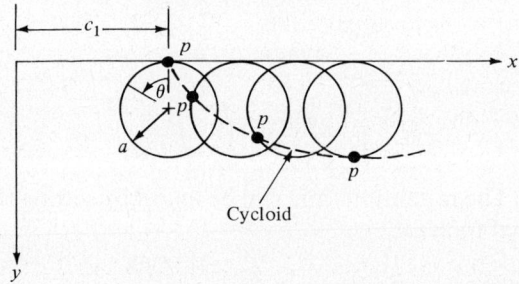
Cycloid

It is left as an exercise for the reader to show that this is a minimum value path.

The cycloid itself has interesting properties. If the circle starts at the origin and is allowed to continue for a full cycle and then return, the curve can be drawn as shown below.

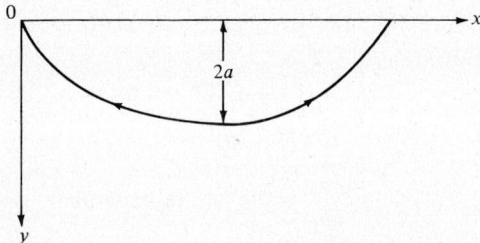

The kinetic energy along the path is expressed by
$$T = \frac{mv^2}{2} = ma^2\dot\theta^2(1 - \cos\theta)$$

The particle starts from rest at $t = 0$. Since total energy is conserved,
$$ma^2\dot\theta^2(1 - \cos\theta) = mga(1 - \cos\theta)$$

Then

$$\dot{\theta} = \sqrt{\frac{g}{a}} \quad \text{(a constant)}$$

$$\theta = \sqrt{\frac{g}{a}} t$$

$$y = a\left(1 - \cos\sqrt{\frac{g}{a}} t\right)$$

$$x = a\left(\sqrt{\frac{g}{a}} t - \sin\sqrt{\frac{g}{a}} t\right)$$

When the motion is allowed to reverse itself, as a pendulum would, it is seen that the period of the motion is *independent* of θ_0 and is given by

$$T = 2\pi\sqrt{\frac{a}{g}}$$

This is the period a simple pendulum of length a would have for *small* angles, but no such approximation is needed here. It was for this reason that in the early days of clock making, several attempts were made to build a clock whose bob was constrained to move along a cycloidal path. ∎

8.8   THE HAMILTONIAN

Hamilton also suggested a new formulation of mechanics whereby the generalized position coordinates q_j *and* the momentum coordinates p_j are both treated equally as independent variables. Instead of attaining m second-order equations of motion, $2m$ first-order equations result. The m-dimensional configuration space becomes a $2m$-dimensional *phase* space. The fact that it is possible to solve for p_j as a function of q_j by using the resultant equations of motion does not imply that they cannot be treated as independent coordinates to begin with.

The *hamiltonian* function constructed to transform the equation of motion to phase space is given by

$H(p_1, p_2, \ldots, p_m, q_1, q_2, \ldots, q_m, t)$

$$= \sum_{j=1}^{m} \dot{q}_j p_j - L(q_1, q_2, \ldots, q_m, \dot{q}_1, \dot{q}_2, \ldots, \dot{q}_m, t) \qquad (8.80)$$

or simply

$$H = \sum_{j=1}^{m} \frac{\partial L}{\partial \dot{q}_j} \dot{q}_j - L \qquad (8.81)$$

which resembles Routh's function. Though one can obtain this function through a Legendre transformation with the original function L dependent on $\dot{q}_j$ and q_j and the final function H dependent on p_j and q_j, it is also possible to obtain it directly from the lagrangian; namely, for a conservative system with L not an explicit function of t,

$$\frac{dL}{dt} = \sum_{j=1}^{m} \frac{\partial L}{\partial q_j} \dot{q}_j + \sum_{j=1}^{m} \frac{\partial L}{\partial \dot{q}_j} \frac{d}{dt}(\dot{q}_j)$$

$$= \sum_{j=1}^{m} \frac{d}{dt}(p_j)\dot{q}_j + \sum_{j=1}^{m} p_j \frac{d}{dt}(\dot{q}_j)$$

$$= \sum_{j=1}^{m} \frac{d}{dt}(p_j \dot{q}_j) \tag{8.82}$$

Then

$$0 = \frac{d}{dt}\left(\sum_{j=1}^{m} p_j \dot{q}_j - L\right)$$

or

$$\sum_{j=1}^{m} p_j \dot{q}_j - L \equiv H \quad \text{(a constant)} \tag{8.83}$$

Remembering also that

$$T = \sum_{i=1}^{n} m_i \frac{\mathbf{v}_i \cdot \mathbf{v}_i}{2} \tag{8.84}$$

where

$$\mathbf{r}_i = \mathbf{r}_i(q_1, q_2, \ldots, q_m, t) \tag{8.1}$$

$$\mathbf{v}_i = \sum_{j=1}^{m} \frac{\partial \mathbf{r}_i}{\partial q_j} \dot{q}_j + \frac{\partial \mathbf{r}_i}{\partial t} \tag{8.8}$$

then

$$T = \sum_{j=1}^{m} \sum_{k=1}^{m} a_{jk} \dot{q}_j \dot{q}_k + \sum_{j=1}^{m} a_j \dot{q}_j + a \tag{8.85}$$

Here

$$a_{jk} = \sum_{i=1}^{m} \frac{m_i}{2} \frac{\partial \mathbf{r}_i}{\partial q_j} \cdot \frac{\partial \mathbf{r}_i}{\partial q_k} \tag{8.86}$$

$$a_j = \sum_{i=1}^{m} m_i \frac{\partial \mathbf{r}_i}{\partial q_j} \cdot \frac{\partial \mathbf{r}_i}{\partial t} \tag{8.87}$$

$$a = \sum_{i=1}^{m} \frac{m_i}{2} \frac{\partial \mathbf{r}_i}{\partial t} \cdot \frac{\partial \mathbf{r}_i}{\partial t} \tag{8.88}$$

If the constraints on the system are independent of time, then $a_j = a = 0$, and

$$T = \sum_{j=1}^{m} \sum_{k=1}^{m} a_{jk} \dot{q}_j \dot{q}_k \tag{8.89}$$

With the potential not a function of the velocities,

$$p_j = \frac{\partial L}{\partial \dot{q}_j} = \frac{\partial T}{\partial \dot{q}_j} = \sum_{k=1}^{m} 2a_{jk} \dot{q}_k \tag{8.90}$$

$$\sum_{j=1}^{m} p_j \dot{q}_j = \sum_{j=1}^{m} \sum_{k=1}^{m} 2a_{jk} \dot{q}_k \dot{q}_j = 2T \tag{8.91}$$

$$H = 2T - L = 2T - (T - V) = E \quad \text{(energy)} \tag{8.92}$$

The hamiltonian then represents the total energy for a conservative system if no moving constraints are present in the system. The moving constraints produce real work (virtual work produced is zero). To summarize, if L is not an explicit function of t, H is conserved. If in addition, $\mathbf{r}_i$ is not a function of t, H equals the total energy. With moving constraints, $\mathbf{r}_i$ is a direct function of t.

Sample Problem 8.15

The massless pipe shown is constrained to move in a horizontal plane with angular speed ω. Inside the pipe is a mass m attached to a spring having zero unstretched length and spring constant k. If the ensuing motion is friction-free, find the hamiltonian for the system. Is it constant? What is the difference between E and H? What physical significance does it have? Plot the phase space diagram for the system.

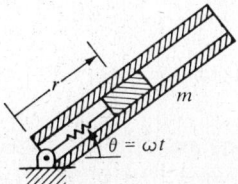

Analysis:

$$H = \sum_{j=1}^{m} p_j \dot{q}_j - L$$

There is only 1 degree of freedom present. Then

$$H = p_r \dot{r} - L$$

$$L = \frac{m}{2}(\dot{r}^2 + r^2\omega^2) - \frac{k}{2}r^2$$

$$\frac{\partial L}{\partial \dot{r}} = p_r = m\dot{r}$$

$$H = \frac{p_r^2}{m} - \left[\left(\frac{p_r^2}{2m} + \frac{mr^2\omega^2}{2}\right) - \frac{k}{2}r^2\right]$$

$$= \frac{p_r^2}{2m} - \frac{mr^2\omega^2}{2} + \frac{k}{2}r^2 \tag{a}$$

Since L is not an explicit function of t, H is constant. H is not equal to E here.

$$E = \frac{p_r^2}{2m} + \frac{mr^2\omega^2}{2} + \frac{k}{2}r^2$$

which is not constant.

$$E - H = mr^2\omega^2$$

This difference between E and H is the work put into the system that affects the constraint $\theta = \omega t$. The force in the θ direction on the mass is

$$F_\theta = m(r\ddot\theta + 2\dot r \dot\theta)$$

or for constant ω,

$$N = m(2\dot r \omega)$$

The necessary moment becomes

$$M_0 = rN = 2mr\dot r\omega$$

and the rotational work is

$$W_{1\to 2} = \int_1^2 M_0\, d\theta = \int_1^2 M_0 \omega\, dt = \int_1^2 2mr\dot r\omega^2\, dt$$
$$= \int_0^r 2mr'\omega^2\, dr' = mr^2\omega^2$$

The hamiltonian then is a constant of the motion and represents the energy of the system in a noninertial frame that is constrained to rotate at an angular rate ω. The constant hamiltonian can also be derived here by integrating the equation of motion for the r coordinate.

For the phase space plot, equation (a) may be rewritten as

$$2H = \frac{p_r^2}{m} + r^2(k - m\omega^2) = \frac{p_r^2}{m} + a^2 r^2 \tag{b}$$

With $a^2 = k - m\omega^2$ and $k - m\omega^2 > 0$, the phase space plot appears as

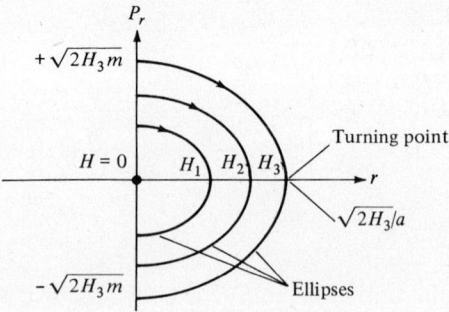

For $k - m\omega^2 = 0$, the phase space plot changes to

ENERGY METHODS 307

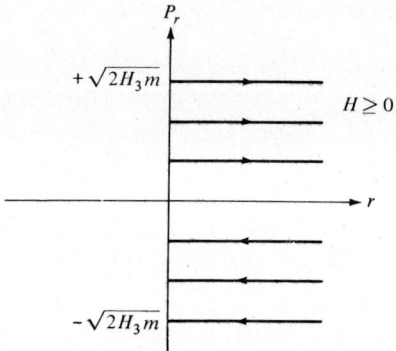

and for $k - m\omega^2 < 0$, where $a^2 = m\omega^2 - k$, the plot becomes

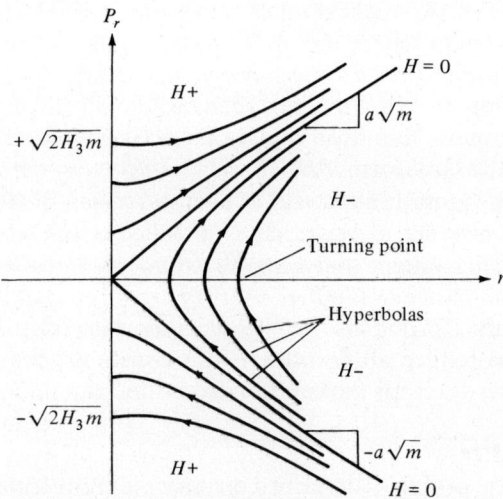

8.9 HAMILTON'S EQUATIONS

The *modified Hamilton's principle* has the integrand in equation (8.62) changed to

$$\delta I = 0 = \delta \int_{t_1}^{t_2} \left(\sum_{j=1}^{n} p_j \dot{q}_j - H \right) dt \tag{8.93}$$

The variation becomes

$$\delta I = 0 = \int_{t_1}^{t_2} \left(\sum_{j=1}^{m} \dot{q}_j \, \delta p_j + \sum_{j=1}^{m} p_j \, \delta \dot{q}_j - \sum_{j=1}^{m} \frac{\partial H}{\partial q_j} \delta q_j - \sum_{j=1}^{m} \frac{\partial H}{\partial p_j} \delta p_j \right) dt \tag{8.94}$$

$$0 = \int_{t_1}^{t_2} \left[\sum_{j=1}^{m} \left(\dot{q}_j - \frac{\partial H}{\partial p_j} \right) \delta p_j - \sum_{j=1}^{m} \left(\dot{p}_j + \frac{\partial H}{\partial q_j} \right) \delta q_j \right] dt \tag{8.95}$$

308 ANALYTICAL MECHANICS

With *both* δp_j and δq_j treated as independent coordinates,

$$\dot{q}_j = \frac{\partial H}{\partial p_j} \quad (j=1, 2, \ldots, m) \tag{8.96}$$

and

$$\dot{p}_j = -\frac{\partial H}{\partial q_j} \quad (j=1, 2, \ldots, m) \tag{8.97}$$

Equations (8.96) and (8.97) are called the *Hamilton equations*. They represent a set of $2m$ first-order equations of motion. From the definition of H, it is also seen that

$$\frac{\partial H}{\partial t} = -\frac{\partial L}{\partial t} \tag{8.98}$$

Note that equations (8.96) and (8.97) are nearly symmetric. Because of this symmetry, one can free the coordinates from any particular physical identity. Routh demonstrated a method of transforming the Lagrange function L to a coordinate system where the cyclic coordinates and their velocities were omitted from consideration. In the same vein, it is also possible to transform Hamilton's function H to a $2m$ phase space where all the new coordinates are constant and the new Hamilton equations derived from the new Hamilton function H' have the same form: $\partial H'/\partial p_j = \dot{Q}_j, \partial H'/\partial Q_j = -\dot{P}_j$. A transformation that transforms the hamiltonian to H' while preserving the form of Hamilton's equations in the new set of coordinates is called a *canonical transformation*. The new canonical moment and generalized coordinates will no longer have any physical resemblance to familiar coordinates or momentum measures. The solution of mechanics problems then becomes a search for the proper canonical transformation to reduce all coordinates to constants, the initial conditions. The *transformation* develops the solution from time equal to zero to time t.

Sample Problem 8.16
For the two-body problem described in Sample Problem 8.12, develop the hamiltonian, and find Hamilton's equations.

Analysis: H equals E here and is constant. Therefore,

$$H = \frac{p_r^2}{2m_2'} + \frac{p_\theta^2}{2m_2' r^2} - \frac{Gm_1' m_2'}{r} \tag{a}$$

$$\dot{q}_j = \frac{\partial H}{\partial p_j}, \quad \dot{r} = \frac{p_r}{m_2'}, \quad \dot{\theta} = \frac{p_\theta}{m_2' r^2}$$

which is not new information. The momentum equations are of more interest.

$$-\dot{p}_j = \frac{\partial H}{\partial q_j}$$

$$-\dot{p}_r = -\frac{p_\theta^2}{m_2' r^3} + \frac{Gm_1' m_2'}{r^2} \tag{b}$$

and $\quad -\dot{p}_\theta = 0 \quad p_\theta = $ constant (c)

Since θ was cyclic in L, it is cyclic in H as well. Note that equation (b) can be developed further to

$$m_2' \ddot{r} - m_2' r \dot{\theta}^2 + \frac{Gm_1' m_2'}{r^2} = 0$$

a previous result.

With p_θ treated as a constant γ, equation (a) becomes

$$p_r^2 + \frac{\gamma^2}{r^2} - \frac{2Gm_1' m_2'^2}{r} = 2m_2' H$$

or $\quad p_r^2 + \dfrac{\gamma^2}{r^2} - \dfrac{\beta}{r} = 2m_2' H$

The phase space plot for equation (d) appears below.

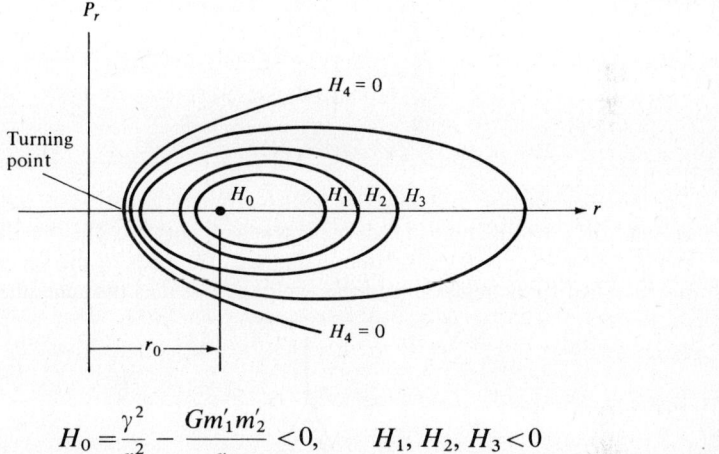

$$H_0 = \frac{\gamma^2}{r_0^2} - \frac{Gm_1' m_2'}{r_0} < 0, \quad H_1, H_2, H_3 < 0 \qquad \blacksquare$$

PROBLEMS

8.1 For the thin homogeneous bar shown of mass M that is bent into a right angle and balanced at the pivot point A, find the angle θ that provides a stable equilibrium position.

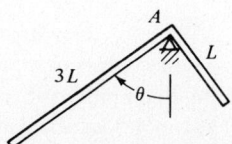

8.2 Derive Lagrange's equations of motion for the system shown in the figure. Stiff wire

$ABCD$ is fixed at points A and D and wound around the smooth pulleys connected to masses m_1 and m_2. The ground is assumed frictionless.

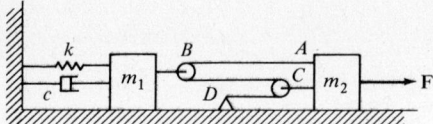

8.3 A particle of mass m is constrained to move on the inside surface of a smooth paraboloid having the equation $z = b^2 r^2$. Derive Lagrange's equations, and use Lagrange multipliers to find the normal force from the surface.

8.4 Write Lagrange's equations of motion for the system shown. The unstretched lengths of the two springs are L_1 and L_2. Note that distance x_2 is measured relative to the cart. Neglect all friction forces.

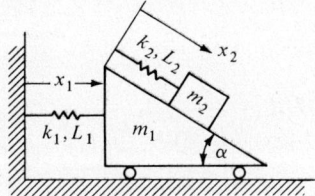

8.5 For the three-member system given, write Lagrange's equations of motion. Members AB and BC are thin uniform rods of mass m and length L. The thin disk at C having mass m_1 and radius R rolls without slipping. The ends of the spring of length L_0 are attached to the midpoints of the two bars. Use θ as the generalized coordinate.

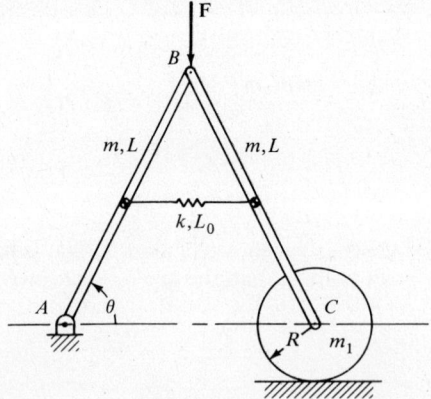

*8.6 For the previous problem, write a computer program that will solve for $\dot{\theta}$ when $\theta = 0°$. Take $F = 100$ N, $m = 2$ kg, $m_1 = 5$ kg, $L = 1$ m, $R = 0.25$ m, $L_0 = 0.5$ m, and $k = 500$ N/m. The system starts from rest when $\theta = 60°$. Compare your numerical results with the exact solution obtained by using energy principles. Use a communication interval of 0.02 s and a final time of 0.95 s.

8.7 The thin homogeneous disk of radius R and mass m rolls (without slipping) on the oscillating cart of mass m_1. Neglecting friction acting between the cart and the floor, write Lagrange's equations of motion for the system.

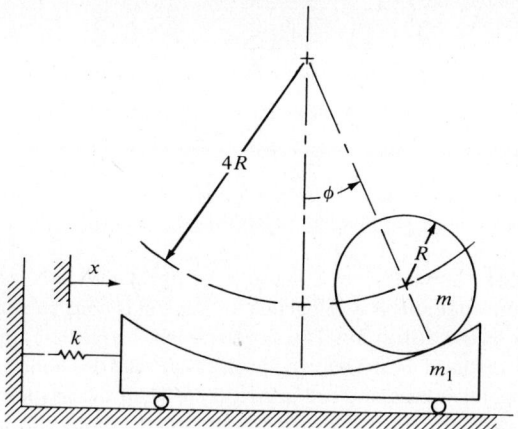

8.8 A simple pendulum of mass m and length L is attached to an oscillating cart of mass m_1. The cart is forced to move so that $x = A \cos \omega t$. Neglecting friction, write Lagrange's equation of motion for the cart and pendulum.

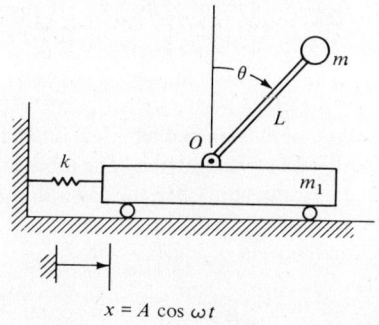

$x = A \cos \omega t$

8.9 Explain physically what the difference is between E and H for the previous problem. Is E or H constant? Write Hamilton's equations for the motion.

8.10 Derive Lagrange's equations of motion for the double pendulum shown.

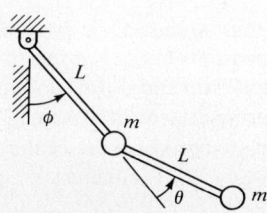

8.11 Derive the hamiltonian for the following 2 degrees of freedom problem. Is H constant? Does H equal to E? What are Hamilton's equations? The unstretched length of the spring is L_0.

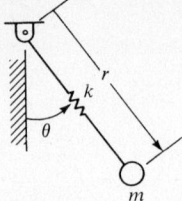

8.12 The thin disk shown rotates about a fixed horizontal axis AB. Its mass moment of inertia about that axis is I. A particle of mass m is constrained to move in the plane of the disk as the disk spins. The slot in which the particle travels makes an angle β with the horizontal axis. The two springs that retard the motion of the particle each have spring constant k and are unstretched when the particle is at the center of the slot. If a constant torque T drives the system about the horizontal axis, derive Lagrange's equations of motion. Neglect friction.

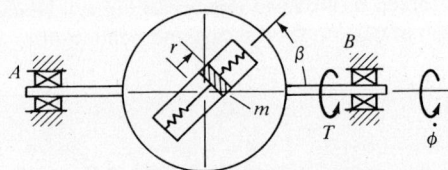

8.13 A thin homogeneous rod of mass m and length L is free to slide outward relative to a tube that rotates about a fixed vertical axis. The moment of inertia of the tube about its axis of rotation is I. For the time when the rod is in the tube, derive the lagrangian. Which coordinates are cyclic? Find the routhian and derive Routh's equation of motion. Ignore frictional effects.

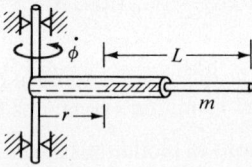

8.14 The T-shaped member of mass m shown has two point masses, each m_1, welded at its ends. The T-shape has dimensions $L \times L$. It is free to twist in a horizontal plane while constrained to move vertically up and down. Both parts of the T can be considered uniform thin rods. The spring attached to one end of the T has spring constant k. z is measured from the unstretched position of the spring. For this problem, derive the lagrangian. Identify the cyclic coordinates. Write Routh's equations of motion. Friction can be neglected.

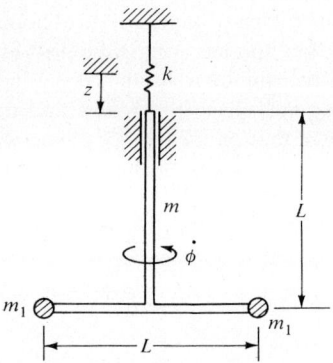

8.15 For the conical pendulum given, identify the cyclic coordinate, and derive Routh's equation of motion.

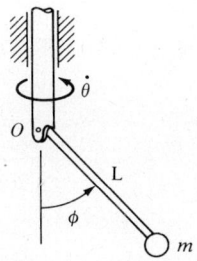

*8.16 If in the previous problem motion was started at $t=0$ with $\phi_0 = 90°$, $\dot{\theta} = 5$ rad/s, and $\dot{\phi}_0 = 0$, write a computer program that will solve for ϕ and θ as functions of time. Using energy and momentum principles, calculate the minimum angle ϕ obtainable, and compare this result with your numerical solution. Take $L = 1$ m and $m = 2$ kg. Use a communication interval of 0.05 s, and terminate the program at $t = 2.0$ s.

8.17 The conical pendulum given, having mass m and length L, has a pivot point O which is free to move vertically. The spring attached to the pivot has a spring constant k and unstretched length L_0. Write Lagrange's equations of motion for the noncyclic coordinates present. Neglect torsional effects on the spring.

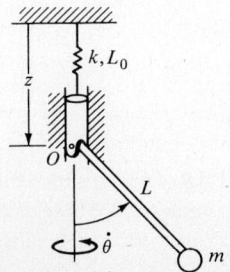

8.18 A particle of mass m travels around a vertical hoop of radius R. The hoop is free to turn about the vertical and has a mass moment of inertia J about that axis. Offset from the center of the ring by a distance d is a fixed spring that is attached to the particle. The spring has spring constant k and unstretched length L_0. Write Lagrange's equations of motion for the hoop-mass system. Use Lagrange multipliers to determine the normal force of the hoop on the particle.

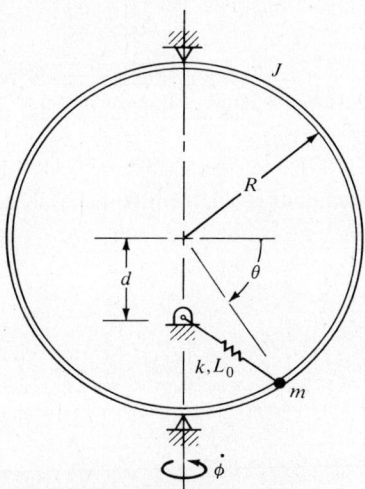

8.19 Prove that Lagrange's equations in the case of impact are $(\partial T/\partial \dot{q}_i)_2 - (\partial T/\partial \dot{q}_i)_1 = \hat{Q}_i$, where $\hat{Q}_i$ is defined to be the *generalized impulse* $\hat{Q}_i = \lim_{\Delta t \to 0} \int_{t_1}^{t_2} Q_i \, dt$.

8.20 Using the results of problem 8.19, find the angular velocity of the top rod immediately after the impulse $\hat{F}$ strikes the bottom rod. Both rods have mass m and length L and can be considered thin and uniform.

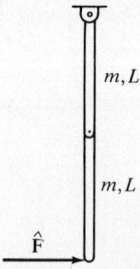

8.21 A thin homogeneous rod AB of mass m and length L is constrained to move so that A slides on a smooth horizontal track while B moves along a smooth vertical one. If the rod is initially at rest in a vertical position, calculate the necessary impulse $\hat{F}$ applied at the rod's center so that the bar will reach a horizontal position.

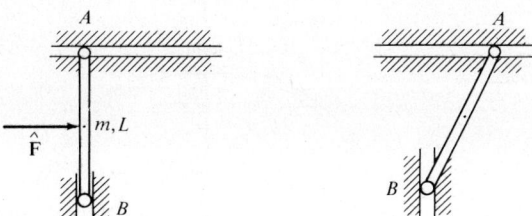

8.22 A thin uniform disk of mass m and radius R rolls (without slipping) down an incline with initial speed v_0. The subsequent impact with the horizontal surface is perfectly plastic. Determine the angular velocity of the disk and the velocity of its center immediately after the impact when the disk is slipping. Use Lagrange's equation with impulse forces acting.

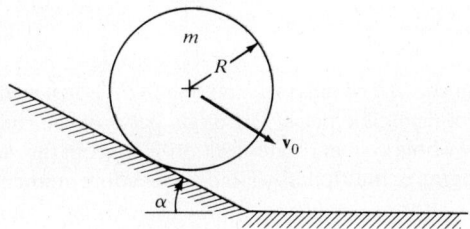

8.23 The uniform thin rod of mass m and length L shown swings in a vertical plane about the pivot point A while the plane of oscillation itself rotates. Write Lagrange's equations of motion for the rod.

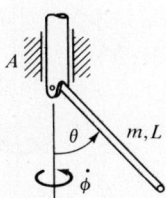

8.24 A circular vertical ring of radius R and mass m is free to rotate about a vertical axis. In the ring is an axle that makes an angle of $60°$ with the vertical and spins relative to the ring at a constant angular speed ω. Fixed to this axle is a body of revolution whose polar mass moment of inertia is J and whose transverse mass moment of inertia is $J/2$. The center of gravity of this body lies on the vertical axis of rotation.
(a) Derive Lagrange's equation of motion, and (b) derive Lagrange's equation of motion for the case when ω is not constant.

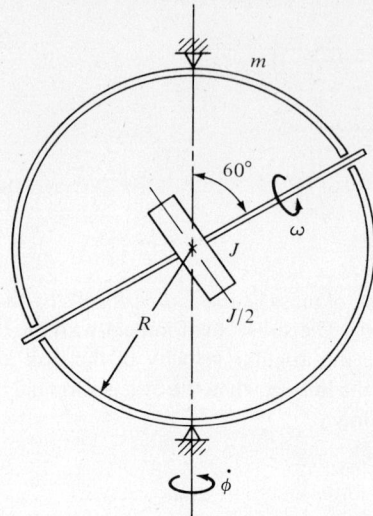

8.25 A uniform thin rod AB of mass m_1 and length $2a$ is fastened at its midpoint O to a vertical shaft of negligible mass. The angle between the rod and shaft is fixed and equal to α. The whole system rotates friction-free about the vertical axis at a constant rate Ω. Use Lagrange multipliers to find the bending moment on the shaft.

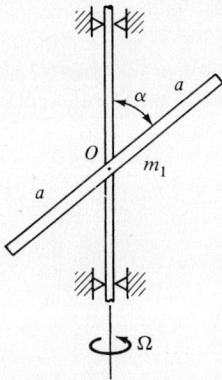

8.26 For problem 8.23 find the routhian, and derive Routh's equation.

8.27 A thin homogeneous ring of mass m_1 and radius R is suspended by an elastic wire of negligible mass whose torsional spring constant is k_ϕ (N · m/rad). The ring is free to turn about the vertical. Mounted on the horizontal axle inside the ring is a thin bar of length L and mass m free to turn relative to the ring. The axle's mass is also negligible and is assumed to stay in a vertical plane. For this friction-free motion, derive the system's hamiltonian. Is H constant? Does H equal E? Find Hamilton's equations of motion.

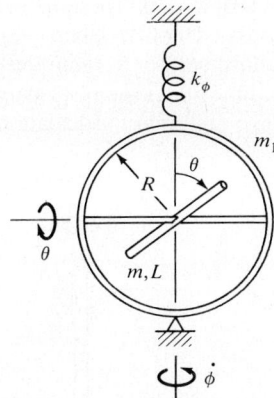

*8.28 Write a computer program that will give numerical results for θ and ϕ as functions of time in problem 8.23. At $t=0$, $\theta_0 = 30°$, $\dot{\theta}_0 = 0$, and $\dot{\phi}_0 = 40$ rad/s. Let mass m equal 2 kg and $L = 0.1$ m. Take your solution past the turning point. Use a communication interval of 0.02 s, and terminate the program at $t = 0.5$ s.

8.29 A thin ring of negligible mass is free to rotate about a vertical axis. Fixed to a horizontal axle along a diameter within the ring is a thin right triangular plate having mass m_1 and dimension $a \times b$. The horizontal axle, also of negligible mass, is welded to the plate and turns freely without friction relative to the ring. The plate itself is mounted on the axle so that its center of mass lies in a plane perpendicular to the axle that contains the vertical axis of rotation. The side of the plate welded to the axle has dimension b. Develop Lagrange's equations of motion for this problem.

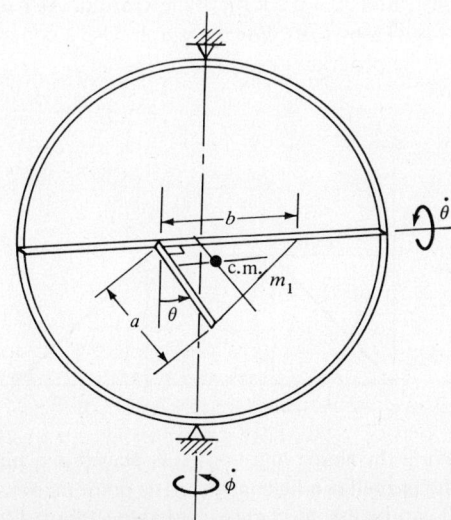

8.30 Shown below is a gyroscopic mill. The thin heavy wheel of mass m_1 and radius R sits at the end of an axle of length $4R$ having negligible mass. The wheel rolls without slipping on the inclined surface, turning freely on its axle. The axle and wheel both rotate at the same time about the vertical axis with constant angular speed Ω. With the aid of Lagrange multipliers, calculate the normal force from the ground on the wheel.

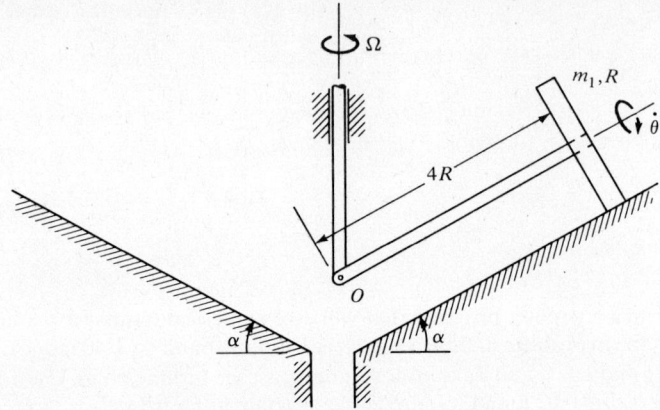

8.31 Derive Lagrange's equations of motion for the rate gyro of Figure 7.17. The spin rate $\dot{\psi}$ is constant, and the torsional spring constant is equal to k_θ with an unstretched position at $\pi/2$. Take the mass polar moment of inertia of the rotor as C and the transverse moment of inertia as A.

8.32 Rework problem 8.31, but find Routh's equations of motion.

8.33 Rework problem 8.31 with $\dot{\phi}$ equal to a given function of time $\dot{\phi}=f(t)$.

*8.34 Write a computer program to obtain results for problem 8.33. Let $C=10\,000$ g·cm², $A=5000$ g·cm², and $k_\theta = 0.2\times 10^5$ dyne·cm/rad. At $t=0$, $\theta=\pi/2$, $\dot{\psi}=2000$ rpm, $\dot{\theta}=0$, and $\dot{\phi}$ is given by the following graph. Use a communication interval of 0.2 s and terminate the program at $t=4$ s.

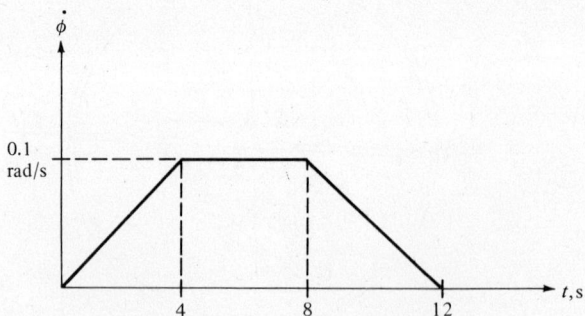

8.35 A satellite having the shape of a barbell is placed in a plane orbit about the earth. The ends of the barbell can be considered as point masses having mass m. The stock of the barbell can be assumed massless having length L. Its length is very small in comparison with the distance from the center of the earth. Derive Lagrange's equation of motion for the barbell.

8.36 A satellite in circular orbit having radius $3R$ about the earth ejects a film capsule so that the capsule's initial motion is parallel to that of the satellite. The capsule is to land on the surface of the earth with a motion almost locally horizontal. Neglecting the effects of air resistance, and assuming the film capsule suitably protected, what must the change in speed of the capsule be when it leaves the satellite?

8.37 On May 15, 1963, a spacecraft commanded by astronaut Cooper was put into an orbit having an altitude of 160.9 km at perigee and 267.3 km at apogee. Find what the eccentricity of the orbit was. How long did it take to orbit the earth? ($GM_e = gR_e^2 = 3.988 \times 10^{14}$ m^3/s^2, $R_e = 6371$ km.)

THEORY OF SMALL OSCILLATIONS

9.1 INTRODUCTION

It is the purpose of this chapter to study the small oscillations of an m degrees of freedom system about a position of static equilibrium and then, later, about a state of steady motion. The reason for limiting the discussion to small oscillations is that otherwise, i.e., for large motion, nonlinear differential equations result. Though these cases are certainly of interest, there is no unique mathematical theory established to treat this class of problems, and only numerical solutions are available. Solely numerical results hinder insight into the common features of oscillatory motion that are present in all vibrating systems.

To develop a linear theory, some assumptions have to be made. The discussion will begin with the case of a *conservative system*, displaced slightly from a static equilibrium position. The coordinates used for the problem will be measured from this equilibrium points, and the potential of the system at that point, a constant, will be arbitrarily set equal to zero. Since only small motion is allowed, only the first nonzero term of any Taylor series expansion about the equilibrium position will be considered significant. Also, constraints that are functions of time will be excluded from the discussion.

The potential V, with the above assumption, can then be expanded about the equilibrium point to yield

$$V = V_0 + \sum_{j=1}^{m} \left(\frac{\partial V}{\partial q_j}\right)_0 q_j + \frac{1}{2} \sum_{\substack{j=1 \\ l=1}}^{m} \left(\frac{\partial^2 V}{\partial q_j \partial q_l}\right)_0 q_j q_l + \cdots \tag{9.1}$$

For equilibrium (see Section 8.3),

$$\left(\frac{\partial V}{\partial q_j}\right)_0 = 0 \tag{8.16}$$

and

$$V_0 \equiv 0 \tag{9.2}$$

Then

$$V \cong \frac{1}{2} \sum_{\substack{j=1 \\ l=1}}^{m} k_{jl} q_j q_l \tag{9.3}$$

where

$$k_{jl} = \left(\frac{\partial^2 V}{\partial q_j \partial q_l}\right)_0 = k_{lj} = \text{constant} \tag{9.4}$$

For a stable equilibrium position the quadratic function for V must be positive definite [equation (8.19)].

The kinetic energy term can now also be approximated for small motion.

$$T = \sum_{i=1}^{n} \frac{m_i \mathbf{v}_i \cdot \mathbf{v}_i}{2} \tag{9.5}$$

$$\mathbf{r}_i = (\mathbf{r}_i)_0 + \sum_{j=1}^{m} \left(\frac{\partial \mathbf{r}_i}{\partial q_j}\right)_0 dq_j + \frac{1}{2} \sum_{\substack{j=1 \\ l=1}}^{m} \left(\frac{\partial^2 \mathbf{r}_i}{\partial q_j \partial q_l}\right)_0 dq_j\, dq_l + \cdots \tag{9.6}$$

With $\partial \mathbf{r}_i/\partial t = 0$ assumed,

$$\mathbf{v}_i \cong \sum_{j=1}^{m} \left(\frac{\partial \mathbf{r}_i}{\partial q_j}\right)_0 \dot{q}_j \tag{9.7}$$

$$T = \sum_{i=1}^{n} \sum_{\substack{j=1 \\ l=1}}^{m} \frac{m_i}{2} \left(\frac{\partial \mathbf{r}_i}{\partial q_j}\right)_0 \cdot \left(\frac{\partial \mathbf{r}_i}{\partial q_l}\right)_0 \dot{q}_j \dot{q}_l \tag{9.8}$$

$$T = \sum_{\substack{j=1 \\ l=1}}^{m} \frac{m_{jl}}{2} \dot{q}_j \dot{q}_l \tag{9.9}$$

where

$$m_{jl} = \sum_{i=1}^{n} m_i \left(\frac{\partial \mathbf{r}_i}{\partial q_j}\right)_0 \cdot \left(\frac{\partial \mathbf{r}_i}{\partial q_l}\right)_0 = m_{lj} = \text{constant} \tag{9.10}$$

Since $T \geq 0$, the quadratic form of equation (9.9) is also positive definite.

For small oscillations about a position of static equilibrium the lagrangian becomes

$$L = \frac{1}{2} \sum_{\substack{j=1 \\ l=1}}^{m} m_{jl} \dot{q}_j \dot{q}_l - \frac{1}{2} \sum_{\substack{j=1 \\ l=1}}^{m} k_{jl} q_j q_l \tag{9.11}$$

The m resultant linear differential equations of motion are, for each j index,

$$\sum_{l=1}^{m} m_{jl} \ddot{q}_l + \sum_{l=1}^{m} k_{jl} q_l = 0 \qquad (j = 1, 2, \ldots, m) \tag{9.12}$$

The mathematical theory concerning the solution of a set of m second-order linear differential equations is well-developed and will be discussed in the following section.

Sample Problem 9.1

Rederive the lagrangian and Lagrange's equation of motion for Sample Problem 8.4. Assume small motion now.

Analysis: Defining y now to be measured from the *equilibrium position*,

$$T = \frac{m v_m^2}{2} = \frac{m}{2} (l^2 \dot{\theta}^2 + \dot{y}^2 - 2 \dot{y} l \dot{\theta} \sin \theta)$$

322 ANALYTICAL MECHANICS

For small motion,
$$T \simeq \frac{m}{2}(l^2\dot{\theta}^2 + \dot{y}^2)$$

$$V = -mg(y + l\cos\theta) + \int_0^y (ky' + mg)\,dy'$$

$$= -mg\left(y + l - l\frac{\theta^2}{2}\right) + \frac{k}{2}y^2 + mgy$$

where
$$\cos\theta = 1 - \frac{\theta^2}{2} + \cdots$$

Then
$$V \simeq -mg\left(l - l\frac{\theta^2}{2}\right) + \frac{k}{2}y^2$$

and with V at the equilibrium point ($\theta = 0$) defined to be zero,
$$V \simeq \frac{mgl\theta^2}{2} + \frac{ky^2}{2}$$

$$T - V = \frac{m}{2}(l^2\dot{\theta}^2 + \dot{y}^2) - \frac{mgl\theta^2}{2} - \frac{ky^2}{2}$$

$$m_{11} = ml^2 \qquad m_{12} = m_{21} = 0 \qquad m_{22} = m$$
$$k_{11} = mgl \qquad k_{12} = 0 = k_{21} \qquad k_{22} = k$$

from
$$T - V = \frac{1}{2}\sum_{\substack{j=1 \\ l=1}}^{m} m_{jl}\dot{q}_j\dot{q}_l - \frac{1}{2}\sum_{\substack{j=1 \\ l=1}}^{m} k_{jl}q_jq_l$$

The linearized equations of motion then become
$$ml^2\ddot{\theta} + mgl\theta = 0 \qquad m\ddot{y} + ky = 0 \qquad \blacksquare$$

9.2 NORMAL MODES OF VIBRATION

Equations (9.11) and (9.12) can be rewritten in matrix form, namely,
$$L = \tfrac{1}{2}\{\dot{q}\}^t[m]\{\dot{q}\} - \tfrac{1}{2}\{q\}^t[k]\{q\} \qquad (9.13)$$

and
$$[m]\{\ddot{q}\} + [k]\{q\} = \{0\} \qquad (9.14)$$

where $[k]$ is called the *stiffness* matrix. $[k]^{-1}$ is called the *flexibility* matrix. If the solution to equation (9.14) is assumed to be of the form
$$q_j = u_j C \cos(\omega t + \phi) \qquad (j = 1, 2, \ldots, m) \qquad (9.15)$$

then substituting equation (9.15) into equation (9.14) yields
$$-\omega^2[m]\{u\} + [k]\{u\} = \{0\} \qquad (9.16)$$

or
$$([k] - \omega^2[m])\{u\} = \{0\} \tag{9.17}$$

Equation (9.17) represents another form of the familiar *eigenvalue* problem. The determinant of the coefficients must be set to zero.

$$|[k] - \omega^2[m]| = 0 \tag{9.18}$$

The *characteristic* equation, or polynomial, that results leads to a solution for the m eigenvalues: $\omega_1^2, \omega_2^2, \ldots, \omega_m^2$. The natural frequencies (in rad/s) of the system are equal to the positive square root of the eigenvalues.

Since both V and T are positive-definite quadratic functions and the matrices $[m]$ and $[k]$ are real and symmetric, it can be shown that all the eigenvalues are real. The system will oscillate with small amplitudes about a state of static equilibrium. Negative eigenvalues are obtained for systems having unstable equilibrium. The resulting exponentially increasing solutions obtained indicate that the regions of small oscillations are soon left and the mathematical model is no longer valid. Eigenvalues with zero magnitude indicate neutral stability. In effect, it represents "rigid body" motion. The whole system moves together at constant speed without relative displacement.

As was done in Section 6.3, assuming the eigenvalues are real and positive or zero, an eigenvalue can be reintroduced into equation (9.17) to solve for its *eigenvector* or *normal mode vector* $\{u\}$. Again, no specific quantities can be solved for. Only a relative relationship between the displacement in each mode is furnished. These relative values at each natural frequency are called the *mode shapes*.

Assuming that all eigenvalues are distinct (in practice this is the usual physical case), one can rewrite equation (9.16) for the ith eigenvalue and mode shape as

$$[k]\{u_i\} = \omega_i^2[m]\{u_i\} \tag{9.19}$$

and for the lth mode

$$[k]\{u_l\} = \omega_l^2[m]\{u_l\} \tag{9.20}$$

Premultiplying equation (9.19) by $[u_l]^t$ yields

$$\{u_l\}^t[k]\{u_i\} = \omega_i^2\{u_l\}^t[m]\{u_i\} \tag{9.21}$$

Premultiplying equation (9.20) by $[u_i]^t$ furnishes

$$\{u_i\}^t[k]\{u_l\} = \omega_l^2\{u_i\}^t[m]\{u_l\} \tag{9.22}$$

Transposing each side of equation (9.22), with $[m]$ and $[k]$ symmetric, results in

$$\{u_l\}^t[k]\{u_i\} = \omega_l^2\{u_l\}^t[m]\{u_i\} \tag{9.23}$$

Subtracting equation (9.23) from equation (9.21) gives

$$0 = (\omega_i^2 - \omega_l^2)\{u_l\}^t[m]\{u_i\} \tag{9.24}$$

324   ANALYTICAL MECHANICS

For distinct eigenvalues this means that

$$\{u_l\}^t[m]\{u_i\} = 0 \tag{9.25}$$

Equation (9.25) is known as the *generalized orthogonality* condition. This relation is used quite often in numerical analysis to "sweep out" unwanted mode shapes and frequencies from iteration attempts. For $l = i$

$$\{u_i\}^t[m]\{u_i\} = M_i \tag{9.26}$$

$$\{u_i\}^t[k]\{u_i\} = K_i \tag{9.27}$$

These terms are defined to be the *generalized mass* and *generalized stiffness* in the ith mode, respectively.

The *modal matrix* is formed by combining the orthogonal eigenvectors as follows:

$$[U] = [\{u_1\} \quad \{u_2\} \quad \cdots \quad \{u_m\}] \tag{9.28}$$

Then the *generalized mass matrix* can be found by

$$[U]^t[m][U] = [M] \tag{9.29}$$

which must be diagonal because of the generalized orthogonality conditions. An equation similar to equation (9.24) can be developed in terms of the stiffness matrix, namely,

$$\left(\frac{1}{\omega_l^2} - \frac{1}{\omega_i^2}\right)\{u_l\}^t[k]\{u_i\} = 0 \tag{9.30}$$

and for $\omega_l^2 \neq \omega_i^2$

$$\{u_l\}^t[k]\{u_i\} = 0 \tag{9.31}$$

The *generalized stiffness matrix* then, which also must be *diagonal* for distinct eigenvalues, is given by

$$[U]^t[k][U] = [K] \tag{9.32}$$

Sample Problem 9.2
For the three-mass system shown, calculate the natural frequencies and the normal mode shapes. Sketch the mode shapes. Find the generalized mass and stiffness matrices. Show that the generalized orthogonality condition holds true.

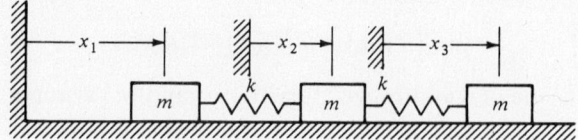

Analysis:

$$L = \frac{m}{2}(\dot{x}_1^2 + \dot{x}_2^2 + \dot{x}_3^2) - k\frac{(x_2 - x_1)^2}{2} - k\frac{(x_3 - x_2)^2}{2}$$

Then the equations of motion are

$$m\ddot{x}_1 - k(x_2 - x_1) = 0$$
$$m\ddot{x}_2 + k(x_2 - x_1) - k(x_3 - x_2) = 0$$
$$m\ddot{x}_3 + k(x_3 - x_2) = 0$$

In matrix form this becomes

$$\begin{bmatrix} m & 0 & 0 \\ 0 & m & 0 \\ 0 & 0 & m \end{bmatrix} \begin{Bmatrix} \ddot{x}_1 \\ \ddot{x}_2 \\ \ddot{x}_3 \end{Bmatrix} + \begin{bmatrix} k & -k & 0 \\ -k & 2k & -k \\ 0 & -k & k \end{bmatrix} \begin{Bmatrix} x_1 \\ x_2 \\ x_3 \end{Bmatrix} = \{0\} \qquad (a)$$

Assuming a solution of the form $x_i = u_i C \cos(\omega t + \phi)$, equation (a) becomes

$$\begin{bmatrix} k - m\omega^2 & -k & 0 \\ -k & 2k - m\omega^2 & -k \\ 0 & -k & k - m\omega^2 \end{bmatrix} \begin{Bmatrix} u_1 \\ u_2 \\ u_3 \end{Bmatrix} = \{0\} \qquad (b)$$

The determinant of the coefficient matrix in equation (b) must equal zero. Proceeding then,

$$(k - m\omega^2)(2k - m\omega^2)(k - m\omega^2) - 2k^2(k - m\omega^2) = 0$$

$$\omega_1^2 = \frac{k}{m}$$

$$(2k - m\omega^2)(k - m\omega^2) - 2k^2 = 0$$

$$m\omega^2(m\omega^2 - 3k) = 0$$

$$\omega_2^2 = 0 \qquad \omega_3^2 = \frac{3k}{m}$$

The eigenvalue $\omega_2 = 0$ corresponds to the rigid body mode.

In the first mode with $\omega_1^2 = k/m$,

$$\frac{u_{21}}{u_{11}} = \frac{k-k}{k} = 0 \qquad \frac{u_{31}}{u_{11}} = -1$$

Similarly in the second mode with $\omega_2^2 = 0$,

$$\frac{u_{22}}{u_{12}} = 1 \qquad \frac{u_{32}}{u_{12}} = 1$$

and in the third mode with $\omega_3^2 = 3k/m$,

$$\frac{u_{23}}{u_{13}} = -2 \qquad \frac{u_{33}}{u_{13}} = 1$$

Then $\{u_1\} = \begin{Bmatrix} 1 \\ 0 \\ -1 \end{Bmatrix} \quad \{u_2\} = \begin{Bmatrix} 1 \\ 1 \\ 1 \end{Bmatrix} \quad \{u_3\} = \begin{Bmatrix} 1 \\ -2 \\ 1 \end{Bmatrix}$

326 ANALYTICAL MECHANICS

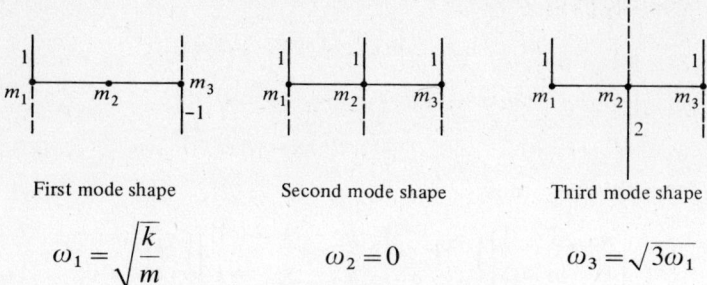

First mode shape　　Second mode shape　　Third mode shape

$$\omega_1 = \sqrt{\frac{k}{m}} \qquad \omega_2 = 0 \qquad \omega_3 = \sqrt{3\omega_1}$$

The modes can be *normalized* so that the mode vector has a magnitude of 1.

$$\{u_1\} = \frac{1}{\sqrt{2}} \begin{Bmatrix} 1 \\ 0 \\ -1 \end{Bmatrix} \qquad \{u_2\} = \frac{1}{\sqrt{3}} \begin{Bmatrix} 1 \\ 1 \\ 1 \end{Bmatrix} \qquad \{u_3\} = \frac{1}{\sqrt{6}} \begin{Bmatrix} 1 \\ -2 \\ 1 \end{Bmatrix}$$

The modal matrix is

$$[U] = [\{u_1\} \quad \{u_2\} \quad \{u_3\}]$$

or

$$[U] = \begin{bmatrix} 1 & 1 & 1 \\ 0 & 1 & -2 \\ -1 & 1 & 1 \end{bmatrix}$$

The generalized mass matrix becomes

$$[U]^t[m][U] = \begin{bmatrix} 1 & 0 & -1 \\ 1 & 1 & 1 \\ 1 & -2 & 1 \end{bmatrix} \begin{bmatrix} m & 0 & 0 \\ 0 & m & 0 \\ 0 & 0 & m \end{bmatrix} \begin{bmatrix} 1 & 1 & 1 \\ 0 & 1 & -2 \\ -1 & 1 & 1 \end{bmatrix}$$

$$= \begin{bmatrix} 2m & 0 & 0 \\ 0 & 3m & 0 \\ 0 & 0 & 6m \end{bmatrix} = [M]$$

The generalized stiffness matrix is

$$[U]^t[k][U] = \begin{bmatrix} 1 & 0 & -1 \\ 1 & 1 & 1 \\ 1 & -2 & 1 \end{bmatrix} \begin{bmatrix} k & -k & 0 \\ -k & 2k & -k \\ 0 & -k & k \end{bmatrix} \begin{bmatrix} 1 & 1 & 1 \\ 0 & 1 & -2 \\ -1 & 1 & 1 \end{bmatrix}$$

$$= \begin{bmatrix} 2k & 0 & 0 \\ 0 & 0 & 0 \\ 0 & 0 & 18k \end{bmatrix} = [K]$$

The generalized orthogonality condition between modes one and two yields

$$\{u_1\}^t[m]\{u_2\} = \{1 \quad 0 \quad -1\} \begin{bmatrix} m & 0 & 0 \\ 0 & m & 0 \\ 0 & 0 & m \end{bmatrix} \begin{Bmatrix} 1 \\ 1 \\ 1 \end{Bmatrix} = \{1 \quad 0 \quad -1\} \begin{Bmatrix} m \\ m \\ m \end{Bmatrix} = 0$$

The same holds true for

$$\{u_1\}^t[m]\{u_3\} \quad \text{and} \quad \{u_2\}^t[m]\{u_3\}$$

The homogeneous solution must be of the form

$$q_1 = u_{11}C_1 \cos(\omega_1 t + \phi_1) + u_{13}C_3 \cos(\omega_3 t + \phi_3) + (C_2 + C_4 t)$$
$$q_2 = u_{21}C_1 \cos(\omega_1 t + \phi_1) + u_{23}C_3 \cos(\omega_3 t + \phi_3) + (C_2 + C_4 t)$$
$$q_3 = u_{31}C_1 \cos(\omega_1 t + \phi_1) + u_{33}C_3 \cos(\omega_3 t + \phi_3) + (C_2 + C_4 t)$$

where $u_{12} = u_{22} = u_{32} = 1$.

$$\{q\} = \{u_1\}C_1 \cos(\omega_1 t + \phi_1) + \{u_3\}C_3(\omega_3 t + \phi) + \{u_2\}(C_2 + C_4 t)$$

Note that the last factor, for the second mode shape $\{u_2\}$, represents rigid body motion. If this mode, and only this mode, is excited, all three masses will move together at a constant speed.

To "drive" one mode alone requires that the initial conditions be such that only that mode is excited. For example, if the three masses are released from a rest position with relative displacements corresponding to the first mode, then only the first mode will subsequently appear when the system vibrates. The general case is a combination of all three mode shapes. Keep in mind that the mode vector shows magnitude ratios at any instant. For example, for mode shape 1, as time progresses, masses 1 and 3 oscillate at the natural frequency ω_1, while mass 2 remains stationary. Their displacements maintain the constant relative ratio given by $\{u_1\}$. When displacement of mass 1 becomes negative, the displacement of mass 3 becomes positive, etc. ∎

9.3 REPEATED ROOTS

When the characteristic equation, equation (9.18), has a repeated root, the associated eigenvectors are not unique. Any linear combination of these eigenvectors must satisfy the state equation, equation (9.17); hence, no unique mode exists. However, using the nonunique eigenvectors, it is possible to build mutually orthogonal vectors that are at the same time perpendicular to the rest of the system eigenvectors. The complete set of eigenvectors will now be linearly independent and may be used in an ordinary manner for further system treatment.

9.4 NORMAL COORDINATES

If the mass matrix of equation (9.13) is not diagonal, the system is said to be *dynamically coupled*. One speaks of *static coupling* if the stiffness matrix is not diagonal. With any form of coupling present, the system equations are difficult to solve since each equation of motion will affect the other. To uncouple the equations for conservative systems, it is possible to transform the original generalized coordinates of the problem $[q]$ to a set of *normal coordinates* $[y]$

328 ANALYTICAL MECHANICS

for which each developed equation of motion now becomes independent of the others. In these new coordinates, the system equations integrate at once, and constants of integration are readily determined for both the homogeneous and forced cases. By this means, a general m degrees of freedom system is in effect converted to a system of m simple independent oscillators. *To affect the transformation, the mode vector must first be known.* Assume a solution to the differential equations of the form

$$\{q\} = [U]\{y\} \tag{9.33}$$

where $[y]$ represents the new *normal coordinates*. Then substitution into equation (9.14) yields

$$[m][U]\{\ddot{y}\} + [k][U]\{y\} = \{0\} \tag{9.34}$$

and, for the nonhomogeneous case,

$$[m][U]\{\ddot{y}\} + [k][U]\{y\} = \{Q\} \tag{9.35}$$

Premultiplying the equation for the homogeneous case by $[U]^t$

$$[U]^t[m][U]\{\ddot{y}\} + [U]^t[k][U]\{y\} = \{0\} \tag{9.36}$$

and reducing it by means of equation (9.29) furnishes

$$[M]\{\ddot{y}\} + [U]^t[k][U]\{y\} = \{0\} \tag{9.37}$$

where $[M]$ is the diagonal generalized mass matrix.

Recalling that for the ith mode

$$[k]\{u_i\} = \omega_i^2[m]\{u_i\} \tag{9.19}$$

and that

$$\{u_l\}^t[k]\{u_i\} = \omega_i^2\{u_l\}^t[m]\{u_i\} \tag{9.21}$$

the generalized orthogonality condition has

$$\{u_i\}^t[k]\{u_i\} = \omega_i^2 M_i \quad \text{for } i=l$$

or

$$\{u_l\}^t[k]\{u_i\} = 0 \quad \text{for } i \neq l$$

or

$$[U]^t[k][U] = [K] = [M][\omega^2] \tag{9.38}$$

where $[\omega^2]$ is a *diagonal* matrix having elements ω_i^2. Then equation (9.37) simplifies further to

$$[M]\{\ddot{y}\} + [M][\omega^2]\{y\} = \{0\} \tag{9.39}$$

or

$$[M]\{\ddot{y}\} + [M]\{\omega^2\}\{y\} = [U]^t\{Q\} \tag{9.40}$$

for the nonhomogeneous case.

Since both $[M]$ and $[\omega^2]$ are diagonal, equation (9.39) represents a set of m independent uncoupled linear differential equations of the form

$$M_i \ddot{y}_i + \omega_i^2 M_i y_i = 0 \tag{9.41}$$

THEORY OF SMALL OSCILLATIONS 329

which have the general solution

$$y_i = C_i \cos \omega_i t + D_i \sin \omega_i t \qquad (9.42)$$

Sample Problem 9.3
For sample Problem 9.2 find the normal coordinates present. If $x_1(0) = \dot{x}_1(0) = x_2(0) = \dot{x}_3(0)$, and $\dot{x}_2(0) = 1$ while $x_3(0) = 2$, solve for $\{y\}$ as a function of time. What is $\{x\}$ as a function of time?

Analysis: From equation (9.33) one can write

$$\{q\} = \{x\} = [U]\{y\}$$
$$[m]\{q\} = [m][U]\{y\}$$
$$\{U\}^t[m]\{q\} = \{U\}^t[m][U]\{y\} = [M]\{y\} \qquad (a)$$

From the previous sample problem,

$$[U]^t = \begin{bmatrix} 1 & 0 & -1 \\ 1 & 1 & 1 \\ 1 & -2 & 1 \end{bmatrix} \qquad [M] = m \begin{bmatrix} 2 & 0 & 0 \\ 0 & 3 & 0 \\ 0 & 0 & 6 \end{bmatrix}$$

and

$$[m] = m \begin{bmatrix} 1 & 0 & 0 \\ 0 & 1 & 0 \\ 0 & 0 & 1 \end{bmatrix}$$

Equation (a) becomes

$$m \begin{bmatrix} 1 & 0 & -1 \\ 1 & 1 & 1 \\ 1 & -2 & 1 \end{bmatrix} \begin{Bmatrix} q_1 \\ q_2 \\ q_3 \end{Bmatrix} = m \begin{Bmatrix} 2y_1 \\ 3y_2 \\ 6y_3 \end{Bmatrix} \qquad (b)$$

or

$$y_1 = \tfrac{1}{2}(q_1 - q_3)$$
$$y_2 = \tfrac{1}{3}(q_1 + q_2 + q_3)$$
$$y_3 = \tfrac{1}{6}(q_1 - 2q_2 + q_3) \qquad (c)$$

At $t = 0$, and with $\{q\} = \{x\}$,

$$\begin{Bmatrix} 2y_1 \\ 3y_2 \\ 6y_3 \end{Bmatrix} = \begin{bmatrix} 1 & 0 & -1 \\ 1 & 1 & 1 \\ 1 & -2 & 1 \end{bmatrix} \begin{Bmatrix} 0 \\ 0 \\ 2 \end{Bmatrix} = \begin{Bmatrix} 2C_1 \\ 3C_2 \\ 6C_3 \end{Bmatrix}$$

and

$$\begin{Bmatrix} 2\dot{y}_1 \\ 3\dot{y}_2 \\ 6\dot{y}_3 \end{Bmatrix} = \begin{bmatrix} 1 & 0 & -1 \\ 1 & 1 & 1 \\ 1 & -2 & 1 \end{bmatrix} \begin{Bmatrix} 0 \\ 1 \\ 0 \end{Bmatrix} = \begin{Bmatrix} 2D_1\omega_1 \\ 3D_2 \\ 6D_3\omega_3 \end{Bmatrix}$$

where from equation (9.41),

$$y_i = \begin{cases} C_i \cos \omega_i t + D_i \sin \omega_i t & \text{for } \omega_i \neq 0 \\ C_i + D_i t & \text{for } \omega_i = 0 \end{cases}$$

330 ANALYTICAL MECHANICS

Therefore

$$C_1 = -1 \quad C_2 = \tfrac{2}{3} \quad C_3 = \tfrac{1}{3} \quad D_1 = 0 \quad D_2 = \tfrac{1}{3} \quad D_3 = -\frac{1}{3\omega_3}$$

$$y_1 = -\cos\sqrt{\frac{k}{m}}\,t$$

$$y_2 = \tfrac{2}{3} + \tfrac{1}{3}t \tag{d}$$

$$y_3 = \frac{1}{3}\left(\cos\sqrt{\frac{3k}{m}}\,t - \sqrt{\frac{m}{3k}}\sin\sqrt{\frac{3k}{m}}\,t\right)$$

Then
$$\{x\} = [U]\{y\} = \begin{bmatrix} 1 & 1 & 1 \\ 0 & 1 & -2 \\ -1 & 1 & 1 \end{bmatrix}\begin{Bmatrix} y_1 \\ y_2 \\ y_3 \end{Bmatrix}$$

or

$$x_1 = -\cos\sqrt{\frac{k}{m}}\,t + \frac{2}{3} + \frac{t}{3} + \frac{1}{3}\left(\cos\sqrt{\frac{3k}{m}}\,t - \sqrt{\frac{m}{3k}}\sin\sqrt{\frac{3k}{m}}\,t\right)$$

$$x_2 = \frac{2}{3} + \frac{t}{3} - \frac{2}{3}\left(\cos\sqrt{\frac{3k}{m}}\,t - \sqrt{\frac{m}{3k}}\sin\sqrt{\frac{3k}{m}}\,t\right)$$

$$x_3 = \cos\sqrt{\frac{k}{m}}\,t + \frac{2}{3} + \frac{t}{3} + \frac{1}{3}\left(\cos\sqrt{\frac{3k}{m}}\,t - \sqrt{\frac{m}{3k}}\sin\sqrt{\frac{3k}{m}}\,t\right)$$

It is left as an exercise for the reader to directly develop $\{x\}$ as a function of time in Sample Problem 9.2 for the same initial conditions. ∎

Sample Problem 9.4
Shown in the figure is a two-mass system driven by a harmonic force $F\sin\omega t$ applied to mass 1. Neglecting the transient solution, develop the steady-state response of the system using normal coordinates. If $\omega = \sqrt{k/m}$, what is x_1 as a function of time?

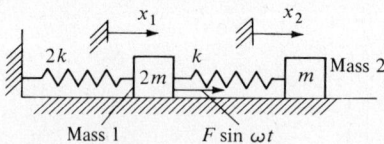

Analysis: The lagrangian of the system is given by

$$L = \frac{2m\dot{x}_1^2}{2} + \frac{m\dot{x}_2^2}{2} - \frac{2kx_1^2}{2} - \frac{k(x_2 - x_1)^2}{2}$$

Lagrange's equations become

$$2m\ddot{x}_1 + 3kx_1 - kx_2 = Q_1 = F\sin\omega t$$

$$m\ddot{x}_2 + kx_2 - kx_1 = Q_2 = 0$$

THEORY OF SMALL OSCILLATIONS

$$\begin{bmatrix} 2m & 0 \\ 0 & m \end{bmatrix} \begin{Bmatrix} \ddot{x}_1 \\ \ddot{x}_2 \end{Bmatrix} + \begin{bmatrix} 3k & -k \\ -k & k \end{bmatrix} \begin{Bmatrix} x_1 \\ x_2 \end{Bmatrix} = \begin{Bmatrix} F \sin \omega t \\ 0 \end{Bmatrix}$$

To find the normal modes, set $x_i = q_i = u_i \cos(\hat{\omega} t + \phi)$,

$$\begin{bmatrix} 3k - 2m\hat{\omega}^2 & -k \\ -k & k - m\hat{\omega}^2 \end{bmatrix} \begin{Bmatrix} u_1 \\ u_2 \end{Bmatrix} = \{0\}$$

The characteristic equation becomes

$$(3k - 2m\hat{\omega}^2)(k - m\hat{\omega}^2) - k^2 = 0$$

$$\left[3 - 2\left(\frac{\hat{\omega}}{\omega_0}\right)^2\right]\left[1 - \left(\frac{\hat{\omega}}{\omega_0}\right)^2\right] - 1 = 0$$

where
$$\omega_0^2 = \frac{k}{m}$$

Then
$$2 - 5\left(\frac{\hat{\omega}}{\omega_0}\right)^2 + 2\left(\frac{\hat{\omega}}{\omega_0}\right)^4 = 0$$

$$\left(\frac{\hat{\omega}}{\omega_0}\right)_{1,2} = \sqrt{0.5}, \sqrt{2}$$

For $(\hat{\omega}_1/\omega_0)^2 = 0.5$, with $u_{11} = 1$, $u_{21} = 2$, and for $(\hat{\omega}_2/\omega_0)^2 = 2$, with $u_{12} = 1$, $u_{22} = -1$, the modal matrix becomes

$$[U] = \begin{bmatrix} 1 & 1 \\ 2 & -1 \end{bmatrix} \qquad [U]^t = \begin{bmatrix} 1 & 2 \\ 1 & -1 \end{bmatrix}$$

With
$$\{q\} = \{x\} = [U]\{y\}$$

then
$$[m][U]\{\ddot{y}\} + [k][U]\{y\} = \{Q\}$$

$$[U]^t[m][U]\{\ddot{y}\} + [U]^t[k][U]\{y\} = [U]^t\{Q\}$$

$$[M]\{\ddot{y}\} + [M][\omega^2]\{y\} = [U]^t\{Q\} \qquad \text{(a)}$$

Equation (a) expanded is

$$m\begin{bmatrix} 6 & 0 \\ 0 & 3 \end{bmatrix}\begin{Bmatrix} \ddot{y}_1 \\ \ddot{y}_2 \end{Bmatrix} + m\begin{bmatrix} 6 & 0 \\ 0 & 3 \end{bmatrix}\begin{bmatrix} \hat{\omega}_1^2 & 0 \\ 0 & \hat{\omega}_2^2 \end{bmatrix}\begin{Bmatrix} y_1 \\ y_2 \end{Bmatrix} = \begin{bmatrix} 1 & 2 \\ 1 & -1 \end{bmatrix}\begin{Bmatrix} F \sin \omega t \\ 0 \end{Bmatrix}$$

This becomes

$$\ddot{y}_1 + \hat{\omega}_1^2 y_1 = \frac{F \sin \omega t}{6m}$$

$$\ddot{y}_2 + \hat{\omega}_2^2 y_2 = \frac{F \sin \omega t}{3m}$$

or
$$\ddot{y}_1 + \frac{\omega_0^2}{2} y_1 = \frac{F \sin \omega t}{6m}$$

$$\ddot{y}_2 + 2\omega_0^2 y_2 = \frac{F \sin \omega t}{3m} \qquad \text{(b)}$$

The steady-state solution for y_1 and y_2 immediately follows due to the uncoupled nature of the transformed equations of motion. Namely,

$$y_1 = \frac{F \sin \omega t}{3m[1 - 2(\omega/\omega_0)^2]\omega_0^2}$$

$$y_2 = \frac{F \sin \omega t}{3m[2 - (\omega/\omega_0)^2]\omega_0^2}$$

$$\begin{Bmatrix} x_1 \\ x_2 \end{Bmatrix} = \begin{bmatrix} 1 & 1 \\ 2 & -1 \end{bmatrix} \begin{Bmatrix} y_1 \\ y_2 \end{Bmatrix}$$

$$x_1 = \frac{F \sin \omega t}{m\omega_0^2} \left(\frac{1 - (\omega/\omega_0)^2}{[2 - (\omega/\omega_0)^2][1 - 2(\omega/\omega_0)^2]} \right)$$

and

$$x_2 = \frac{F \sin \omega t}{m\omega_0^2} \left(\frac{1}{[2 - (\omega/\omega_0)^2][1 - 2(\omega/\omega_0)^2]} \right)$$

If $\omega^2 = k/m = \omega_0^2$, $x_1 = 0$! The vibration has been *absorbed*. The second mass, proportioned properly, has canceled the motion of the first mass. The second mass in the system is called a *vibration absorber*.

The second mass, for $\omega^2 = k/m$, has

$$x_2 = -\frac{F \sin \omega t}{k}$$

The force in the second spring then is

$$x_2 k = -F \sin \omega t$$

which *cancels* the force on the first mass. A sketch of the frequency response for x_1 follows (dashed lines indicate a negative value for x_1):

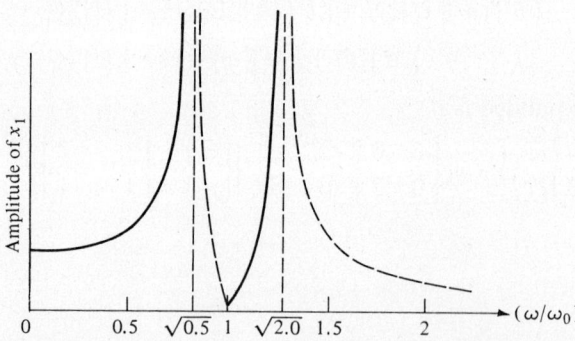

Before proceeding to the damped case, it should also be pointed out that using normal mode analysis allows the engineer to neglect frequency components that have little influence. Solutions to the system equations, though then approximate, are easier to obtain. A numerical method that only integrates the equations of motion with respect to time does not leave room for simplification (or for system understanding). ∎

9.5 FORCED VIBRATIONS WITH DAMPING PRESENT

A system of equations representing forced vibration with damping present would be written in matrix form as

$$[m]\{\ddot{q}\} + [c]\{\dot{q}\} + [k]\{q\} = \{Q\} \tag{9.43}$$

where $[c]$ is assumed symmetric and real. In this case the coordinates cannot be uncoupled directly because of the presence of the first derivative terms that cause a phase change. The eigenvalues and eigenvectors are no longer all real but are complex. For stable motion, all the eigenvalues must have negative or zero real parts. One way to simplify the damping problem is to transform the generalized coordinates to *state space*. In state space the velocities act as independent variables as well as the position coordinates, thus forming a $2m$-dimensional space. (Recall Hamilton's phase space.)

In state space equation (9.43) becomes with

$$\{\dot{q}\} = \{v\} \tag{9.44}$$

$$[m]\{\dot{v}\} + [c]\{v\} + [k]\{q\} = \{Q\} \tag{9.45}$$

Rewriting this in terms of the $2m$ *state variables* z_i and their derivatives, equation (9.45) becomes

$$[m]\{\dot{z}\} = -[c]\{z\} - [k]\{z\} + \{Q\} \tag{9.46}$$

or finally,

$$\{\dot{z}\} = [A]\{z\} + [B]\{r\} \tag{9.47}$$

The matrix $[A]$ is called the *system matrix*, and the matrix $[B]$ is called the *input matrix*. Keep in mind that the vector $\{z\}$ is composed of both generalized velocities and coordinates.

To illustrate the theory, assume a 1 degree of freedom problem of the form

$$m\ddot{x} + c\dot{x} + kx = F(t)$$

Then

$$\ddot{x} = -\frac{c}{m}\dot{x} - \frac{k}{m}x + \frac{F(t)}{m}$$

Defining $z_1 = x$ and $z_2 = v$, one has

$$\dot{z}_1 = \dot{x} = v = z_2$$

$$\dot{z}_2 = -\frac{c}{m}z_2 - \frac{k}{m}z_1 + \frac{F(t)}{m}$$

$$\begin{Bmatrix} \dot{z}_1 \\ \dot{z}_2 \end{Bmatrix} = \begin{bmatrix} 0 & 1 \\ -\frac{k}{m} & -\frac{c}{m} \end{bmatrix} \begin{Bmatrix} z_1 \\ z_2 \end{Bmatrix} + \begin{bmatrix} 0 & 0 \\ 0 & \frac{1}{m} \end{bmatrix} \begin{Bmatrix} 0 \\ F(t) \end{Bmatrix}$$

$$[A] = \begin{bmatrix} 0 & 1 \\ -\frac{k}{m} & -\frac{c}{m} \end{bmatrix} \qquad [B] = \begin{bmatrix} 0 & 0 \\ 0 & \frac{1}{m} \end{bmatrix}$$

334 ANALYTICAL MECHANICS

The solution to equation (9.47) may be developed by using Laplace transforms; i.e.,

$$(S[1]-[A])\{Z(S)\} - \{Z(0+)\} = [B]\{R(S)\} \tag{9.48}$$

Ignoring the transient part of the solution,

$$\{Z(S)\} = (S[1]-[A])^{-1}[B]\{R(S)\} \tag{9.49}$$

and

$$\{z(t)\} = \int_0^t [\phi(t-T)][B]\{r(T)\}\, dT \tag{9.50}$$

where the *transition matrix* $[\phi]$ is defined by

$$[\phi] = L^{-1}(S[1]-[A])^{-1} \tag{9.51}$$

To continue with the 1 degree of freedom problem previously discussed,

$$S[1]-[A] = \begin{bmatrix} S & -1 \\ \dfrac{k}{m} & S+\dfrac{c}{m} \end{bmatrix}$$

$$(S[1]-[A])^{-1} = \frac{1}{S^2 + (c/m)S + k/m} \begin{bmatrix} S+\dfrac{c}{m} & 1 \\ -\dfrac{k}{m} & S \end{bmatrix}$$

$$\begin{Bmatrix} Z_1(S) \\ Z_2(S) \end{Bmatrix} = \frac{1}{S^2 + (c/m)S + k/m} \begin{bmatrix} 0 & \dfrac{1}{m} \\ 0 & \dfrac{S}{m} \end{bmatrix} \begin{Bmatrix} 0 \\ F(S) \end{Bmatrix}$$

$$\begin{Bmatrix} Z_1(S) \\ Z_2(S) \end{Bmatrix} = \frac{1}{(S+c/2m)^2 + k/m - (c/2m)^2} \begin{Bmatrix} \dfrac{F(S)}{m} \\ \dfrac{SF(S)}{m} \end{Bmatrix}$$

For simplicity, assume $F(t)$ is an impulse function; then $F(S)=1$ and

$$z_1(t) = \frac{e^{-(c/2m)t}\sin\sqrt{k/m-(c/2m)^2}\,t}{m\sqrt{k/m-(c/2m)^2}}$$

$$z_2(t) = \frac{e^{-(c/2m)t}\sin\sqrt{k/m-(c/2m)^2}\,t}{m\sqrt{k/m-(c/2m)^2}}\left(-\frac{c}{2m}\right) + \frac{e^{-(c/2m)t}}{m}\cos\sqrt{k/m-(c/2m)^2}\,t$$

The concept of *separating the coordinates* can be used to advantage here as well. For the moment, assume a homogeneous problem.

$$\{\dot{z}\} = [A]\{z\} \tag{9.52}$$

A trial solution of the form

$$z_j = u_j e^{\lambda t} \tag{9.53}$$

yields

$$([A] - \lambda_i[1])\{u_i\} = \{0\} \tag{9.54}$$

for the ith mode, where the characteristic equation is

$$\text{Det}\,([A] - \lambda[1]) = 0 \tag{9.55}$$

Then

$$[A]\{u_i\} = \lambda_i\{u_i\} \tag{9.56}$$

$$[A][U] = [U][\lambda] \tag{9.57}$$

where $[U]$ is the modal matrix and $[\lambda]$, the *spectral matrix*, is a diagonal matrix with elements λ_i. By premultiplying by $[U]^{-1}$, a diagonal matrix results:

$$[U]^{-1}[A][U] = [\lambda] \tag{9.58}$$

Suppose now, as before, a solution to the system of equations is attempted by transforming to a new set of coordinates by using

$$\{z\} = [U]\{y\} \tag{9.59}$$

Then

$$\{\dot{z}\} = [A]\{z\} + [B]\{r\} \tag{9.47}$$

becomes

$$[U]\{\dot{y}\} = [A][U]\{y\} + [B]\{r\} \tag{9.60}$$

Premultiplying by $[U]^{-1}$ yields

$$\{\dot{y}\} = [\lambda]\{y\} + [U]^{-1}[B]\{r\} \tag{9.61}$$

and the coordinates have once more been uncoupled. The solution to the system of equations for $\{y\}$ is now direct. The transformation $\{z\} = [U]\{y\}$ transforms the variables back to state space.

Sample Problem 9.5
For the 1 degree of freedom problem previously described, find the eigenvalues of the system matrix. Find the modal matrix. Discuss the stability of the system. Uncouple the system equations, and solve for $\{y\}$ and $\{z\}$ assuming that $F(t)$ is a unit impulse.

Analysis: From the illustrative problem discussed in the text, it is known that

$$\begin{Bmatrix} \dot{z}_1 \\ \dot{z}_2 \end{Bmatrix} = \begin{bmatrix} 0 & 1 \\ -\dfrac{k}{m} & -\dfrac{c}{m} \end{bmatrix} \begin{Bmatrix} z_1 \\ z_2 \end{Bmatrix} + \begin{bmatrix} 0 & 0 \\ 0 & \dfrac{1}{m} \end{bmatrix} \begin{Bmatrix} 0 \\ F(t) \end{Bmatrix}$$

336 ANALYTICAL MECHANICS

The eigenvalues can be found from the equation

$$\begin{bmatrix} -\lambda & 1 \\ -\dfrac{k}{m} & -\left(\dfrac{c}{m}+\lambda\right) \end{bmatrix} \{u\} = \{0\}$$

The determinant of the coefficients must equal zero, which leads to the characteristic equation

$$\lambda^2 + \frac{c}{m}\lambda + \frac{k}{m} = 0$$

$$\lambda_{1,2} = -\frac{c}{2m} \pm \sqrt{\left(\frac{c}{2m}\right)^2 - \frac{k}{m}}$$

The system is stable since both eigenvalues have a negative real part. It is oscillatory if

$$\frac{k}{m} > \left(\frac{c}{2m}\right)^2$$

Assuming this to be the case,

$$\lambda_{1,2} = -\frac{c}{2m} \pm i\omega_d$$

where

$$\omega_d = \sqrt{\frac{k}{m} - \left(\frac{c}{2m}\right)^2}$$

For λ_1,

$$u_{11} = 1 \qquad u_{21} = \lambda_1$$

and for λ_2,

$$u_{12} = 1 \qquad u_{22} = \lambda_2$$

The modal matrix is

$$[U] = \begin{bmatrix} 1 & 1 \\ \lambda_1 & \lambda_2 \end{bmatrix}$$

Further,

$$[\lambda] = \begin{bmatrix} \lambda_1 & 0 \\ 0 & \lambda_2 \end{bmatrix}$$

$$[U]^{-1} = \frac{-1}{2i\omega_d} \begin{bmatrix} \lambda_2 & -1 \\ -\lambda_1 & 1 \end{bmatrix}$$

$$[U]^{-1}[B]\{r\} = \frac{F(t)}{2i\omega_d m} \begin{Bmatrix} 1 \\ -1 \end{Bmatrix}$$

Then
$$\dot{y}_1 - \lambda_1 y_1 = \frac{F(t)}{2i\omega_d m} \quad \text{(a)}$$

$$\dot{y}_2 - \lambda_2 y_2 = -\frac{F(t)}{2i\omega_d m} \quad \text{(b)}$$

and the equations have been uncoupled.

Solving equations (a) and (b) yields for $F(S) = 1$

$$y_1 = \frac{e^{\lambda_1 t}}{2i\omega_d m}, \qquad y_2 = -\frac{e^{\lambda_2 t}}{2i\omega_d m}$$

Since $\{z\} = [U]\{y\}$,

$$z_1 = y_1 + y_2 = x$$

$$x = \frac{e^{-(c/2m)t}}{m\omega_d} \sin \omega_d t$$

and
$$z_2 = \lambda_1 y_1 + \lambda_2 y_2 = v$$

or
$$v = -\frac{c}{2m} \frac{e^{-(c/2m)t}}{m\omega_d} \sin \omega_d t + \frac{e^{-(c/2m)t} \cos \omega_d t}{m}$$

which agrees with previous results. For computer calculations, there is no advantage gained in using this method when damping is present. It is faster to proceed with the direct solution of the differential equations. ∎

9.6 NUMERICAL METHODS

For a system with many degrees of freedom, solutions for the natural frequencies (eigenvalues) and the normal modes (eigenvectors) are obtainable only through numerical means. There are many proven methods available to solve for the eigenvalues and eigenvectors of a matrix using a digital computer. They fall into three categories. In the first category, the numerical analysis begins by solving for the coefficients in the characteristic equation derived from the system matrix. The solution then uses another procedure to solve for the roots of this polynomial. If eigenvectors are desired, further processing is required using a Gauss-Jordan algorithm specially tailored to treat linear simultaneous equations that are homogeneous. The second method of approach transforms the original matrix, through a series of orthogonal transformations to a reduced form that is more amenable to numerical work. For example, Jacobi's method transforms the original matrix through a succession of plane coordinate system rotations until the given matrix is diagonalized (see Sample Problem 6.6). The new diagonal elements of the matrix are the eigenvalues. The eigenvector matrix is the product of the successive rotation matrices. The method produces all the eigenvalues and eigenvectors. It is an inefficient procedure though, since off-diagonal elements that once were reduced to zero in one transformation may inadvertently assume values again after a new transformation. Householder's method employs another approach to matrix reduction. It avoids the slowness of the Jacobi method by transforming the original matrix through a series of orthogonal transformations that reduce most of the elements in a row to zero, row by row. This technique precludes elements returning to nonzero values. The final form of the matrix is tridiagonal. Once the matrix is in this form, several algorithms are available to solve for the eigenvalues and eigenvectors. Since Householder's method is very efficient and accurate, it is considered to be one of the best means available for treating large matrices (50 eigenvalues or more).

The third common method of approach is the power method using *matrix iteration*. This method is popular because it is simple to employ and to program. After the method converges to the lowest eigenvalues and its associated eigenvector, this eigenvector is "swept out" from successive calculations. When the method converges again, it is to the next higher eigenvalue and its eigenvector. Unfortunately, accuracy is increasingly diminished for each new eigenvector found. The method then is restricted to smaller problems. Even with this limitation, the method of matrix iteration is very practical for engineering work. This section will confine its discussion to this method's operation and implementation.

To start with, it will be shown that straight matrix iteration converges to the highest eigenvalue and its eigenvector. From previous results it is known that

$$[k]\{u\} = \omega^2 [m]\{u\} \tag{9.17}$$

Premultiplying by $[m]^{-1}$ yields

$$[D]\{u\} = \omega^2 \{u\} \tag{9.62}$$

where $[D] = [m]^{-1}[k]$.

A first trial vector $\{u\}^1$ can be written as a function of assumed distinct and independent eigenvectors

$$\{u\}^1 = c_1\{u_1\} + c_2\{u_2\} + \cdots + c_n\{u_n\} \tag{9.63}$$

Then with $\{u\}^2$ defined to be the second trial vector,

$$[D]\{u\}^1 = \{u\}^2 = [D](c_1\{u_1\} + c_2\{u_2\} + \cdots + c_n\{u_n\}) \tag{9.64}$$

$$\{u\}^2 = c_1\omega_1^2\{u_1\} + c_2\omega_2^2\{u_2\} + \cdots + c_n\omega_n^2\{u_n\} \tag{9.65}$$

since

$$[D]\{u_i\} = \omega_i^2\{u_i\} \tag{9.66}$$

Rewriting equation (9.65) as

$$\{u\}^2 = \omega_1^2 \left\{ c_1\{u_1\} + c_2\left(\frac{\omega_2}{\omega_1}\right)^2 \{u_2\} + c_3\left(\frac{\omega_3}{\omega_1}\right)^2 \{u_3\} + \cdots + c_n\left(\frac{\omega_n}{\omega_1}\right)^2 \{u_n\} \right\} \tag{9.67}$$

this vector can now be used as the next trial vector. If this procedure is repeated m times, then after the mth iteration,

$$\{u\}^m = \omega_1^{2m}\left[c_1\{u_1\} + c_2\left(\frac{\omega_2}{\omega_1}\right)^{2m}\{u_2\} + \cdots + c_n\left(\frac{\omega_n}{\omega_1}\right)^{2m}\{u_n\}\right] \quad (9.68)$$

In equation (9.68), the choices for ω_1 and $\{u_1\}$ was left completely open. If ω_1 is now taken to stand for the *largest* eigenvalue, it is seen that the fractions present in the expression will decrease rapidly to negligible quantities. One can conclude that if the method converges, it will converge to the largest eigenvector multiplied by a constant. If before each iteration pass, it is made a practice to *normalize* the trial vector by factoring out a constant so that the remaining vector has unit length, or has its first component or its largest component equal to 1, etc., convergence can be ascertained when two succeeding eigenvectors have equal components.

$$\{u_j\}^{m+1} = \{u_j\}^m \quad (9.69)$$

The last constant extracted in the normalization process must be the largest eigenvalue.

$$[D]\{u_j\}^m = \omega_j^2\{u_j\}^{m+1} \quad (9.70)$$

To converge to the *lowest eigenvalue* and its associated *eigenvector*, one can rewrite equation (9.17) as

$$\frac{1}{\omega^2}\{u\} = [k]^{-1}[m]\{u\} \quad (9.71)$$

$[k]^{-1}$ is recognized as the flexibility matrix. Then

$$\frac{1}{\omega_2}\{u\}^2 = [E]\{u\}^1 \quad (9.72)$$

after m iterations becomes

$$\{u\}^m = \left(\frac{1}{\omega_1}\right)^{2m}\left[c_1\{u_1\} + c_2\left(\frac{\omega_1}{\omega_2}\right)^{2m}\{u_2\} + \cdots + c_n\left(\frac{\omega_1}{\omega_n}\right)^{2m}\{u_n\}\right] \quad (9.73)$$

It is seen that convergence is to the lowest-magnitude eigenvalue and corresponding eigenvector since the other terms diminish more rapidly, in comparison, with each iteration.

When convergence has occurred

$$\{u_j\}^{m+1} = \{u_j\}^m \quad (9.74)$$

and

$$[E]\{u_j\}^m = \left(\frac{1}{\omega_j}\right)^2\{u_j\}^{m+1} \quad (9.75)$$

In both methods the rate of convergence depends on the numerical separation of the natural frequencies.

Sample Problem 9.6
Use the matrix iteration procedure to solve for the highest natural frequency present in Sample Problem 9.2. Assume a trial vector of $[1 \quad 0 \quad 0]$.

Analysis: From Sample Problem 9.2,

$$\begin{bmatrix} k & -k & 0 \\ -k & 2k & -k \\ 0 & -k & k \end{bmatrix} \{u\} = \omega^2 \begin{bmatrix} m & 0 & 0 \\ 0 & m & 0 \\ 0 & 0 & m \end{bmatrix} \{u\}$$

$$[m]^{-1}[k] = \frac{k}{m} \begin{bmatrix} 1 & -1 & 0 \\ -1 & 2 & -1 \\ 0 & -1 & 1 \end{bmatrix} = [D]$$

$$\{u\}^1 = \begin{Bmatrix} 1 \\ 0 \\ 0 \end{Bmatrix}$$

$$[D]\{u\}^1 = \frac{k}{m} \begin{Bmatrix} 1 \\ -1 \\ 0 \end{Bmatrix}$$

In this example let the first component of the trial vector always equal 1 (normalization). Then the next trial vector is

$$\{u\}^2 = \begin{Bmatrix} 1 \\ -1 \\ 0 \end{Bmatrix}$$

$$[D]\{u\}^2 = \frac{k}{m} \begin{Bmatrix} 2 \\ -3 \\ 1 \end{Bmatrix} = \frac{2k}{m} \begin{Bmatrix} 1 \\ -1.5 \\ 0.5 \end{Bmatrix}$$

$$\{u\}^3 = \begin{Bmatrix} 1 \\ -1.5 \\ 0.5 \end{Bmatrix}$$

$$[D]\{u\}^3 = \frac{k}{m} \begin{Bmatrix} \frac{5}{2} \\ -\frac{9}{2} \\ \frac{4}{2} \end{Bmatrix} = \frac{5k}{2m} \begin{Bmatrix} 1 \\ -\frac{9}{5} \\ \frac{4}{5} \end{Bmatrix}$$

$$[D]\{u\}^4 = \frac{k}{m} \begin{Bmatrix} \frac{14}{5} \\ -\frac{27}{5} \\ \frac{13}{5} \end{Bmatrix} = \frac{14k}{5m} \begin{Bmatrix} 1 \\ -\frac{27}{14} \\ \frac{13}{14} \end{Bmatrix}$$

$$[D]\{u\}^5 = 2.92 \frac{k}{m} \begin{Bmatrix} 1 \\ -1.98 \\ 0.98 \end{Bmatrix}$$

This converges to

$$[D]\{u\}^m \to 3 \frac{k}{m} \begin{Bmatrix} 1 \\ -2 \\ 1 \end{Bmatrix}$$

The highest eigenvalue then is $\omega^2 = 3k/m$, and the corresponding eigenvector is

$$\{u\} = \begin{Bmatrix} 1 \\ -2 \\ 1 \end{Bmatrix}$$

Compare these results with those previously attained.

As pleasant as the method seems, the reader is cautioned here that there could be complications developing when repeated roots are present or when inappropriate first guesses are taken. ∎

Sample Problem 9.7
For Sample Problem 9.4 solve for the lowest eigenvalue and eigenvector present using matrix iteration. Use a first trial guess of $[1 \quad 1]$.

Analysis: From equation (9.71)

$$\frac{1}{\omega^2}\{u\} = \{k\}^{-1}[m]\{u\}$$

$$[k] = \begin{bmatrix} 3k & -k \\ -k & k \end{bmatrix}$$

$$[k]^{-1} = \frac{1}{2k}\begin{bmatrix} 1 & 1 \\ 1 & 3 \end{bmatrix}$$

$$[E] = [k]^{-1}[m] = \frac{m}{2k}\begin{bmatrix} 2 & 1 \\ 2 & 3 \end{bmatrix}$$

$$[E]\{u\}^1 = \frac{m}{2k}\begin{bmatrix} 2 & 1 \\ 2 & 3 \end{bmatrix}\begin{Bmatrix} 1 \\ 1 \end{Bmatrix} = \frac{m}{2k}\begin{Bmatrix} 3 \\ 5 \end{Bmatrix} = \frac{3m}{2k}\begin{Bmatrix} 1 \\ \frac{5}{3} \end{Bmatrix}$$

$$\{u\}^2 = \begin{Bmatrix} 1 \\ \frac{5}{3} \end{Bmatrix}$$

$$[E]\{u\}^2 = \frac{m}{2k}\begin{Bmatrix} \frac{11}{3} \\ \frac{21}{3} \end{Bmatrix} = \frac{11}{3}\frac{m}{2k}\begin{Bmatrix} 1 \\ \frac{21}{11} \end{Bmatrix}$$

$$\{u\}^3 = \begin{Bmatrix} 1 \\ \frac{21}{11} \end{Bmatrix}$$

$$[E]\{u\}^3 = \frac{m}{2k}\begin{Bmatrix} \frac{43}{11} \\ \frac{85}{11} \end{Bmatrix} = \frac{43}{11}\frac{m}{2k}\begin{Bmatrix} 1 \\ \frac{85}{43} \end{Bmatrix}$$

$\{u\}^4 = \begin{Bmatrix} 1 \\ 1.98 \end{Bmatrix}$, which approaches $\begin{Bmatrix} 1 \\ 2 \end{Bmatrix}$, the answer previously calculated. Also, $1/\omega^2 = \frac{43}{11}(m/2k)$, and $\omega^2 = 0.51k/m$ (compare with the earlier result of $0.5k/m$).

∎

Sample Problem 9.8
Write a Fortran subroutine that will find the inverse of a given matrix; e.g., the flexibility matrix is the inverse of the stiffness matrix.

Analysis: One way of solving a set of simultaneous linear equations $[A]\{X\} = \{C\}$ is to use the Guass-Jordan reduction. In this method the coefficient matrix $[A]$ is reduced to the identity matrix $[1]$ while the $\{C\}$ matrix is converted to $\{d\}$.

$$[A]\{X\} = \{C\} \to [1]\{X\} = \{d\}$$

The solution matrix is $\{d\}$. It is seen that in effect both sides of the matrix equation have been premultiplied by the inverse matrix $[A]^{-1}$. If the Gauss-Jordan method were now used on the matrix $[A]$ augmented with the identity matrix,

$$\begin{bmatrix} a_{11} & a_{12} & a_{13} & 1 & 0 & 0 \\ a_{21} & a_{22} & a_{23} & 0 & 1 & 0 \\ a_{31} & a_{32} & a_{33} & 0 & 0 & 1 \end{bmatrix}$$

the method would convert the augmented matrix to

$$\begin{bmatrix} 1 & 0 & 0 & b_{11} & b_{12} & b_{13} \\ 0 & 1 & 0 & b_{21} & b_{22} & b_{23} \\ 0 & 0 & 1 & b_{31} & b_{32} & b_{33} \end{bmatrix}$$

It follows that the inverse of $[A]$ must therefore be $[B]$. Viewed another way, the Gauss-Jordan method is used to solve three sets of problems:

$$[A]\{X_1\} = \begin{Bmatrix} 1 \\ 0 \\ 0 \end{Bmatrix}$$

$$[A]\{X_2\} = \begin{Bmatrix} 0 \\ 1 \\ 0 \end{Bmatrix}$$

$$[A]\{X_3\} = \begin{Bmatrix} 0 \\ 0 \\ 1 \end{Bmatrix}$$

where ultimately

$$[A]^{-1} = [\{X_1\}\{X_2\}\{X_3\}]$$

The Gauss-Jordan program shown in Figure 9.1 does not actually store the augmented identity matrix but generates it. It saves storage space by replacing the elements of the $[A]$ matrix with $[A]^{-1}$. It also uses partial pivoting whereby rows are interchanged so that the largest elements will always appear on the diagonal. This reduces round-off error and also prevents inadvertent division by zero. Interchanging two row numbers in $[A]$ requires the two corresponding *column* elements of the answer to be interchanged in reverse order. ∎

```
00100 SUBROUTINE RMATIN (A,B,N,M,DET)
00110CTHIS PROGRAM CAN SOLVE SIMULTANEOUS EQUATIONS, FIND THE DETERMINANT
00120COF A MATRIX,OR FIND THE INVERSE OF MATRIX A.  MATRIX B
00130CIS THE MATRIX OF CONSTANT VECTORS WHEN SIMULTANEOUS EQUATIONS ARE
00140C TO BE SOLVED (M>0).  IF M=0, ONLY INVERSION IS SOUGHT.DET=
00150C DETERMINANT OF A.  N.LE.50.
00160 DIMENSION A(N,N),B(N,M),INDEX(50,3)
00170 EQUIVALENCE (IROW,JROW),(ICOLUM,JCOLUM)
00180C INITIALIZATION
00190 DET=1.
00200 DO 2 J=1,N
00210 INDEX(J,3)=0
00220 2 CONTINUE
00230 DO 989 I=1,N
00240C SEARCH FOR LARGEST PIVOT ELEMENTS
00250 AMAX=0.
00260 DO 3 J=1,N
00270 IF (INDEX(J,3).EQ.1)GO TO 3
00280 DO 4 K=1,N
00290 IF (INDEX(K,3)-1)5,4,999
00300 5IF (ABS(AMAX).GE.ABS(A(J,K))) GO TO 4
00310 IROW=J
00320 ICOLUM=K
00330 AMAX=A(J,K)
00340 4 CONTINUE
00350 3 CONTINUE
00360 INDEX(ICOLUM,3)=INDEX(ICOLUM,3)+1
00370 INDEX(I,1)=IROW
00380 INDEX(I,2)=ICOLUM
00390C PLACE LARGEST ELEMENT ON DIAGONAL BY ROW INTERCHANGE
00400 IF(IROW.EQ.ICOLUM) GO TO 6
00410 DET=-DET
00420 DO 7 L=1,N
00430 T=A(IROW,L)
00440 A(IROW,L)=A(ICOLUM,L)
00450 A(ICOLUM,L)=T
00460 7 CONTINUE
00470 IF (M.LE.0) GO TO 6
00480 DO 8 L=1,M
00490 T=B(IROW,L)
00500 B(IROW,L)=B(ICOLUM,L)
00510 B(ICOLUM,L)=T
00520 8 CONTINUE
00530 6 CONTINUE
```

Figure 9.1

```
00540 PIVOT=A(ICOLUM,ICOLUM)
00550 DET=DET*PIVOT
00560C DIVIDE PIVOT ROW BY PIVOT ELEMENT
00570 A(ICOLUM,ICOLUM)=1.
00580 DO 9 L=1,N
00590 A(ICOLUM,L)=A(ICOLUM,L)/PIVOT
00600 9 CONTINUE
00610 IF (M.LE.0) GO TO 16
00620 DO 10 L=1,M
00630 B(ICOLUM,L)=B(ICOLUM,L)/PIVOT
00640 10 CONTINUE
00650 16 CONTINUE
00660C REDUCE NON-PIVOT ROWS
00670 DO 11 L1=1,N
00680 IF(L1.EQ.ICOLUM) GO TO 11
00690 T=A(L1,ICOLUM)
00700 A(L1,ICOLUM)=0.
00710 DO 12 L=1,N
00720 A(L1,L)=A(L1,L)-A(ICOLUM,L)*T
00730 12 CONTINUE
00740 IF (M.LE.0) GO TO 11
00750 DO 13 L=1,M
00760 B(L1,L)=B(L1,L)-B(ICOLUM,L)*T
00770 13 CONTINUE
00780 11 CONTINUE
00790 989 CONTINUE
00800C INTERCHANGE COLUMNS BACK
00810 DO 14 I=1,N
00820 L=N-I+1
00830 IF(INDEX(L,1).EQ.INDEX(L,2)) GO TO 14
00840 JROW=INDEX(L,1)
00850 JCOLUM=INDEX(L,2)
00860 DO 15 K=1,N
00870 T=A(K,JROW)
00880 A(K,JROW)=A(K,JCOLUM)
00890 A(K,JCOLUM)=T
00900 15 CONTINUE
00910 14 CONTINUE
00920 999 RETURN
00930 END
```

Figure 9.1 (continued)

THEORY OF SMALL OXCILLATIONS

To get *intermediate modes*, one can use the generalized orthogonality condition to sweep out the mode that has already been solved for, and the matrix iteration method will then converge to the next highest/lowest mode. Suppose, for example, that the highest mode vector $\{u_j\}$ has already been solved for Taking any arbitrary guess vector $\{u\}^1$,

$$\{u\}^1 = c_1\{u_1\} + c_2\{u_2\} + \cdots + c_j\{u_j\} + \cdots + c_n\{u_n\} \quad (9.63)$$

The generalized orthogonality condition states [see equations (9.25) and (9.26)] that

$$\{u_i\}^t[m]\{u_l\} = M_i \delta_{il} \quad (9.76)$$

where $\delta_{il} =$ Kroneker delta function

$$= \begin{cases} 1 & \text{for } i = l \\ 0 & \text{for } i \neq 1 \end{cases} \quad (9.77)$$

Then for eigenvector $\{u_j\}$,

$$\{u_j\}^t[m]\{u_j\} = M_j \quad (9.78)$$

and

$$\{u_j\}^t[m]\{u\}^1 = M_j c_j \quad (9.79)$$

To sweep out $\{u_j\}$ from $\{u\}^1$, one can subtract $c_j\{u_j\}$ from $\{u\}^1$ so that a modified trial vector results.

$$\{u\}_s^1 = \{u\}^1 - c_j\{u_j\} \quad (9.80)$$

This modified trial vector $[u]_s^1$ is used in the iteration process to form a new trial vector, and the process then repeats itself until convergence is assured. To sweep out two or more vectors, one has

$$\{u\}_s^1 = \{u\}^1 - c_j\{u_j\} - c_k\{u_k\} - \cdots \quad (9.81)$$

Sample Problem 9.9
In Sample Problem 9.6 find the next largest eigenvalue and eigenvector by sweeping out the mode vector already established.

Analysis: The previous results had for the largest eigenvalue $\omega^2 = 3k/m$ and the eigenvector

$$\{u\} = \begin{Bmatrix} 1 \\ -2 \\ 1 \end{Bmatrix}$$

Taking as a trial guess

$$\{u\}^1 = \begin{Bmatrix} 1 \\ 0 \\ 0 \end{Bmatrix}$$

346 ANALYTICAL MECHANICS

then $\quad M_j = \{u_j\}^t[m]\{u_j\} = \{1 \quad -2 \quad 1\} \begin{bmatrix} m & 0 & 0 \\ 0 & m & 0 \\ 0 & 0 & m \end{bmatrix} \begin{Bmatrix} 1 \\ -2 \\ 1 \end{Bmatrix} = 6m$

$$\{u_j\}^t[m]\{u\}^1 = \{1 \quad -2 \quad 1\} \begin{bmatrix} m & 0 & 0 \\ 0 & m & 0 \\ 0 & 0 & m \end{bmatrix} \begin{Bmatrix} 1 \\ 0 \\ 0 \end{Bmatrix} = m$$

Therefore $c_j = \frac{1}{6}$, and

$$\{u\}_s^1 = \begin{Bmatrix} 1 \\ 0 \\ 0 \end{Bmatrix} - \frac{1}{6} \begin{Bmatrix} 1 \\ -2 \\ 1 \end{Bmatrix} = \frac{1}{6} \begin{Bmatrix} 5 \\ 2 \\ -1 \end{Bmatrix} = \frac{5}{6} \begin{Bmatrix} 1 \\ \frac{2}{5} \\ -\frac{1}{5} \end{Bmatrix} \rightarrow \begin{Bmatrix} 1 \\ \frac{2}{5} \\ -\frac{1}{5} \end{Bmatrix}$$

where the first component is arbitrarily selected to be unity. Continuing,

$$[D]\{u\}_s^1 = \frac{k}{m} \begin{bmatrix} 1 & -1 & 0 \\ -1 & 2 & -1 \\ 0 & -1 & 1 \end{bmatrix} \begin{Bmatrix} 1 \\ \frac{2}{5} \\ -\frac{1}{5} \end{Bmatrix} = \frac{3}{5} \frac{k}{m} \begin{Bmatrix} 1 \\ 0 \\ -1 \end{Bmatrix}$$

Taking the next trial vector as

$$\{u\}^2 = \begin{Bmatrix} 1 \\ 0 \\ -1 \end{Bmatrix}$$

$$\{u_j\}^t[m]\{u\}^2 = \{1 \quad -2 \quad 1\} \begin{bmatrix} m & 0 & 0 \\ 0 & m & 0 \\ 0 & 0 & m \end{bmatrix} \begin{Bmatrix} 1 \\ 0 \\ -1 \end{Bmatrix} = 0$$

$$\{u\}_s^2 = \begin{Bmatrix} 1 \\ 0 \\ -1 \end{Bmatrix} - 0 \begin{Bmatrix} 1 \\ -2 \\ 1 \end{Bmatrix} = \begin{Bmatrix} 1 \\ 0 \\ -1 \end{Bmatrix}$$

$$[D]\{u\}_s^2 = \frac{k}{m} \begin{Bmatrix} 1 \\ 0 \\ -1 \end{Bmatrix}$$

Therefore, the solution has already converged, and the next largest eigenvalue is found to be

$$\omega^2 = \frac{k}{m}$$

with the corresponding eigenvector

$$\{u\} = \begin{Bmatrix} 1 \\ 0 \\ -1 \end{Bmatrix}$$

These results of course agree with those found previously. ∎

Sample Problem 9.10

Write a Fortran program that will find the lowest eigenvalue and its corresponding eigenvector for an oscillatory system. Use equation (9.75). Proceed then to sweep out s successive eigenvalues and their associated eigenvectors. Assume that the mass matrix $[m]$ and the stiffness matrix $[k]$ have been given and that the mass matrix is diagonal. (The system is dynamically uncoupled.) Use the results of Sample Problem 9.8 to find the flexibility matrix. The procedure should *normalize each trial vector to a unit length.*

Analysis: The solution is shown in Figure 9.2. Special thanks are offered to Mr. James Richter, a student at California State University, Long Beach, who wrote the program. ∎

The sweeping method, though powerful, decreases in accuracy as more modes are swept out. The last values obtained should be checked by some other means to guarantee their correctness.

9.7 SMALL OSCILLATIONS ABOUT STEADY MOTION

In the previous sections, the small oscillations of an m degree of freedom system about a position of static equilibrium was analyzed. In this section, small oscillations about steady motion will be studied. The importance of this work lies in the fact that the character of the subsequent *small* vibrations established about a perturbed steady motion indicates whether the system's motion will be stable in that configuration. If after the small disturbance, the generalized coordinates remain small, the system is said to be *stable*. If any one coordinate becomes large, the motion is *unstable*. In his famous work on the stability of motion, Routh showed that the roots of the characteristic equation established for the small oscillations determine system stability. Further, he was able to establish criteria for the coefficients of the characteristic equation that ensured negative real parts for the eigenvalues and thus system stability. The *Routh-Hurwitz criteria* allow for the study of the stability of motion *without* having to solve for the actual eigenvalues.

When motion is "steady," it will be characterized in state space as a path where all the noncyclic generalized coordinates and all the generalized velocities for the cyclic coordinates remain constant. To determine stability, slight perturbations of the state variables about a known steady motion could be introduced into the routhian. The resulting equations of motion obtained from the routhian, after subtracting out the steady motion, would be in terms of the perturbed variables only. The Routh-Hurwitz criteria can be applied to the coefficients of these equations to investigate the nature of the stability of the original unperturbed state, or one can introduce the perturbed variables into the equations of motion themselves, subtract out the steady motion, and solve the new equations of motion for the perturbed variables. The latter approach will be demonstrated here in the two following sample problems.

```
00100      PROGRAM EIGEN(INPUT,OUTPUT)
00110C     THIS PROGRAM WILL FIND THE LOWEST EIGENVALUE AND SUCCESSIVELY
00120C     LARGER EIGENVALUES IF THE STIFFNESS AND MASS MATRICES ARE KNOWN.
00130C     THE CORRESPONDING EIGENVECTORS ARE ALSO FOUND USING A SWEEPING
00140C     METHOD.  IT WILL SWEEP OUT A MAXIMUM OF 9 EIGENVECTORS AND MUST
00150C     BE DIMENSIONED TO CORRESPOND TO THE ORDER OF THE STIFFNESS MATRIX.
00160      REAL LAMBDA,MASS
00170      INTEGER S,SP1
00180      DIMENSION FMASS(4,4),F(4,4),X(4,1),D1(4,1),X1NEW(10,4),XS(10,4),
00190+     COEF(9,1),G(9),SUM2(4,1),B(4,1),X1S(10,4),D2(4,1),X2NEW(10,4),
00200+     LAMBDA(10),MASS(4,4),E(4,3),SUM1(1,1),A(1,4),X1NEWA(1,4),XS1(4,1),
00210+     X1NEWB(4,1),B(1,4),C(1,1)
00220      PRINT 4
00230 4    FORMAT('ENTER: ORDER OF STIFFNESS MATRIX, # OF EIGENVECTORS TO BE SWE
00240+PT,EPSILON')
00250      READ *, N,S,EPSI
00260      PRINT 5
00270 5    FORMAT(/,'ENTER: STIFFNESS MATRIX BY ROWS')
00280      READ *, ((F(I,J),J=1,N),I=1,N)
00290      PRINT 6
00300 6    FORMAT(/,'ENTER: MASS MATRIX BY ROWS')
00310      READ *, ((MASS(I,J),J=1,N),I=1,N)
00320      PRINT 7
00330 7    FORMAT(//,15X,'THE STIFFNESS MATRIX IS:'/)
00340      DO 8 I=1,N
00350 8    PRINT *, (F(I,J),J=1,N)
00360      M1=0
00370      M2=1
00380      CALL RMATIN(F,E,N,M1,DET)
00390      PRINT 9
00400 9    FORMAT(//,15X,'THE FLEXIBILITY MATRIX IS:'/)
00410      DO 10 I=1,N
00420 10   PRINT *, (F(I,J),J=1,N)
00430      PRINT 11
00440 11   FORMAT(//,15X,'THE MASS MATRIX IS:',/)
00450      DO 12 I=1,N
00460 12   PRINT *, (MASS(I,J),J=1,N)
00470      PRINT 13
00480 13   FORMAT(/)
00490      CALL MULT(F,MASS,FMASS,N,N,N)
00500      DO 14 I=1,N
00510 14   X(I,1)=1.
00520      IT=0
00530 15   CALL MULT(FMASS,X,D1,N,M2,N)
00540      IT=IT+1
```

Figure 9.2

```
00550      DD1=0.
00560      DO 16 I=1,N
00570 16   DD1=DD1+D1(I,1)**2
00580      SDD1=SQRT(DD1)
00590      DO 17 I=1,N
00600 17   X1NEW(1,1)=D1(I,1)/SDD1
00610      DO 18 I=1,N
00620      DIFF=X(1,1)-X1NEW(1,I)
00630      IF (ABS(DIFF)-EPSI*ABS(X1NEW(1,I))) 18,19,19
00640 18   CONTINUE
00650      GO TO 23
00660 19   DO 20 I=1,N
00670 20   X(I,1)=X1NEW(1,I)
00680      IF (IT.GE.50) GO TO 21
00690      GO TO 15
00700 21   PRINT 22
00710 22   FORMAT('50 ITERATIONS HAVE PASSED FOR LAMBDA(1)')
00720 23   LAMBDA(1)=1./SDD1
00730      DO 43 I=1,S
00740      IP1=I+1
00750      DO 24 L=1,N
00760 24   XS(IP1,L)=1.
00770      DO 25 L=1,N
00780      X1NEWA(1,L)=X1NEW(I,L)
00790      X1NEWB(L,1)=X1NEW(1,L)
00800 25   CONTINUE
00810      CALL MULT(X1NEWA,MASS,B,M2,N,N)
00820      CALL MULT(B,X1NEWB,C,M2,M2,N)
00830      G(1)=C(1,1)
00840      IT=0
00850 26   DO 28 I1=1,I
00860      DO 27 L=1,N
00870      XS1(L,1)=XS(IP1,L)
00880      X1NEWA(1,L)=X1NEW(II,L)
00890 27   CONTINUE
00900      CALL MULT(X1NEWA,MASS,A,M2,N,N)
00910      CALL MULT(A,XS1,SUM1,M2,M2,N)
00920 28   COEF(II,1)=SUM1(1,1)/G(II)
00930      II=IT+1
00940      DO 29 L=1,N
00950 29   SUM2(L,1)=0.
00960      DO 31 K=1,I
00970      DO 30 L=1,N
00980      R(L,1)=COEF(K,1)*X1NEW(K,L)
00990 30   SUM2(L,1)=SUM2(L,1)+R(L,1)
```

Figure 9.2 (continued)

```
01000 31  CONTINUE
01010     DO 32 L=1,N
01020 32  X1S(IP1,L)=XS(IP1,L)-SUM2(L,1)
01030     DO 33 II=1,N
01040 33  X1S(II,1)=X1S(IP1,II)
01050     CALL MULT(FMASS,X1S,D2,N,M2,N)
01060     DD2=0.
01070     DO 34 II=1,N
01080 34  DD2=DD2+D2(II,1)**2
01090     SDD2=SQRT(DD2)
01100     DO 35 L=1,N
01110 35  X2NEW(IP1,L)=D2(L,1)/SDD2
01120     DO 36 L=1,N
01130     DIFF=XS(IP1,L)-X2NEW(IP1,L)
01140     IF (ABS(DIFF)-EPST*ABS(X2NEW(IP1,L))) 36,37,37
01150 36  CONTINUE
01160     GO TO 41
01170 37  DO 38 L=1,N
01180 38  XS(IP1,L)=X2NEW(IP1,L)
01190     IF (IT.GE.50) GO TO 39
01200     GO TO 26
01210 39  PRINT 40, IP1
01220 40  FORMAT('50 ITERATIONS HAVE PASSED FOR LAMBDA(',I1,')')
01230 41  LAMBDA(IP1)=1./SDD2
01240     DO 42 L=1,N
01250 42  X1NEW(IP1,L)=X2NEW(IP1,L)
01260 43  CONTINUE
01270     SP1=S+1
01280     DO 48 I=1,SP1
01290     PRINT 44
01300 44  FORMAT(///)
01310     PRINT 45, I,LAMBDA(I)
01320 45  FORMAT('LAMBDA(',I1,')=',E14.7,//)
01330     PRINT 46
01340 46  FORMAT('THE ASSOCIATED EIGENVECTOR COMPONENTS ARE',/)
01350     PRINT 47, (X1NEW(I,L),L=1,N)
01360 47  FORMAT(13X,E14.7)
01370 48  CONTINUE
01380     STOP
01390     END
01400     SUBROUTINE MULT(A,B,C,L,N,M)
01410     DIMENSION A(L,M),B(M,N),C(L,N)
01420     DO 2 I=1,L
01430     DO 2 J=1,N
01440     C(I,J)=0.
```

Figure 9.2 (continued)

```
01450          DO 1 K=1,M
01460          C(I,J)=C(I,J)+A(I,K)*B(K,J)
01470       1  CONTINUE
01480       2  CONTINUE
01490          RETURN
01500          END
01510          SUBROUTINE RMATIN(A,B,N,M,DET)
01520C         THIS PROGRAM CAN SOLVE SIMULTANEOUS EQUATIONS, FIND THE DETERMINANT
01530C         OF A MATRIX, OR FIND THE INVERSE OF MATRIX A.  MATRIX B IS THE
01540C         MATRIX OF CONSTANT VECTORS WHEN SIMULTANEOUS EQUATIONS ARE TO BE
01550C         SOLVED(M>0).  IF M=0, ONLY INVERSION IS SOUGHT.  DET=DETERMINANT
01560C         OF A.  N.LE.50
01570          DIMENSION A(N,N),B(N,M),INDEX(50,3)
01580          EQUIVALENCE (IROW,JROW),(ICOLUM,JCOLUM)
01590          DET=1.
01600          DO 1 J=1,N
01610          INDEX(J,3)=0
01620       1  CONTINUE
01630          DO 14 I=1,N
01640C         SEARCH FOR LARGEST PIVOT ELEMENTS
01650          AMAX=0.
01660          DO 4 J=1,N
01670          IF (INDEX(J,3).EQ.1) GO TO 4
01680          DO 3 K=1,N
01690          IF ((INDEX(K,3)-1) 2,3,999
01700       2  IF (ABS(AMAX).GE.ABS(A(J,K))) GO TO 3
01710          IROW=J
01720          ICOLUM=K
01730          AMAX=A(J,K)
01740       3  CONTINUE
01750       4  CONTINUE
01760          INDEX(ICOLUM,3)=INDEX(ICOLUM,3)+1
01770          INDEX(I,1)=IROW
01780          INDEX(I,2)=ICOLUM
01790C         PLACE LARGEST ELEMENT ON DIAGONAL BY ROW INTERCHANGE
01800          IF (IROW.EQ.ICOLUM) GO TO 7
01810          DET=-DET
01820          DO 5 L=1,N
01830          T=A(IROW,L)
01840          A(IROW,L)=A(ICOLUM,L)
01850          A(ICOLUM,L)=T
01860       5  CONTINUE
01870          IF (M.LE.0) GO TO 7
01880          DO 6 L=1,M
01890          T=B(IROW,L)
```

Figure 9.2 (continued)

```
01900       B(IROW,L)=B(ICOLUM,L)
01910       B(ICOLUM,L)=T
01920  6    CONTINUE
01930  7    CONTINUE
01940       PIVOT=A(ICOLUM,ICOLUM)
01950       DET=DET*PIVOT
01960C      DIVIDE PIVOT ROW BY PIVOT ELEMENT
01970       A(ICOLUM,ICOLUM)=1.
01980       DO 8 L=1,N
01990       A(ICOLUM,L)=A(ICOLUM,L)/PIVOT
02000  8    CONTINUE
02010       IF (M.LE.0) GO TO 10
02020       DO 9 L=1,M
02030       B(ICOLUM,L)=B(ICOLUM,L)/PIVOT
02040  9    CONTINUE
02050 10    CONTINUE
02060C      REDUCE NON-PIVOT ROWS
02070       DO 13 L1=1,N
02080       IF (L1.EQ.ICOLUM) GO TO 13
02090       T=A(L1,ICOLUM)
02100       A(L1,ICOLUM)=0.
02110       DO 11 L=1,N
02120       A(L1,L)=A(L1,L)-A(ICOLUM,L)*T
02130 11    CONTINUE
02140       IF (M.LE.0) GO TO 13
02150       DO 12 L=1,M
02160       B(L1,L)=B(L1,L)-B(ICOLUM,L)*T
02170 12    CONTINUE
02180 13    CONTINUE
02190 14    CONTINUE
02200C      INTERCHANGE COLUMNS BACK
02210       DO 16 I=1,N
02220       L=N-I+1
02230       IF ((INDEX(L,1).EQ.INDEX(L,2)) GO TO 16
02240       JROW=INDEX(L,1)
02250       JCOLUM=INDEX(L,2)
02260       DO 15 K=1,N
02270       T=A(K,JROW)
02280       A(K,JROW)=A(K,JCOLUM)
02290       A(K,JCOLUM)=T
02300 15    CONTINUE
02310 16    CONTINUE
02320 999   RETURN
02330       END
```

Figure 9.2 (continued)

Sample Problem 9.11
Investigate the stability of motion in a circular orbit about the earth.

Analysis: The lagrangian of the system is

$$L = \frac{m_2'}{2}(\dot{r}^2 + r^2\dot{\theta}^2) + \frac{Gm_1'm_2'}{r}$$

The cyclic coordinate is θ. The equations of motion are

$$m_2'(\ddot{r} - r\dot{\theta}^2) + \frac{Gm_1'm_2'}{r^2} = 0 \tag{a}$$

and

$$\frac{d}{dt}(m_2'r^2\dot{\theta}) = 0 \tag{b}$$

Equations (a) and (b) can be expanded in a Taylor series about the reference conditions for a circular orbit. With

$$\ddot{r} = \ddot{r}_0 + \delta\ddot{r} \qquad \dot{\theta} = \dot{\theta}_0 + \delta\dot{\theta} \qquad r = r_0 + \delta r$$

one obtains

$$m_2'(\ddot{r}_0 - r_0\dot{\theta}_0^2) + m_2'(\delta\ddot{r} - \dot{\theta}_0^2\,\delta r - 2r_0\dot{\theta}_0\,\delta\dot{\theta}) + \frac{Gm_1'm_2'}{r_0^2} - \frac{2Gm_1'm_2'}{r_0^3}\delta r = 0 \tag{c}$$

$$m_2'(r_0^2\delta\ddot{\theta} + 2\dot{r}_0\dot{\theta}_0\,\delta r + 2r_0\dot{\theta}_0\,\delta\dot{r}) + \frac{d}{dt}(m_2'r_0^2\dot{\theta}_0) = 0 \tag{d}$$

after discarding higher-order terms.

For the steady motion in a circular orbit though

$$\ddot{r}_0 = \dot{r}_0 = 0$$

or

$$-m_2'(r_0\dot{\theta}_0^2) + \frac{Gm_1'm_2'}{r_0^2} = 0 \tag{e}$$

$$\frac{d}{dt}(m_2'r_0\dot{\theta}_0) = 0 \tag{f}$$

Equations (e) and (f) describe the motion in the given reference state. Then equations (c) and (d) become

$$\delta\ddot{r} - \dot{\theta}_0^2\,\delta r - 2r_0\dot{\theta}_0\,\delta\dot{\theta} - \frac{2Gm_1'\,\delta r}{r_0^3} = 0 \tag{g}$$

$$r_0\,\delta\ddot{\theta} + 2\dot{\theta}_0\,\delta\dot{r} = 0 \tag{h}$$

From equation (e) one has

$$\dot{\theta}_0^2 = \frac{Gm_1'}{r_0^3}$$

and so equation (g) becomes

$$\delta\ddot{r} - 3\dot{\theta}_0^2\,\delta r - 2r_0\dot{\theta}_0\,\delta\dot{\theta} = 0 \tag{i}$$

354 ANALYTICAL MECHANICS

From equation (h),
$$r_0 \,\delta\dot\theta + 2\dot\theta_0\, \delta r = c \tag{j}$$
but
$$c = r_0\, \delta\dot\theta(0) + 2\dot\theta_0\, \delta r(0)$$
Then
$$r_0\, \delta\dot\theta = r_0\, \delta\dot\theta(0) + 2\dot\theta_0\, \delta r(0) - 2\dot\theta_0\, \delta r \tag{k}$$
Substituting equation (k) into equation (i) yields
$$\delta\ddot r + \dot\theta_0^2\, \delta r = 2\dot\theta_0 r_0\, \delta\dot\theta(0) + 4\dot\theta_0^2\, \delta r(0)$$
or
$$\delta\ddot r + \dot\theta_0^2\, \delta r = 2\dot\theta_0 c$$
Then
$$\delta r = A \cos \dot\theta_0 t + B \sin \dot\theta_0 t + \frac{2c}{\dot\theta_0} \tag{l}$$
where A and B can be found from $\delta r(0)$ and $\delta\dot r(0)$.

Similarly, from equation (j)
$$\delta\dot\theta = -\frac{3c}{r_0} - \frac{2\dot\theta_0}{r_0}(A \cos \dot\theta_0 t + B \sin \dot\theta_0 t)$$

Since the perturbations remain small, they vary harmonically, the system is said to be stable. ∎

Sample Problem 9.12
Investigate the stability of rotational motion about a principal axis for the moment-free motion of a rigid body. See Section 7.5.

Analysis: The equations of motion for the moment-free case are
$$I_1\dot\omega_1 + \omega_2\omega_3(I_3 - I_2) = 0$$
$$I_2\dot\omega_2 + \omega_1\omega_3(I_1 - I_3) = 0$$
$$I_3\dot\omega_3 + \omega_1\omega_2(I_2 - I_1) = 0$$
If $\omega_1(0) = \beta$ and $\omega_2(0) = \omega_3(0)$, then for the perturbed equations
$$I_1\, \delta\dot\omega_1 = 0 \tag{a}$$
$$I_2\, \delta\dot\omega_2 + \beta\, \delta\omega_3(I_1 - I_3) = 0 \tag{b}$$
$$I_3\, \delta\dot\omega_3 + \beta\, \delta\omega_2(I_2 - I_1) = 0 \tag{c}$$
From equation (a) it is seen that $\delta\omega_1$ remains constant. Equations (b) and (c) can be reduced to
$$\delta\ddot\omega_2 - \beta^2 \frac{(I_2 - I_1)(I_1 - I_3)}{I_2 I_3} \delta\omega_2 = 0 \tag{d}$$
or
$$\delta\ddot\omega_3 - \beta^2 \frac{(I_2 - I_1)(I_1 - I_3)}{I_2 I_3} \delta\omega_3 = 0 \tag{e}$$

For the motion to be stable,

$$I_1 > I_2 \quad \text{and} \quad I_1 > I_3$$

or $\quad I_3 > I_1 \quad \text{and} \quad I_2 > I_1$

This means that only rotation about the largest or smallest principal axis will lead to stable motion. Initial turning of the body about an intermediate principal axis is not stable. ∎

PROBLEMS

9.1 It is desired to substitute for a pendulum in a clock the mechanism shown. A thin wire is wrapped around a wheel at A having mass $M = 6$ kg and radius $r = 5$ cm. At each end of the wire is fastened a mass of 2 kg, and attached to the two small masses are springs having equal spring constants k. What must be the value of k if the system is to have a period of 2 s? Assume no slipping of the wire on the wheel.

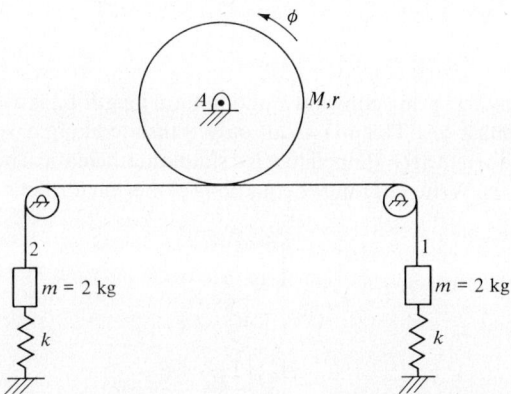

9.2 A weight having mass m is lowered with a speed v_0 by unwinding cable from a revolving drum. The cable has cross-sectional area A and modulus of elasticity E. If the motion is suddenly stopped when a cable length L has been lowered, what will be the

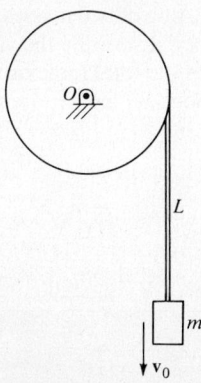

value of the maximum stress the cable experiences? Take $m = 500$ kg, $E = 10^{11}$ N/m², $v_0 = 1$ m/s, $A = 13$ cm², and $L = 20$ m.

9.3 A jet motor with mass m and mass moment of inertia I about its center of mass is suspended from an airplane wing. A simple ideal model of the jet motor and its support is shown in the figure. Two springs affect the motion. The spring constant k_z is used for the motor's vertical translation, and the spring constant k_θ for the rotation of the motor in a vertical plane. The center of gravity of the motor lies a distance d behind the support point at A. If a constant thrust force P always acts along the jet motor axis, derive Lagrange's equations for this 2 degrees of freedom problem.

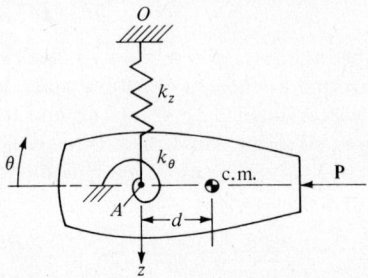

9.4 A spring having spring constant k and natural length L_0 is attached to a small mass at A with mass αm. This mass can only translate along a vertical line that passes through the origin at O. Pinned to it is a simple mathematical pendulum having length a and mass m. Write Lagrange's equations of motion for the system.

9.5 For the system shown consisting of two masses m_1 and m_2 connected by springs having sring constants k_1 and k_2, determine the value of the spring constant k_3 for the third spring that will reduce the total force on the wall to zero. The applied force is $P = a \sin \omega_e t$. No friction acts. (x_1 and x_2 are measured from the unstretched positions.)

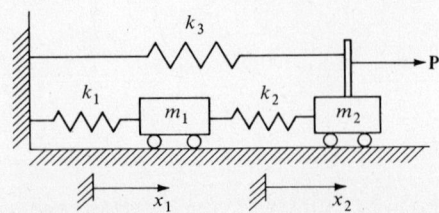

THEORY OF SMALL OSCILLATIONS

9.6 A small circular cylinder having mass m and radius r rolls, without sliding, inside a thin cylindrical shell having the same mass m and inner radius R. The shell is free to roll, also without sliding, along the floor. Derive Lagrange's equations for the motion.

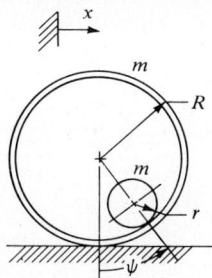

9.7 Two identical cylindrical rollers of mass m and radius r are fixed at their centers. They are free to turn, without friction, in their bearings, as a beam, also having mass m, moves over them without sliding. Attached to the midpoint of the beam at A is a mathematical pendulum of length a and mass $6m$. Set up Lagrange's equations for the motion. Solve for x and ϕ as a function of time if at $t=0$ the system starts from rest with $x=0$ and $\phi=\phi_0$. Assume small motion for x and ϕ so that the equations of motion can be linearized.

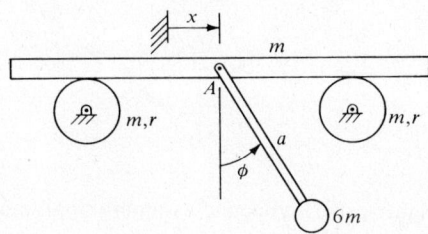

9.8 The disk of mass m and radius R rolls, without slipping, on a horizontal plane. Moving with the disk is a point mass m_1 that is free to slide without friction along an attached massless radial track. Restraining the mass is the spring shown. It has unstretched length L_0 and spring constant k. Derive Lagrange's equations of motion for the system.

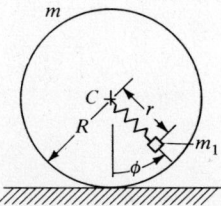

9.9 As a first trial approximation to the suspension characteristics of an automobile, a 2 degrees of freedom system can be used as a model. The automobile may be thought of as a thin rigid beam of mass *m* and length *L*, and the tires can be treated as massless springs having spring constants k_1 and k_2. Damping (for example, from the shock absorbers) can be neglected. For this elementary model, derive Lagrange's equations of motion. The distance *y* is measured from the static equilibrium position.

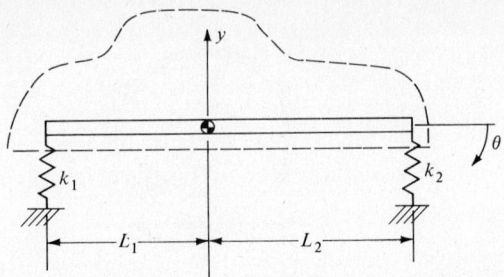

9.10 Assuming small motions and neglecting higher-order terms, linearize the equations of motion for the double pendulum shown.

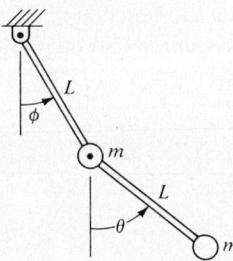

9.11 By neglecting higher-order terms and assuming small motions, linearize the equations of motion previously found for the jet motor of problem 9.3.

9.12 Derive the linear equations of motion for the two-mass system given. Mass *m* is restrained by a spring having spring constant *k*. Mass m_1 acts as the bob for a simple pendulum that has length *L*. The pendulum is pivoted at mass *m*.

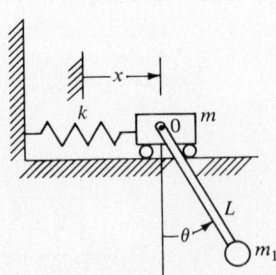

9.13 Find the natural frequencies for the double pendulum in problem 9.10.

9.14 Determine the natural frequencies for the system shown. Point mass B is twice the size of point mass A. The springs are made of the same material, but segment BA has twice the unstretched length of segment OB. Neglect the weight of the springs and friction.

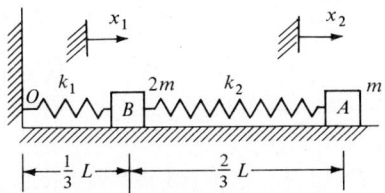

9.15 The thin beam of mass m and length L is supported at a distance $L/4$ from each end by two identical springs having spring constant k. The system is assumed to have only 2 degrees of freedom. If the downward motion is measured from the static equilibrium position, find the natural frequencies of the system.

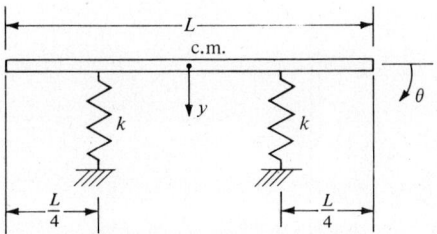

9.16 A thin beam of mass m and length $4a$ is attached to pivot point A and rotates freely in the vertical plane. Attached at the midway point and at the far end of the beam are two identical springs having spring constant k. The springs are connected to each other by a thin massless line that passes over a fixed pulley having mass $3m$ and radius a. There is sufficient friction on the pulley surface to prevent slipping of the line. For this system, determine the natural frequencies present. The springs are unstretched when the beam is horizontal.

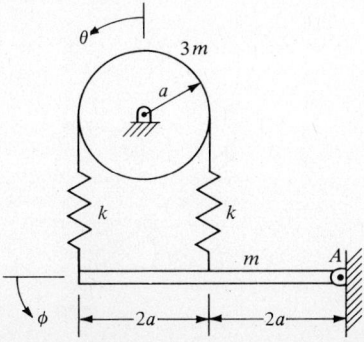

9.17 Repeat problem 9.12 for the case where there is no restraining spring. Solve for the natural frequencies of the motion.

9.18 Two ideal pendulums having mass $5m$ and m can swing about pivot point A and B, respectively. Their motion is restrained by the three springs shown having spring constants $2k, k,$ and k. Determine the natural frequencies of the system. Let $k = mg/L$.

9.19 A thin rod of mass m and length L is free to swing in a vertical plane about point A. Also fastened to point A is a weightless spring that slides over the rod and is attached to a slider having mass m_1. The slider's motion is also friction-free along the rod. In the static equilibrium position the spring's length is known to be r_0, and it has a spring constant k. Determine the natural frequencies of the system for small motions.

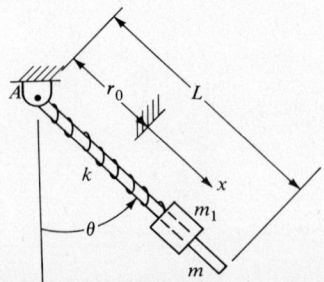

9.20 Determine the natural frequencies and normal mode shapes for problem 8.7 assuming only small motions occur. Let $k/(m_1 + m) = g/2R$ and $m_1 = m$.

9.21 What are the normal modes of vibration for the system shown?

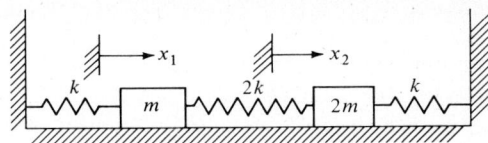

9.22 If the simple pendulum in problem 9.12 is replaced by a rod of length $2a$ having the same mass as the cart m, find the normal modes of vibration. Let $k = 3mg/a$, and assume small motions.

9.23 Shown below is a schematic drawing of a church bell. With the improper selection of dimensions, the bell will not ring. An example of this is the famous "Kaiser glocke" in Köln, Germany. Assuming small motions, determine the criteria that must be avoided if the bell is to function. The mass moments of inertia of the bell casing about the pivot point O is given as J_B, and the mass moment of inertia of the clapper is given as J_C, about its pivot point O'. The mass of the casing and the clapper are M and m, respectively.

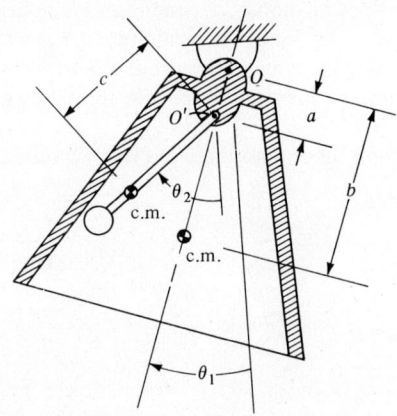

9.24 If a system's equations of motion have a mass matrix given by

$$[m] = m \begin{bmatrix} 2 & 0 \\ 0 & 3 \end{bmatrix}$$

and a stiffness matrix given by

$$[k] = k \begin{bmatrix} 2 & -\sqrt{1.5} \\ -\sqrt{1.5} & 3 \end{bmatrix}$$

find the normalized eigenvectors and eigenvalues of the vibration.

9.25 For the system shown, determine the elements of the flexibility matrix.

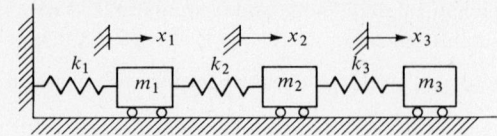

9.26 Determine the normal modes of the torsional system shown. k_1 and k_2 are torsional spring constants. Let $J_1 = J_2$ and $k_1 = 2k_2$.

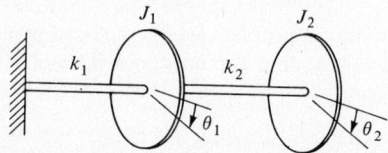

9.27 Demonstrate that the mode vectors found in problem 9.24 are orthogonal with respect to the mass matrix.

9.28 Two identical simple pendulums having mass m and length L are connected by a massless spring having spring constant k. The ends of the spring lie a distance d from the fixed pivot points along the pendulum's length. The unstretched length of the spring is selected so that when the pendulums are in a vertical position, the spring is not deformed. Derive Lagrange's equations of motion for the system. Solve for the motion of the system as a function of time for the case where the system starts from rest at $t = 0$ with $\theta_1 = \theta_0$ and $\theta_2 = 0$. Show that for a weak spring, i.e., k is very small, a "beating" phenomenon takes place between the two masses.

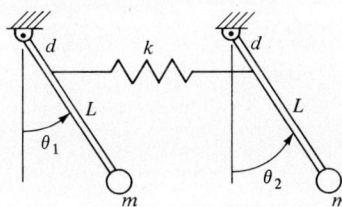

9.29 In problem 9.24 it is known that at $t = 0$, $x_1 = x_2 = 0$ and $\dot{x}_1 = 0$. Determine the time until the kinetic energy from the second mass of $3m$ totally transfers to the first mass of $2m$.

9.30 In problem 9.22 determine the free vibrations of the system if at $t = 0$, $x = 2$, $\theta = 2/a$, and the system starts from rest.

9.31 In Sample Problem 9.2 determine the free vibrations of the system if at $t = 0$, $x_1 = 2$, $x_2 = -2$, $x_3 = 0$, and $\dot{x}_1 = \dot{x}_2 = \dot{x}_3 = 1$.

9.32 Without transforming to normal coordinates, solve for the time response of the given system shown. Deflections are measured from the static equilibrium position.

THEORY OF SMALL OSCILLATIONS

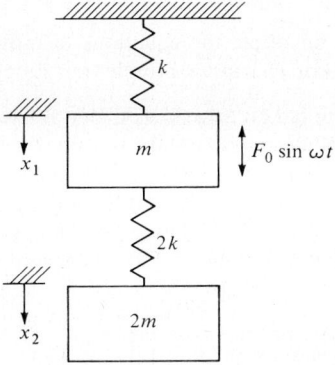

9.33 Repeat problem 9.32 with damping elements now acting. Let $m = 2$ kg, $c = c_1 = 50$ N·s/m, $k = 2000$ N/m, and $\omega = 5$ rad/s.

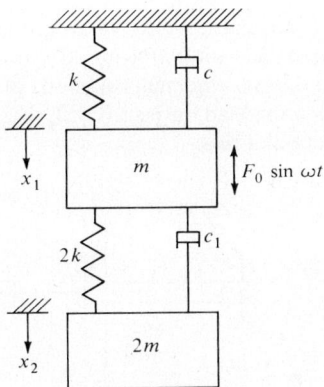

9.34 Repeat problem 9.32 using normal coordinates.

9.35 A wheel of mass m and radius r is free to roll, without slipping, on a horizontal floor. It is coupled to a translating mass of equal weight through a spring having constant k. In addition, the wheel is attached to a fixed wall by a spring having spring constant $2k$. If a driving force $F(t)$ is applied to the translating mass, derive the uncoupled equations of motion for the system. Neglect friction acting on the translating mass.

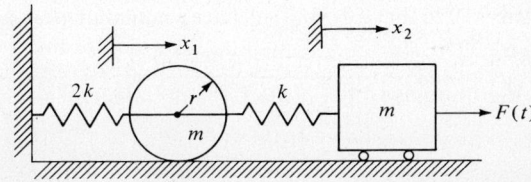

364 ANALYTICAL MECHANICS

9.36 If in problem 9.35 $F(t)$ is replaced by an impulse $\hat{F}$, find x_1 and x_2 as functions of time.

9.37 For problem 9.22, uncouple the equations of motion using normal coordinates. Assume a driving force $F(t)$ applied to the translating carriage.

9.38 Two wheels are mounted on an axle ABC that is fixed at end A. The first wheel at B has mass moment of inertia J_1, and the second wheel has moment of inertia J_2. It is also known that the torsional stiffness of the shaft portion AB is k_1 (N·m/rad) and that the portion BC has k_2 (N·m/rad) torsional stiffness. It is desired to tune the system by selecting a new J_2 so that, under an applied moment at B of magnitude $M_0 \cos \omega t$, wheel B will not vibrate. Find the magnitude of J_2 needed.

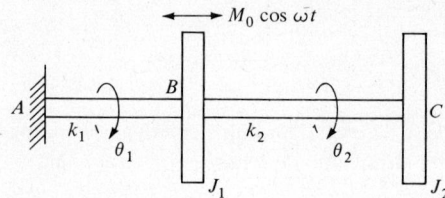

9.39 Shown below is a schematic sketch of an untuned viscous vibration damper. The torsional stiffness of the shaft is k(N·m/rad), D is the damping constant (N·m·s/rad), and J_2 and J_1 are the mass moments of inertia of the load and a free rotational mass that is sealed inside the load, respectively. Using complex notation, find the output angle θ_2 as a function of time.

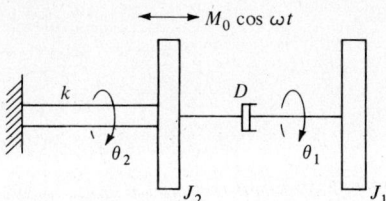

*9.40 One type of nonlinear vibration that increases for small oscillation and decreases when they become too large, reaching a stable limit cycle, is described by Van der Pol's equation

$$\ddot{x} - \mu \dot{x}(1 - x^2) + x = 0$$

Write a digital computer program that will output x and $\dot{x}$ with increasing time. Plot $\dot{x}$ versus x. Let $\mu = 0.5$ and $x = \dot{x} = 1$ at $t = 0$. Repeat the computer run with $x = 1$ and $\dot{x} = 3$ at $t = 0$. Enough values should be found to establish the limit cycle. Use a communication interval of 0.1 s and a final time of 20 s.

*9.41 Write a computer program that will find the optimum value of the damping ratio ζ in problem 9.39 so that $|k\theta_2/M_0|$ will have a minimum peak amplitude.

$$\left|\frac{k\theta_2}{M_0}\right| = \left[\frac{\alpha^2(\omega/\omega_n)^2 + 4\zeta^2}{\alpha^2(\omega/\omega_n)^2[1-(\omega/\omega_n)^2]^2 + 4\zeta^2\{\alpha(\omega/\omega_n)^2 - [1-(\omega/\omega_n)^2]\}^2}\right]^{1/2}$$

$$\alpha = \frac{J_1}{J_2} \qquad \zeta = \frac{D}{2J_2\omega_n} \qquad \omega_n^2 = \frac{k}{J_2}$$

Take $\alpha=1$, and let ω/ω_n vary from 0.5 to 1.1 in steps of 0.1. Trial values of ζ should run from $\zeta=0.1$ to $\zeta=0.9$ in steps of 0.2. Plot your results.

*9.42 A simplified linear model of part of an automobile suspension system is shown below. The mass m_1 is one-fourth of that of the vehicle's chassis. The spring constant of the main spring is k_1, and c is the damping coefficient for the shock absorber. k_2 and m_2 are the spring constant and mass of the tire, respectively. With input from the ground, x_3, the system's motion can be described by the following equations:

$$350\ddot{x}_1 + c(\dot{x}_1 - \dot{x}_2) + 15\,000(x_1 - x_2) = 0$$

$$25\ddot{x}_2 + c(\dot{x}_2 - \dot{x}_1) + 15\,000(x_2 - x_1) + 75\,000(x_2 - x_3) = 0$$

Write a digital computer program that tries to find the optimum value of c to minimize the peak displacement of m_1. Vary c from 300 to 30 300 in increments of 5000 N·s/m. The input x_3 is a pulse of amplitude 0.1 m starting at $t=0$ and lasting 0.2 s. Take $t=2.0$ s as the final time. What are your recommendations?

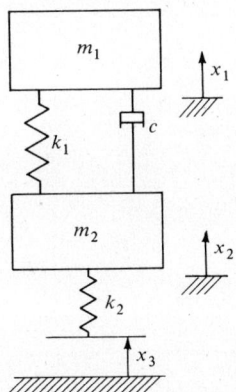

*9.43 Write a digital computer program to solve for the eigenvalues and eigenvectors of the following system dynamic matrix:

$$\begin{bmatrix} -\omega^2 + 6 & -3 & 0 \\ -3 & -\omega^2 + 6 & -3 \\ 0 & -3 & -\omega^2 + 4 \end{bmatrix} \{u\} = \{0\}$$

*9.44 Write a digital computer program to solve for the eigenvalues and eigenvectors of the following system dynamic matrix:

$$\begin{bmatrix} -\omega^2 + 18 & -9 & 0 & 0 & 0 & 0 \\ -9 & -\omega^2 + 18 & -9 & 0 & 0 & 0 \\ 0 & -9 & -\omega^2 + 18 & -9 & 0 & 0 \\ 0 & 0 & -9 & -\omega^2 + 18 & -9 & 0 \\ 0 & 0 & 0 & -9 & -\omega^2 + 18 & -9 \\ 0 & 0 & 0 & 0 & -9 & -\omega^2 + 18 \end{bmatrix} \{u\} = \{0\}$$

*9.45 If in problem 9.44 the mass matrix is given by $m[1]$, where m equals 10 kg, find the generalized mass matrix $[m]$ using the computer. Find the transformed force matrix for the normal coordinates if $F_{x_1} = 500 \sin 10t$ and F_{x_2} through F_{x_6} equal to 0. Discuss how you could approximate the total time solution for x using only the two lowest modes of vibration.

9.46 Investigate the criteria for the stability of a sleeping top.

9.47 Investigate the stability of motion for the conical pendulum of problem 8.15 about its steady motion positions if $\dot{\theta}$ is a constant equal to Ω.

9.48 Investigate the stability of motion for mass m_2 of problem 6.18 about its steady motion state.

DIGITAL COMPUTER SIMULATION

10

10.1 INTRODUCTION

In the not-so-distant past, most engineering practice was limited to the analysis of linear systems. If a system had large damping terms, was nonlinear in nature, or had discontinuous loading functions, approximations were made to linearize the system so that classical mathematical theory could be employed. Only a few researchers worked with the problem of nonlinear mechanics, and only a few did computer calculations. Today, all that has changed; with the introduction of the inexpensive minicomputer, engineers and scientists now have universal access to digital computers. The dynamic problems that were ignored or linearized in the past are now commonly investigated in detail. The computer has produced a fundamental shift in attitude and in practice. All these changes are coming at an exponentially increasing rate. Because the engineer or scientist dealing with complex dynamical systems has frequent need of the computer, it is the intent of this chapter to introduce to the reader the broad subject of digital computer simulation and to study in detail a versatile simulation language, CSSL-IV. Though analog simulation will first be discussed, it is only done to establish background introductory material for the main topic.

10.2 THE ANALOG COMPUTER

The first computers were based on analogs between the behavior of different physical systems. For example, before World War II, engineers used mathematical gear trains and linkages to analyze differential equations. The solution, the mechanical output, was analogous to the studied differential equation's dependent variable. The mechanical input of the unit was analogous to the driving function. The gear trains and linkages represented the scaled system equations. For instance, it was possible for the sighting mechanism for an antiaircraft gun on a ship to have gear trains compute the firing angle after having airspeed, boat speed, altitude, etc., cranked in. The output was instantaneous, performing in *real time*.

Next, from the laboratories came specially designed circuits, whose voltage output was analogous to the mechanical output of a specific physical problem. The compact, inexpensive, easy to manufacture and assemble, and fairly accurate

368 ANALYTICAL MECHANICS

electronic component's circuit had the same form of differential equation governing its behavior as that of the mechanical system. (Analogous forms of governing equations hold true throughout all branches of the physical sciences.) In dynamics, the similarity between the equation for the flow of charge q in an RLC circuit having an impressed voltage $e(t)$,

$$L\frac{d^2q}{dt^2} + R\frac{dq}{dt} + \frac{q}{C} = e(t) \tag{10.1}$$

and the equation of motion for a spring-mass system with damping present,

$$m\frac{d^2x}{dt^2} + c\frac{dx}{dt} + kx = F(t) \tag{10.2}$$

was of great practical interest.

Ultimately these specially designed circuits were replaced by the general-purpose analog computer that is still used today. The building blocks of the analog computer are high-gain dc operational amplifiers. They are characterized by amplitude gains in the order of 10^8. Their input grid voltage e_g and grid current i_g are negligible. Voltage outputs are usually ± 10 V or ± 100 V.

The dc operational amplifier in combination with resistors and capacitors can perform the function of differentiation, integration, addition, etc. Shown in Figure 10.1 is a circuit designed to add inputs. The triangular symbol used is the usual representation for an operational amplifier.

Kirchoff's current law applied to the circuit of Figure 10.1 yields

$$i_1 + i_2 + i_3 + i_g + i_f = 0 \tag{10.3}$$

but since $i_g \simeq 0$ and $e_g \simeq 0$, then

$$\frac{e_1}{R_1} + \frac{e_2}{R_2} + \frac{e_3}{R_3} = -\frac{e_0}{R_f} \tag{10.4}$$

$$e_0 = -R_f\left(\frac{e_1}{R_1} + \frac{e_2}{R_2} + \frac{e_3}{R_3}\right) \tag{10.5}$$

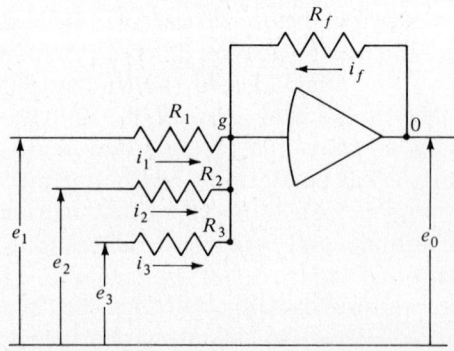

Figure 10.1

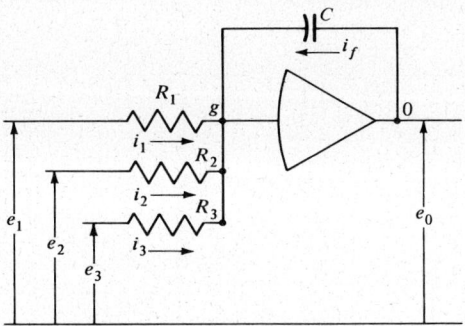

Figure 10.2

Figure 10.2 shows a circuit that performs *integration*. The output voltage of this circuit equals

$$e_0 = -\frac{1}{C}\int_0^t \left(\frac{e_1}{R_1}+\frac{e_2}{R_2}+\frac{e_3}{R_3}\right)dt' + e_0(0) \tag{10.6}$$

where $e_0(0)$ is the initial voltage across the capacitor.

The engineer using these units never deals directly with the electronics. He physically wires a *patch panel* located on the front of the analog computer. This internally connects the various *standard* resistors, capacitors, and amplifiers together. For example, shown in Figure 10.3 is a simple spring-mass problem having $x(0)$ and $\dot{x}(0)$ as given initial conditions. The equation of motion governing the system is

$$m\ddot{x} + c\dot{x} + kx = F(t) \tag{10.7}$$

or

$$\ddot{x} = \frac{F(t)}{m} - \frac{c}{m}\dot{x} - \frac{k}{m}x \tag{10.8}$$

Figure 10.4 shows the analog computer circuit diagram for the solution of this equation. Note that each operational amplifier changes the sign of the input. The symbol I will be used to denote an integrator; i.e., the operational amplifier is wired to perform integration. The symbol $(-)$ stands for sign inversion. The small circle represents a potentiometer (multiplication by a constant less than 1). Possible multiplication by a factor of 10 is available and built into the patch panel at each amplifier input G_i. Note also the signs on the initial conditions.

Because the amplifiers are limited to voltages of ± 10 V or ± 100 V, the user must be careful that at no stage of the solution are these values exceeded.

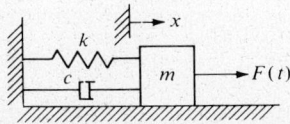

Figure 10.3

370 ANALYTICAL MECHANICS

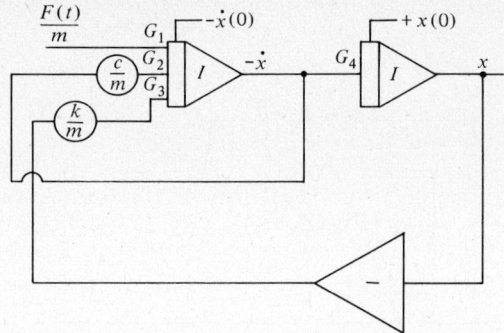

Figure 10.4

To avoid this he or she must estimate the maximum values each variable will experience and then *scale* the terms to fit the limitations of the analog computer. Time scaling may also be necessary if the period of the motion is too small or too large. To see how this operates, assume that the previous example had $F(t) = 100$, $c = 1, k = 2000, m = 0.2$. Ignoring damping and assuming zero initial conditions,

$$x \sim \frac{F}{k} + A \cos \sqrt{\frac{k}{m}} t + B \sin \sqrt{\frac{k}{m}} t$$

$$x \sim 0.05 (1 - \cos 100t)$$

where

$$\omega_n = \sqrt{\frac{k}{m}} = 100 \text{ rad/s}$$

With damping ignored, the maximum values estimated are

$$x_{max} \sim 0.1 \text{ m}$$

$$\dot{x}_{max} \sim \omega_n x_{max} = 10 \text{ m/s}$$

$$\ddot{x}_{max} \sim \omega_n^2 x_{max} = 1000 \text{ m/s}^2$$

$$\frac{F_{max}}{m} = 500 \text{ m/s}^2$$

With a $+100$-V computer, let the scale factors s_i be as follows:

$$s_1 \leqslant \frac{100}{x_{max}}, \quad s_1 \leqslant 1000 \text{ V/m}$$

Arbitrarily limiting the maximum output voltage to 80 V,

$$s_1 = 800 \text{ V/m}$$

$$s_2 \leqslant \frac{100}{\dot{x}_{max}}, \quad s_2 = 8 \text{ V} \cdot \text{s/m}$$

$$s_3 \leqslant \frac{100}{F_{max/m}}$$

DIGITAL COMPUTER SIMULATION

Take

$$s_3 = 0.1 \text{ V} \cdot \text{s}^2/\text{m}$$

$$s_4 \leq \frac{100}{\ddot{x}_{max}}, \quad s_4 = 0.08 \text{ V} \cdot \text{s}^2/\text{m}$$

Since $\ddot{x}$ is not formed explicitly in the computer diagram, s_4 will not be needed. For the scaled *computer variables* $\dot{x}_c$ and x_c one can now write

$$\dot{x}_c = s_2 \dot{x} = \int_0^t s_2 \left(\frac{F}{m} - \frac{c}{m} \dot{x} - \frac{k}{m} x \right) dt$$

and

$$x_c = s_1 x = \int_0^t s_1 \dot{x} \, dt = \int \frac{s_1}{s_2} \dot{x}_c \, dt$$

The computer diagram will then appear as shown in Figure 10.5.

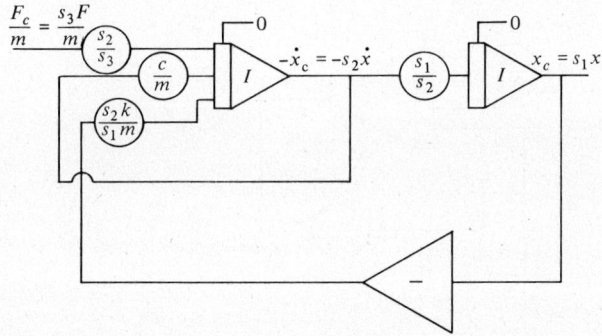

Figure 10.5

The values for the potentiometer settings would be

$$\frac{s_1}{s_2} = 100 \quad \frac{c}{m} = 5 \quad \frac{s_2 k}{s_1 m} = 100 \quad \frac{s_2}{s_3} = 80$$

which is far too high for the computer. The problem lies in the large natural frequency present. (Even the computer's graph and pen apparatus could not follow at this frequency.) What is further needed is time scaling. Defining the computer time as τ,

$$\tau \equiv \alpha t \quad \frac{d}{d\tau} = \frac{1}{\alpha} \frac{d}{dt} \quad \frac{d^2}{d\tau^2} = \frac{1}{\alpha^2} \frac{d^2}{dt^2}$$

The differential equation transformed to scaled computer time is now

$$\alpha^2 \frac{d^2 x}{d\tau^2} + \alpha \frac{c}{m} \frac{dx}{d\tau} + \frac{k}{m} x = \frac{F}{m}$$

or

$$\frac{d^2 x}{d\tau^2} = \frac{1}{\alpha^2} \left(\frac{F}{m} - \alpha \frac{c}{m} \frac{dx}{d\tau} - \frac{k}{m} x \right)$$

372 ANALYTICAL MECHANICS

On the computer

$$\frac{dx}{d\tau} = \frac{1}{\alpha} \int \left(\frac{F}{m} - \alpha \frac{c}{m} \frac{dx}{d\tau} - \frac{k}{m} x \right) d\tau$$

$$x = \frac{1}{\alpha} \int \frac{dx}{d\tau} d\tau$$

It is apparent that time scaling changes each integrator gain by a factor of $1/\alpha$. The computer diagram with time scaling now appears as shown in Figure 10.6. For a value of $\alpha = 100$ (operating in slowed-down time),

$$\frac{s_2}{\alpha s_3} = 0.8 \qquad \frac{s_1}{\alpha s_2} = 1$$

$$\frac{c}{\alpha m} = 0.05 \qquad \frac{s_2 k}{\alpha s_1 m} = 1$$

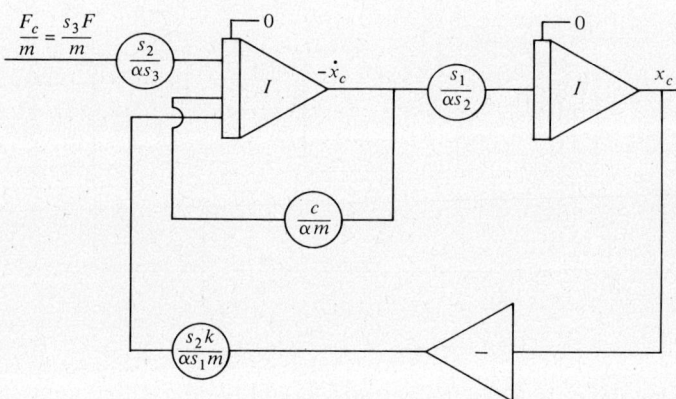

Figure 10.6

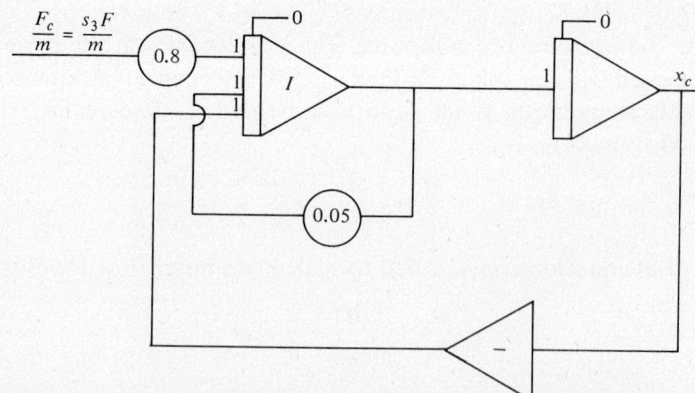

Figure 10.7

and the problem can be easily implemented. The final patch wiring diagram would now appear as shown in Figure 10.7. Note that several alternative wiring solutions are possible.

10.3 DIGITAL COMPUTER SIMULATION

In the early 1960s, with the introduction of the first commercially available digital computers, engineers who were responsible for analog computations began to check their results with numerical analyses. It soon became apparent that the digital computer had much to offer. It could be programmed. It could make logical decisions. It had a memory. It had the capability of function generation and nonlinear operation. It was extremely accurate. Simulation runs could be programmed to loop back and repeat trials with different constants automatically if resulting values were not satisfactory. With the introduction of *timesharing*, interactive control was made possible, and programs could be changed easily at the terminal. The analog computer, on the other hand, was limited in accuracy. It was physically restricted by the number of integrators that could be placed in the machine. It took time to scale. It was tedious to wire and required extensive checking of the patching. No memory or programming capability was available. Nonlinear elements and special function generation required special-purpose circuit design. However, as fast as the digital computer was, it could not operate in *real time*, as the analog machine could. An analog machine provides instantaneous output upon signal input. Because of this, most calculations were still done on analog machines until recent years. Today, the cost of digital computing units, chips, random access memory (RAM), etc., has gone down phenomenally, and the computer is so ubiquitous that the analog machine now needs justification for its use. (*Hybrid computation* takes advantage of the best features of both machines. It uses the analog computer's speed to calculate and the digital computer's ability to process and control the experiment. Its technology and hardware have advanced enormously in the years since its promising introduction. Since the state of the art is still limited to special-purpose machines custom designed for a specific computing facility and product, hybrid computation lacks universality.)

To solve differential equations on a digital computer, one should have knowledge of the many numerical techniques available for solution approximation. Stepwise integration can be carried out by Euler integration, Heun's formula, Adam-Moulton fixed- or variable-step algorithms, Runge-Kutta algorithm, Milne's method, etc. One must also be familiar with the common computer languages Fortran, Basic, and the newer Pascal that are used to program the numerical analysis. Alternatively, the latest special-purpose languages, *continuous system simulation languages*, write the desired program for the user and thus eliminate the need for computer expertise. Their development started with the early use of the digital computer for checking analog results. General programs were developed with subroutine construction and terminology that modeled the analog computer. These programs were essential, for they eliminated all the time wasted writing and debugging individual programs by the technical staff that basically performed the same integration task. They also allowed the nonexpert easy access to the computer. Later, it was decided that for computational efficiency it would be better to write a new translator. This new translator would accept an entirely new language whose words would be commands or tasks similar to those performed by the analog computer, and it would generate an operable Fortran program. These new languages, the continuous system simulation languages, would be able to take a command from the keyboard,

$$X = \text{INTEG}(\text{INTEG}(F/M - C/M*XD - K/M*X, XD0), X0)$$

for instance, and construct the numerical analysis program needed to solve the indicated differential equations (a) and (b), simultaneously. The computer variable XD would correspond to v.

$$\dot{v} = \frac{F}{m} - \frac{c}{m}v - \frac{k}{m}x \qquad v = \int_{v_0}^{v} \dot{v}\, dt \qquad \text{(a)}$$

$$\dot{x} = v \qquad x = \int_{x_0}^{x} \dot{v}\, dt \qquad \text{(b)}$$

By combining this feature with the interactive capability of cathode-ray-tube (CRT) terminal timesharing and by incorporating many built-in subroutine blocks, i.e., Pulse, Lead-Lag, Hysteresis, and Delay, with which the analog computer user is familiar, a very powerful tool was formed. Extensive capability for report output in the forms of built-in plotting routines, table forming, and labeling was also included. The names of some of the better known languages that have been developed over the years are Midas, Mimic, Dare, ACSL, CSL, CSMP (IBM's software), and CSSL. It is the purpose of the following sections to describe, in abbreviated form, the functioning of CSSL, version 4. It is a language owned, distributed, and fully supported by Simulation Services, Chatsworth, California. It is one of the simulation languages included in the California State University's timesharing computer user's library. *Continuous System Simulation Language—version four* (1976), a user's guide and reference manual, is also available for Nilsen Associates of Chatsworth, California.

Discussion of discrete-time simulation languages will not be presented here.

10.4 CSSL, VERSION 4

A. Introduction

Most simulation languages perform approximately in the same manner. Though CSSL, version 4 (CSSL-IV or CSSLV4), will be discussed here, a sample problem programmed in CSMP will also be included for comparison. It might also be mentioned that a senior-level student at California State University, Long Beach, easily fitted his own simulation language, written in Basic, to his home microcomputer.

An overview of the CSSL-IV language shows that it has many useful features. For one, it accepts Fortran statements and has access to the Fortran function library. It has its own library of (39) simulation operators for performing integration, function generation, stochastic analysis, linear transfer function modeling, digital switching, etc. The user can also write his or her own library function, which is called a MACRO (it is similar to a subroutine or function subprogram in Fortran). Further, the user can plot his or her results or have them reported in tabular output. Up to seven different numerical integration algorithms are provided. Ten directives control numerical analysis; they provide for the specification of algorithms, truncation interval (stepsize), relative and absolute error limits, name and initial conditions of the independent variable, and iteration control. The language has 15 run-time directives that can be used interactively to change variables, parameters, input/output options, etc. Up to three independent variables may be used to define a tabular function. The language also allows for logical control variables, vectors, and arrays. In the DERIVATIVE section, the translator sorts the order for processing equations unless specifically proscribed. The translator writes the Fortran program to be compiled and run, which is also accessible to the user. If problems arise, CSSL-IV provides some diagnostic help. The user also has a DEBUG command. CSSL-IV can be presently implemented using Fortran on the CDC 6000, UNIVAC 1108, PDP-10, and IBM 370 computers.

To introduce the CSSL-IV language, the example of Section 10.2 will be repeated here. A typical CSSL-IV program written to solve this problem is shown in Figure 10.8. The values of k, m, c, F, x_0, v_0, and the final time are given as 2000, 0.2, 1.0, 1000, 0.0, 0.0, and 10, respectively.

Note that the program has three main sections. Constant values are defined or calculated in the INITIAL region. The problem solution is performed in the DYNAMIC region; each pass through this section advances the solution by an amount specified by the communication interval, CINTERVAL. Included in the DYNAMIC section is a section titled DERIVATIVE in which the system dynamic equations are defined. The equations are integrated until the independent variable reaches the communication interval value. When this happens, control is passed to the DYNAMIC section, and information can be printed out using an OUTPUT directive or, as done here, stored in an array for future plotting purposes (PREPAR). In the DERIVATIVE section, ultimate control is transferred to the operating translator system that determines the proper sorting and sequencing of statements that appear. For instance, XDD appears last in the listing of this section but will actually be processed first. The TERMT statement sets the condition(s) that the independent variable T must meet before the program moves to the TERMINAL phase. TFIN was defined in the INITIAL section. In the TERMINAL region final calculations are made, if necessary, and then the program ends. Note that in CSSL-IV all variables or constants, such as M, are assumed *real*. Integer variables must be defined by a data

```
*** CONTINUOUS SYSTEM SIMULATION LANGUAGE-VERSION FOUR (CSSL-IV) ***
    *** SUPPORTED BY SIMULATION SERVICES, CHATSWORTH, CA. ***

PROGRAM FIRST
INITIAL
CONSTANT M=0.2,K=2000.,C=1.0,F=1000.,XD=0.,...
XDD=0.,TFIN=10.
END
DYNAMIC
DERIVATIVE SPRING
CINTERVAL CI=0.5
XD=INTEG(XDD,XD0)
X=INTEG(XD,X0)
XDD=(F-C*XD-K*X)/M
END
PREPAR XDD,XD,X
TERMT(T.GT.TFIN)
END
TERMINAL
END
END
```

Figure 10.8

declaration directive. The character T is reserved for the independent variable. Variable names such as ZXXX, where XXX stands for numeric characters, are not permitted. Variable names are limited to six characters.

In the interactive mode, upon successful compilation of the program, the computer will ask

ENTER NEXT RUN-TIME DIRECTIVE

(abbreviated in this text as ENRTD). The user then can employ any of the 15 run-time directives available. Here, one might elect to change the value of c to 0.5, for example, using an INPUT command. The dialogue with the computer might appear as shown in Figure 10.9. At this point the computer will commence

```
        ENTER NEXT RUN-TIME DIRECTIVE
? INPUT C=.5
        ENTER NEXT RUN-TIME DIRECTIVE
? OUTPUT X,XD,XDD
        ENTER NEXT RUN-TIME DIRECTIVE
? START
        DO YOU WISH A STATIC CHECK?
? NO
```

Figure 10.9

to numerically solve the problem. The OUTPUT command causes the variables listed to be printed in tabular form. Proper headings, page spacing, and format control will be selected automatically by the processor. An HDR statement can be used to place a title on each page of results generated by the OUTPUT statement. The static check capability allows for a system check of initial conditions very much like that provided for on the analog computer. The program output will appear as shown in Figure 10.10. After completion of this task the computer will then request from the user additional commands (see Figure 10.11).

T	X	XD	XDD
0.	0.	0.	.500000E+04
.500000E+00	.242546E+00	-.711127E+01	.259232E+04
.100000E+01	.377536E+00	-.734853E+01	.124301E+04
.150000E+01	.447286E+00	-.555933E+01	.541039E+03
.200000E+01	.480727E+00	-.363421E+01	.201817E+03
.250000E+01	.495227E+00	-.217164E+01	.531543E+02
.300000E+01	.500647E+00	-.120353E+01	-.345963E+01
.350000E+01	.502097E+00	-.625387E+00	-.194101E+02
.400000E+01	.502067E+00	-.304787E+00	-.199101E+02
.450000E+01	.501596E+00	-.134506E+00	-.156253E+02
.500000E+01	.501076E+00	-.487015E-01	-.106368E+02
.550000E+01	.500690E+00	-.248582E-02	-.689132E+01
.600000E+01	.500379E+00	.102657E-01	-.381734E+01
.650000E+01	.500180E+00	.145745E-01	-.183767E+01
.700000E+01	.500003E+00	.169309E-01	-.734498E-01
.750000E+01	.499860E+00	.390089E-02	.139305E+01
.800000E+01	.499963E+00	-.125305E-01	.406181E+00
.850000E+01	.500113E+00	-.392236E-02	-.112711E+01
.900000E+01	.500038E+00	.101259E-01	-.402645E+00
.950000E+01	.499909E+00	.384593E-02	.903721E+00
.100000E+02	.499963E+00	-.814254E-02	.386714E+00

Figure 10.10

```
         ENTER NEXT RUN-TIME DIRECTIVE
       ?
         LABEL C=0.5
              ENTER NEXT RUN-TIME DIRECTIVE
           ? PLOT T,X
```

Figure 10.11

The output of the computer to the responses given in Figure 10.11 are now

378 ANALYTICAL MECHANICS

```
C=0.5

    PLOT SYMBOLS ARE:  X     = *

 .600E+00 +    +    +    +    +    +    +    +    +    +    +




 .525E+00 +    +    +    +    +    +    +    +    +    +    +

                        * * * * * * * * * * * * * * * * * *
                   *
 .450E+00 +    +  * +    +    +    +    +    +    +    +    +




 .375E+00 +  *  +    +    +    +    +    +    +    +    +    +




 .300E+00 +    +    +    +    +    +    +    +    +    +    +


             *
 .225E+00 +    +    +    +    +    +    +    +    +    +    +




 .150E+00 +    +    +    +    +    +    +    +    +    +    +




 .750E-01 +    +    +    +    +    +    +    +    +    +    +
```

Figure 10.12

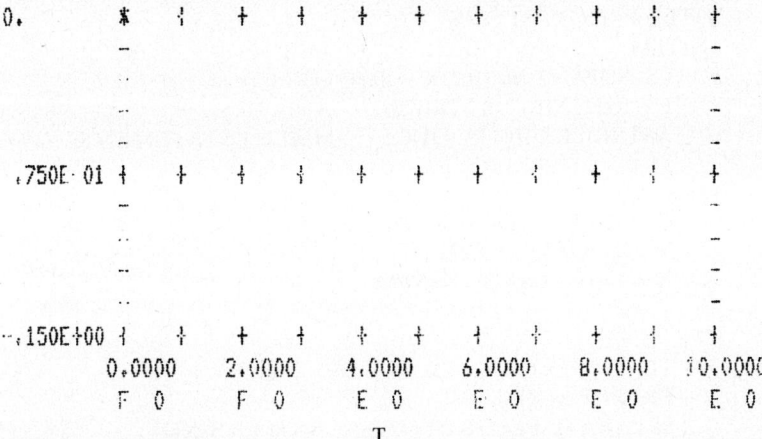

Figure 10.12 (continued)

shown in Figure 10.12. The command LABEL places a heading in the upper left corner of the plot. PLOT, as can be seen, automatically determines the range of variable output, scales and determines symbols for each variable, and plots the output (provided they were first stored in an array through a PREPAR statement). The first variable in the list, for this example T, becomes the abscissa on any plot. It is possible to take direct control of the plotting by specifying the user's own scales for the axis.

After the plotting has been completed, the following dialogue may occur.

ENRTD
? HALT

The command HALT stops the program and returns control from the CSSL-IV processor back to the standard operating system.

Since an integration algorithm was not specified, a default option was implemented. The CSSL-IV default integration option is the Adams-Moulton variable-step algorithm. It has a truncation error approximately equal to h^5, where h is the variable step size. Error control is provided by both relative and absolute error checks (which are also specified by default). It has a wide range of stability and good accuracy. It uses a Runge-Kutta starting routine. The solution advances by first predicting Y_{n+1},

$$Y_{n+1}^P = Y_n + \frac{h}{24}(55\dot{Y}_n - 59\dot{Y}_{n-1} + 37\dot{Y}_{n-2} - 9\dot{Y}_{n-3}) \qquad (10.9)$$

and then corrects this value with the equation

$$Y_{n+1}^C = Y_n + \frac{h}{24}(9\dot{Y}_{n+1}^P + 19\dot{Y}_n - 5\dot{Y}_{n-1} + \dot{Y}_{n-2}) \qquad (10.10)$$

A similar program written for continuous system modeling program (CSMP)

```
        TITLE PROGRAM ONE
        INITIAL
            CONSTANT M=0.2, K=2000., C=1.0,...
            F=1000., X0=0.0, XD0=0.0
        *   ASTERISKS INTRODUCE COMMENT STATEMENTS

        DYNAMIC
            XD=INTGRL(XD0, XDD)
            X=INTGRL(X0, XD)
            XDD=(F-C*XD-K*X)/M

        TERMINAL
            TIMER FINTIME=1.0, OUTDEL=0.125
            PRTPLT X, XD
        *   SEPARATE PLOTS OF T VS. X AND T VS. XD
        *   WILL BE MADE, CSMP-III ALLOWS FOR
        *   SIMULTANEOUS PLOTS
        END
        STOP
        ENDJOB
```

Figure 10.13

might appear as shown in Figure 10.13. The default integration algorithm used for CSMP is a variable-step Runge-Kutta method.

Sample Problem 10.1
Write a CSSLV4 computer program for Sample Problem 5.12.

Analysis: The program is shown in Figure 10.14. ∎

Sample Problem 10.2
Write a CSSLV4 computer program for Sample Problem 8.7.

Analysis: The program is shown in Figure 10.15. ∎

Sample Problem 10.3
Write a CSSLV4 computer program for Sample Problem 8.9.

Analysis: The program is shown in Figure 10.16. ∎

Sample Problem 10.4
Write a CSSLV4 computer program for Sample Problem 8.13.

Analysis: The program is shown in Figure 10.17. ∎

```
PROGRAM RLAND
INITIAL
CONSTANT OM=7.3E-5,XDO=0.,YDO=0.,ZDO=600.,...
X0=0.,Y0=0.,Z0=0.,TFIN=124.,LAM=50.
A=COS(LAM/57.3)
B=SIN(LAM/57.3)
C=2.*OM
END
DYNAMIC
CINTERVAL CI=2.
DERIVATIVE ROCKET
XD=INTEG(XDD,XDO)
YD=INTEG(YDD,YDO)
ZD=INTEG(ZDD,ZDO)
X=INTEG(XD,X0)
Y=INTEG(YD,Y0)
Z=INTEG(ZD,Z0)
XDD=C*YD*B
YDD=-C*(XD*B+ZD*A)
ZDD=-9.81+C*YD*A
END
IF(T.GT.TFIN.OR.Z.LT.0.) GO TO FIN
END
TERMINAL
FIN..CONTINUE
END
END
```

Figure 10.14

```
PROGRAM ROD
INITIAL
CONSTANT PSID=10.,THO=30.,TFIN=2.1,DTHO=0.
THO=THO/57.3
PSID=PSID*.1047
END
DYNAMIC
CINTERVAL CI=.08
DERIVATIVE ROD
DDTH=PSID*PSID*SIN(TH)*COS(TH)
DTH=INTEG(DDTH,DTHO)
TH=INTEG(DTH,THO)
END
IF (T.GE.TFIN.OR.TH.GE.1.571)GO TO FIN
END
TERMINAL
FIN..CONTINUE
END
END
```

Figure 10.15

```
PROGRAM TOP
INITIAL
CONSTANT MO=.07,IXX=23.1E-5,IZZ=1.25E-5,DPSI=2000.,...
THO=90.,DTHO=0.,TFIN=0.2.
R=DPSI*.01047
A=IZZ*R/IXX
THO=THO/57.29578
END
DYNAMIC
CINTERVAL CI=.005
DERIVATIVE FIRST
TH=INTEG(DTH,THO)
DTH=INTEG(DDTH,DTHO)
DPH=-A/(1.-COS(TH))
DDTH=(DPH*COS(TH)-A )*DPH*SIN(TH)+MO*SIN(TH))/IXX
END
IF(TH.GE.3.1415.OR.T.GT.TFIN)GO TO FIN
PREPAR TH
END
TERMINAL
FIN..CONTINUE
END
END
```

Figure 10.16

```
PROGRAM ORBIT
INITIAL
CONSTANT H=1.819,RO=6.614,DRO=.2750,DTHO=.0419,...
THDO=135.,OM=.058,TFIN=107.22
COMMENT REAL TIME EQUALS T*0.2237
THO=THDO/57.29578
THA=THO
A=H*H
END
DYNAMIC
CINTERVAL CI=1.375
DERIVATIVE EARTH
DDR=(A/R-1.)/(R*R)
DDTH=-2.*DR*DTH/R
DR=INTEG(DDR,DRO)
DTH=INTEG(DDTH,DTHO)
R=INTEG(DR,RO)
TH=INTEG(DTH,THO)
END
THA=THA+OM*1.375
IF(THA.GT.(6.2832+THO).OR.T.GT.TFIN)GO TO FIN
END
TERMINAL
FIN..CONTINUE
END
END
```

Figure 10.17

Sample Problem 10.5
Write a CSSLV4 computer program to simulate the free vibrations of a spring-mass system with Coulomb friction acting.

Analysis: The applicable equations of motion are
$$m\ddot{x} = \begin{cases} -kx - \mu_k mg & \text{for } \dot{x} > 0 \\ -kx + \mu_k mg & \text{for } \dot{x} < 0 \end{cases}$$

```
PROGRAM COULOM
INITIAL
CONSTANT M=0.5,K=1000.,UK=0.3,XD0=0.,X0=.015,...
TFIN=.3125,G=9.81
XCM=X0*100.
XDCM=XD*100.
F=UK*M*G
END
DYNAMIC
CINTERVAL CI=.01171875
DERIVATIVE COULOM
ALGORITHM IA=5,JA=5
PROCEDURAL (FR=F,XD)
FR=0.
IF(XD.GE.0.) GO TO L1
FR=-F
GO TO L2
L1..IF(XD.EQ.0.) GO TO L2
FR=F
L2..CONTINUE
END
XDD=-(K*X+FR)/M
XD=INTEG(XDD,XD0)
X=INTEG(XD,X0)
END
XCM=X*100.
XDCM=XD*100.
PREPAR XCM,XDCM
IF(T.GT.TFIN) GO TO L3
END
TERMINAL
L3..CONTINUE
END
END
```

Figure 10.18

384 ANALYTICAL MECHANICS

The CSSLV4 program is shown in Figure 10.18. Though this problem could have been solved in another way using the simulation operator FCNSW, the intent was to demonstrate how a PROCEDURAL statement is used. The purpose of a PROCEDURAL block is to preserve the order of statements in the DERIVATIVE section that would otherwise be sorted, i.e., rearranged or separated. Within the PROCEDURAL and END PROCEDURAL statements, the logic and sequencing of the assignment statements are unaltered. The whole block is free to be moved as a unit in the DERIVATIVE section. The form of the directive is

$$\text{PROCEDURAL (output list = input list)}$$

Note also how the self-starting Runge-Kutta integration algorithm was called out. (Algorithm IA = 5, JA = 5.)

B. CSSL-IV Directives

A complete list of CSSL-IV translator directives appears in Table 10.1.
 A summary of the CSSL-IV interactive run-time control directives is shown in Table 10.2. Table 10.3 lists the input/output control directives available.

C. CSSL-IV Simulation Operators

As mentioned before, one of the chief advantages of using CSSL-IV is that a library of 39 simulation operators have already been preprogrammed as MACRO subroutines and are available to the user. They can be called upon and used as easily as one would call upon SIN X from the Fortran library. Provision is made for the user to write additional programs of his or her own to be included as a subroutine. Most all of the simulation-operator terms listed below in Table 10.4 trace their origins to the field of atuomatic control.

Sample Problem 10.6
Shown in the figure is a block diagram of the dynamics of a position control system with minor-loop tachometer feedback. Write a CSSL-IV program that will calculate the response of the system, $c(t)$, to a unit-step input, $r(t)$. Vary the value of b from 0 to 0.09 in steps of 0.03. Let the final time be 5 s and the communication interval be 0.1 s.

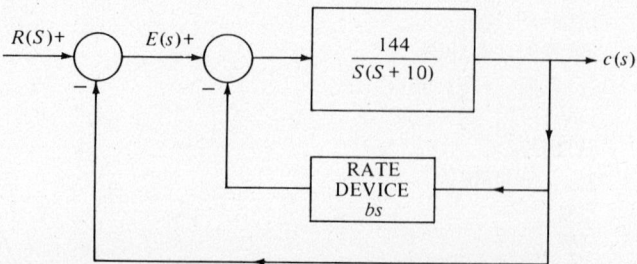

Analysis: The problem should be reformulated to eliminate the derivative

TABLE 10.1   CSSL-IV Translator Directives

Data declaration directives

ARRAY	Variable (J, K, L), ... list, similar to Fortran dimension statement.
CONSTANT	Variable = value, ... list.
INTEGER	Variable, ... list.
TABLE	Name, NVAR, J. K. L, independent variable list (X, Y, Z), dependent variable list (f(X, Y, Z)). For example, TABLE XSQ, 1, 5, 0., 1., 2., 3., .4, .5, 0., 1., 4., 9., 16., 25 establishes a stored table called XSQ with one independent variable, five samples, and five functional values $[Y = F(X) = X^2]$. It can be then called in the DERIVATIVE section for any value of X. Linear interpolation is used for values that fall between those established in the table. NVAR is the number of independent variables, J, K, L are the number of data points in X, Y, Z, respectively. In this example, NVAR = 1; so $K = L = 0$, $J = 5$.

Structure directives

DERIVATIVE	Name.
DYNAMIC	
END	
INITIAL	
PROCEDURAL	(output list = input list)
PROGRAM	Name.
SEGMENT	Name. Used only when multiple simulations are done at once where each simulation has its own INITIAL, DYNAMIC, AND TERMINAL sections.
TERMINAL	

Program control directives

COMMENT	Alphanumeric string.
DROP(-PICKUP)	Stops variables from being entered into a symbol table
LCV	Logical control variable, statements preceded by an LCV will only be performed if the LCV is presently true.
PICKUP	Restarts the storage of variable names in a symbol table
TERMT	Defines conditions for ending program.

TABLE 10.2 CSSL-IV Runtime Directives

CONTIN	Resumes a run after it terminates. Terminate conditions must be updated. Program returns to the beginning of the DYNAMIC section.
DEBUG	NDUMP, TSTART. NDUMP is the number of dumps requested per communication interval. TSTART specifies the time each dump cycle is to start.
GRAPH	Options, SCALEX, SXMIN, SXMAX, SCALEY, SYMIN, SYMAX, X variable, Y variable list. Used with digital plotter. See PLOT.
HDR	*Alphanumeric string, associate with OUTPUT. Up to 131 alphanumeric characters can be used.
INPUT	Variable = value, . . . list.
LABEL	*Alphanumeric string, plot title. Up to 60 alphanumeric characters can be used.
MERROR	Variable = bound, . . . list, relative error limits established.
OUTPUT	*n, variable list, output every n communication intervals. Time is listed automatically. For example, OUTPUT 10, X.

PLOT	Grid options, SCALEX, SXMIN, SXMAX, SCALEY, SYMIN, SYMAX, X variable, Y variable list. Grid options:
	    SQ: square grid
	    NG: no grid
	    OG: open grid
	    NGSQ: square format no grid
	    OGSQ: square format open grid
	Scale X: SXMIN and SXMAX and SXMAX sets scale for X axis with the minimum and maximum values, respectively, of X specified.
	Scale Y: SYMIN and SYMAX sets scale for Y axis with the minimum and maximum values, respectively, of Y specified.
	X variable is the name of the X coordinate.
	Y variable are the names (up to four) of the variables to be plotted.
	Default mode: PLOT X, XD, ... Automatic scaling is performed.
	Variable to be plotted must be listed in a PREPAR directive first.

PREPAR	*Variable, . . . list. Prepares storage table for eventual plotting.
RANGE	*Variable, . . . list. Maximum and minimum values reached in a run for variables listed is provided.
START	Start processing.
HALT	Leave system.
TAG	*Alphanumeric string, plot subtitle. Up to 30 alphanumeric characters are allowed.
XERROR	Variable = bound, . . . list. Absolute error limits established.

Note: Directives marked with an asterisk (*) may also be employed within the program.

TABLE 10.3 CSSL-IV Input/Output Control Directives

CARDS	Variable, ... list. Batch command.
GRAPH	Runtime use only. See Table 10.2.
HDR	Available at runtime also. See Table 10.2.
INPUT	Variable = value, ... list. Runtime use only.
LABEL	Alphanumeric string. Available at runtime. See Table 10.2.
OUT	Variable, ... list, up to six output columns formed. Translator creates Fortran write and format statements. Time will not appear unless called for.
OUTPUT	Preferred use at runtime. See Table 10.2.
PAGE EJECT	New page for experiment. Called for in INITIAL section.
PAGE SKIP	Skip lines.
PLOT	Runtime use only. See Table 10.2.
PREPAR	Variable, ... list. Available at runtime. See Table 10.2.
TAG	Alphanumeric string. Available at runtime. See Table 10.2.
TITLE	Variable, ... list, up to six column headings are provided. Associated with and precedes the OUT directive.

simulation operator needed to represent the rate device. From control theory

$$R(S) - C(S) = E(S) \qquad (a)$$

$$E(S) - bSC(S) = E1(S) \qquad (b)$$

$$E1(S)\left[\frac{144}{S(S+10)}\right] = C(S) \qquad (c)$$

Then
$$R(S) = \left[1 + bS + \frac{S(S+10)}{144}\right] C(S)$$

$$\frac{C(S)}{R(S)} = \frac{144}{S^2 + (10 + 144b)S + 144} \qquad (d)$$

The equivalent block diagram for equation (d) is

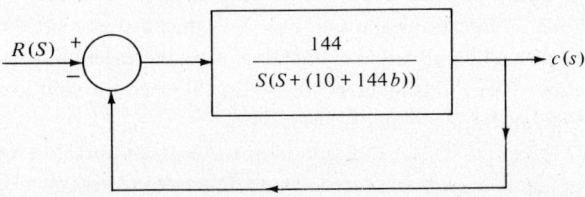

Though equation (d) can be directly implemented, by using CMPXPL, the solution to the problem will be developed with reference to the equivalent block diagram instead. The solution is shown in Figure 10.24. ∎

TABLE 10.4 CSSL-IV Simulation Operators

ANDL, Q = ANDL(X1, X2, ..., XN), logical AND function. Q = 1.0 if all inputs are greater than or equal to 0.5; otherwise, Q = 0.

BOUND, Y = BOUND (LL, UL, X), simulates saturation. It is shown in Figure 10.19. It should not be used to limit the output of an integrator. See LIMINT, Section 10.4D.

CMPXPL, Y = CMPXL (P, Q, X, IC1, IC2), complex pole. It represents the transfer function $1/(PS^2 + QS + 1)$ with X as the input, $\dot{Y}_0 = IC1$, and $Y_0 = IC2$.

COMPAR, Y = COMPAR (X1, X2), analog comparator. Y = 1.0 if X1 ⩾ X2; Y = 0.0 otherwise.

DEAD, Y = DEAD (LL, UL, X), dead zone. The function DEAD has Y = 0 for LL < X < UL, Y = X − UL for X ⩾ UL, and Y = X − LL for X ⩽ LL (see Figure 10.20).

DELAY, Y = DELAY (IC, N, X, TS), multistep delay. Provides fixed time delay of N times the fixed integration step size. Output Y = IC for the first N time steps. After t becomes greater than the delay, Y equals the X value found N time steps ago. TS is the name of a storage array. Each time a DELAY function is specified a new storage name must be given.

DERIVT, Y = DERIVT (IC, X, T, TS), numerical differentiation. As with the analog computer, this function is to be used with extreme caution. IC is Y_0. TS is a storage array which again must have a unique name.

EXPF, Y = EXPF (IC, TAU, T, SW), exponential pulse. IC is the initial condition for Y, either 0.0 or 1.0. TAU is the rise/fall time constant. T is the independent variable. SW is a switch that turns the pulse on when SW ⩾ 0.5 and off when SW < 0.5.

FCNSW, Y = FCNSW (CNTRL, X1, X2, X3), three position function switch. Y = X1 for CNTRL < 0.0; Y = X2 for CNTRL = 0.0; Y = X3 for CNTRL > 0.0.

GAUSS, Y = GAUSS (MEAN, SIGMA), normally distributed random variable generator.

HARM, Y = HARM (TON, FREQ, PHASE, T) harmonic sine-wave generator. TON is the turn-on time. PHASE is the phase in radians at the turn-on time. It is leading for PHASE > 0. T is the name of the independent variable.

HSTRSS, Y = HSTRSS (IC, PL, PR, X) or Y = HSTRSS (IC, PL, PR, SLOPE, X). IC is the value of the output Y at t = 0. X is the input. The two functions are shown in Figure 10.21.

IMPL, Y = IMPL (YO, ERR, NITER, ERFLG, FOFY, YN), implicit equation solver. YO is an initial guess for Y. ERR is the error bound on the solution. NITER is the maximum number of iterations allowed. FOFY is an expression for the function of Y. YN is optional with a default value of 0.0001; it is an incremental perturbation value of the initial guess. ERFLG is an error flag set to 1.0 when the solution does not converge. A Newton-Raphson algorithm is used.

INTEG, Y = INTEG (YDOT, IC), integration operator. Can operate even on vector arrays. All arguments must be declared as one-dimensional arrays in the INITIAL section, each having the same size.

LEDLAG, Y = LEDLAG (P, Q, X, IC), lead-lag transfer function. Y is the solution to $(PS + 1)/(QS + 1)*X(S)$. IC is the value of Y at $T = 0$.

LIMINT, Y = LIMINT (X, IC, LL, UL), limiting integrator. See Section 10.4D.

TABLE 10.4 (continued)

LOGIC, Q = LOGIC (JK, 14, IC, J, K, CLEAR, CLOCK), JK flip-flop. JK is the name of the device; 14 is the dimension of temporary storage for past values; IC is the initial condition of Q; J is the set input; K is the reset input; CLEAR is the automatic clear signal when set to 0.0, and CLEAR = 1.0 otherwise; CLOCK causes flip-flop transition to occur when it goes from 1.0 to 0.0, trailing-edge triggered.

LOGIC, Q = LOGIC (NAND, 7, IC, A, B), NAND gate with history. Use only when a digital circuit with feedback is simulated. See NANDL. Q = 0.0 when both A and B = 1.0; Q = 0.0 otherwise. Q initially is set to IC.

LOGIC, Q = LOGIC (RST, 14, IC, RD, SD, RC, SC, C), RST flip-flop. With RD and SD set to 1.0 (direct asynchronous reset and set lines, respectively), RC and SC and the clock C control the behavior of the flip-flop output synchronously.

MODINT, Y = MODINT (X, IC, SW1, SW2), mode-controlled integrator. Simulates analog computer operation. When switches SW1 and SW2 are both less than or equal to zero or both greater than zero, the computer is in the computer mode. When SW1 is less than or equal to zero while SW2 is greater than zero, it is the hold mode. When SW1 is greater than zero while SW2 is less than or equal to zero, it is in the reset mode.

NANDL, Q = NANDL (X1, X2, ..., XN), logical NAND. Do not use in a feedback circuit. See LOGIC (NAND). The output Q is set equal to 1.0 if at least one input is 0.0. It is set to 0.0 if all inputs have the value 1.0.

ORL, Q = ORL (X1, X2, ..., XN), logical OR. The output Q will be 0.0 if all inputs have the value 0.0. Q will be set to 1.0 if at least one input has the value 1.0.

OU, Y = OU (TAU, T, SIGMA, MEAN). The output Y is the result of passing band-limited white noise having constant power over a finite bandwidth through a first-order lag network. TAU is the time constant of the low-pass filter, and T is the independent variable. Y is the output.

OUTSW, Y1, Y2, = OUTSW (SW, X), two position demultiplexor. For SW < 0, Y1 = X and Y2 = 0.0. For SW ⩾ 0, Y1 = 0.0 and Y2 = X.

PTR, X, Y = PTR (R, THETA), polar-to-rectangular coordinate conversion.

PULSE, Y = PULSE (TO, PERIOD, PW, T, STROBE, STMIN, XERR), pulse synchronized to communication interval. Used with multistep integration algorithm. Y is a pulse train starting at TO, having pulse width PW; PERIOD is the time between pulses; STROBE is the character string specifying synchronization; XERR represents the absolute error per step; default value is 0.001. STMIN is the minimum number of steps to be taken when the pulse appears high; default value is 1.0.

PULSE, Y = PULSE (TO, PERIOD, PW, T), nonsynchronized pulse generator. Used with fixed-step-size integration algorithms.

QNTZR, Y = QNTZR (P, X), digital quantizer. Simulates output of a digital-to-analog converter. See Figure 10.22.

RAMP, Y = RAMP (P, T), ramp generator. See Figure 10.23.

REALPL, Y = REALPL (TAU, X, IC), first-order lag. Solves the equation $1/(TAU*S + 1)*X(S)$ with $Y_0 = IC$.

RTP; R, THETA = RTP (X, Y), rectangular to polar conversion.

TABLE 10.4 (continued)

SDELAY, Y = SDELAY (IC, X, TS), single-step transport delay. TS is a uniquely named storage array. Used with fixed-step integration algorithms.

STEP, Y = STEP (P, T), unit step function. Y = 1.0 for T > P.

SWIN, Y = SWIN (SW, X1, X2), input multiplexor. Y = X1 when SW < 0, and Y = X2 for SW ⩾ 0.0.

TDELAY, Y = TDELAY (X, XO, TD, T, XS, TS, I, J, K), variable transport lag. Used with variable-step integration.

    X = input at time T_n

    XO = Y initially

    TD = Amount of delay required at time T_n

    XS = uniquely named k-dimensional storage array for past values of X

    TS = uniquely named k-dimensional storage array for past values of T

        I = value 1 if delayed quantity corresponds to previous sample; I = 0 if result is from interpolation

        J = remaining stack storage; J must be less than or equal to K

        K = dimension of stack storage for XS and TS

TRANF, Y = TRANF (NN, ND, P, Q, X, TS) linear transfer function.

    NN = order of numerator

    ND = order of denominator

        P = array containing coefficients of numerator polynomial, higher-order terms first

        Q = array containing coefficients of denominator polynomial, higher-order terms first

        X = input variable

        TS = uniquely named storage array of dimension ND

UNIF, Y = UNIF (LB, UB) uniformly distributed random variable. Describes random process with lower and upper bounds LB and UB, respectively.

ZHOLD, Y = ZHOLD (IC, SW, X), zero order hold. IC is the initial output, Y = X whenever SW ⩾ 0.5 but equals the previous value of Y otherwise. Simulates various sampled-data systems.

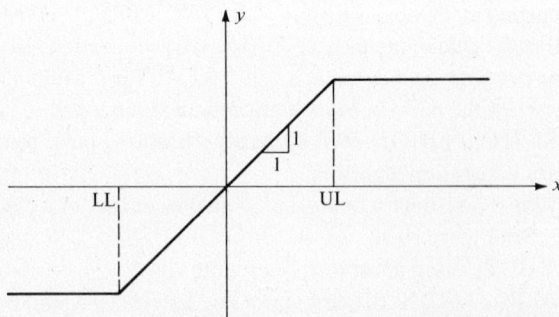

Figure 10.19

DIGITAL COMPUTER SIMULATION 391

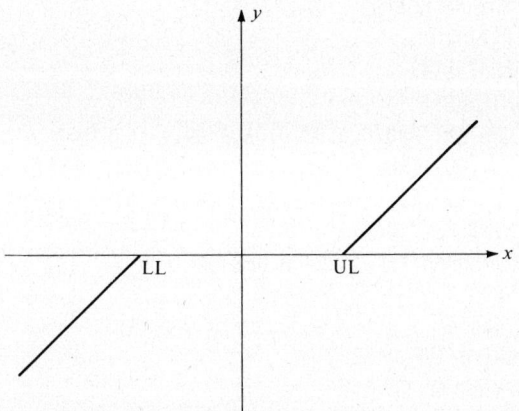

Figure 10.20

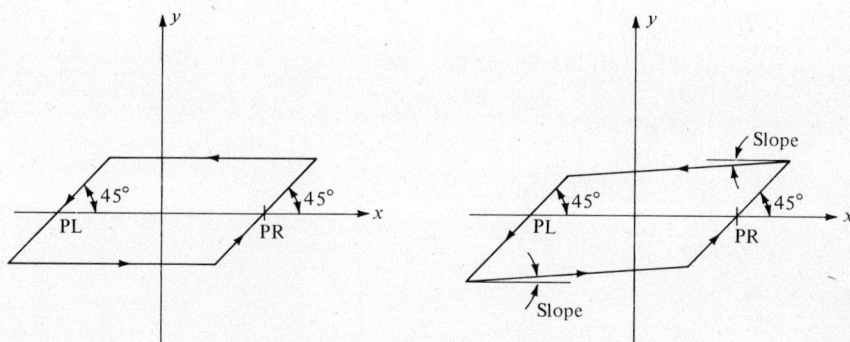

Figure 10.21

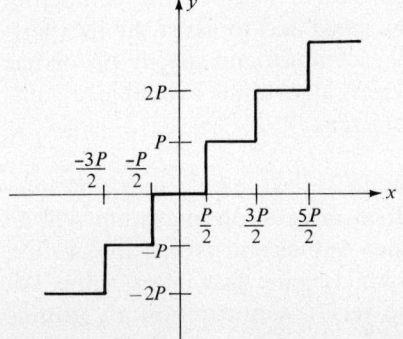

Figure 10.22

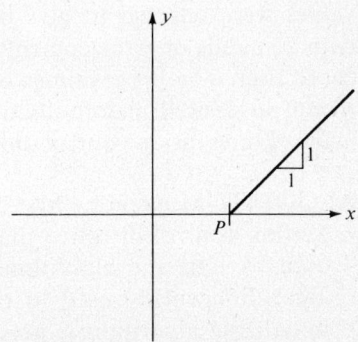

Figure 10.23

```
PROGRAM CONTRL
INITIAL
ARRAY B(4)
INTEGER I
CONSTANT B=0.0,0.03,0.06,0.09,I=1,TFIN=2.
L1..CONTINUE
TAU =1./(10.*144.*B(I))
PRINT 12,I,B(I)
L2..FORMAT(1X,'THE VALUES ARE FOR B(',I1,')=',F7.3)
END
DYNAMIC
DERIVATIVE RATE
CINTERVAL CI=0.1
E=STEP(0.,T)-C
E1=144./TAU*E
X1=REALPL(TAU,E1,0.)
C=INTEG(X1,0.)
END
PREPAR C
IF (T.GT.TFIN) GO TO L3
END
TERMINAL
L3..I=I+1
IF (I.LE.4) GO TO L1
END
END
```

Figure 10.24

D. CSSL-IV Numerical Analysis

CSSL-IV provides seven different integration algorithms that the user can call forth. Each integration algorithm has preprogrammed default values. These values were selected to give optimum compromises between the competing twin demands of error control and execution speed and to assist the inexperienced user. The programmer can override these values and specify his or her own if so desired. Before discussing the different algorithms available, a summary of integration control directives will be given.

ALGORITHM name 1 = first index value, name 2 = last index value is an integration control directive that specifies which integration algorithm will be used to start the algorithm, i.e., first index value, and which integration algorithm will be used to run the program, i.e., the last index value. An initializing algorithm is necessary because very few integration algorithms are self-starting. The Runge-Kutta method and Euler's method are two important exceptions. If no algorithm is specified, the preprogrammed default method for CSSL-IV is a variable-step predictor-corrector algorithm with error control.

CINTERVAL name=value specifies the communication interval from the DERIVATIVE section to the DYNAMIC section described previously. The default value is 1.0.

ERRTAG name is an error flag set equal to 1 when a variable-step integration algorithm tries to set the truncation interval below that specified by the MINTERVAL directive.

ITERATE name=value specifies an upper bound on the number of iterations that an integration algorithm will be performed. Default value equals 2.

MERROR name=value, name=value, ... specifies the maximum relative error per truncation interval each variable is to have when using a variable-step integration routine. The last value in the list provides values for all remaining unlisted variables. A default value of 1.E-5 is called forth if no MERROR is specified.

XERROR name=value, name=value, ... is similar to the MERROR directive but controls absolute error instead. A default value of 1.E-4 is precoded.

MINTERVAL name=value specifies the minimum value to which the truncation interval can be reduced by a variable-step algorithm. See ERRTAG. Default value is 1.E-6.

NISTEPS name=value specifies the number of integration steps taken per communication interval for a fixed-step starting integration algorithm. The default value is 1.

NSTEPS name=value specifies the number of integration steps taken per communication interval for the running algorithm. The default value here is 1.

VARIABLE name=value specifies a unique name and initial value for the independent variable in a DERIVATIVE section. Default value is T=0.0.

The integration algorithms provided by CSSL-IV are

Euler integration: Index value is 1. Euler integration is a fixed-step integration method that is self-starting. It provides a first-order approximation to the solution thus having truncation error $\varepsilon \sim (\Delta t)^2$. It has a region of calculation stability of $-2 < \Delta t \lambda < 0$, where Δt is the step size and λ is the largest time constant of the system. With large function error, Δt must be kept small, but this in turn leads to round-off error accumulation. A compromise Δt is found for each problem usually by trial-and-error experimentation. This algorithm is the recommended method for systems having a discontinuous input function. The numerical algorithm is

$$Y_{n+1} = Y_n + \Delta t \, \dot{Y}_n \qquad (10.11)$$

Trapezoidal integration: Index value is 2. Trapezoidal integration is a fixed-step integration method that requires one previous derivative value to start. It provides a second-order approximation to the solution with truncation error $\varepsilon \sim (\Delta t)^3$. The stability region is limited to $-1 < \Delta t \lambda < 0$. The numerical algorithm used is

$$Y_{n+1} = Y_n + \Delta t \, \dot{Y}_n + \frac{\Delta t}{2}(\dot{Y}_n - \dot{Y}_{n-1}) \tag{10.12}$$

Adams predictor algorithm: Index value is 3. The Adams four-point predictor algorithm is a fixed-step integration method that requires three previous derivative values to start. It is a fourth-order method having truncation error $\varepsilon \sim (\Delta t)^5$. The stability region is very narrow for this method, $-0.3 < \Delta t \, \lambda < 0$, though execution time is fast. Here

$$Y_{n+1} = Y_n + \frac{\Delta t}{24}(55\dot{Y}_n - 59\dot{Y}_{n-1} + 37\dot{Y}_{n-2} - 9\dot{Y}_{n-3}) \tag{10.13}$$

Adams-Moulton algorithm: Index value is 4. The Adams-Moulton predictor-modifier-corrector-modifier algorithm is a fixed-step integration method that also requires three previous derivative values. It is a fourth-order method with a larger stability region than the Adams predictor algorithm; i.e., $-1.3 < \Delta t \, \lambda < 0$. The predictor is

$$P_{n+1} = Y_n + \frac{\Delta t}{24}(55\dot{Y}_n - 59\dot{Y}_{n-1} + 37\dot{Y}_{n-2} - 9\dot{Y}_{n-3}) \tag{10.14}$$

The first modifier is

$$M_{n+1} = P_{n+1} + \frac{251}{270} D_n \tag{10.15}$$

where the corrector is

$$D_{n+1} = \frac{3 \Delta t}{8}(\dot{M}_{n+1} - 4\dot{Y}_n + 6\dot{Y}_{n-1} - 4\dot{Y}_{n-2} + \dot{Y}_{n-3}) \tag{10.16}$$

The final modifier is given by

$$M_{n+1} = P_{n+1} + \frac{251}{270} D_{n+1} = Y_{n+1} \tag{10.17}$$

The Runge-Kutta algorithm: Index value is 5. The Runge-Kutta-Gill algorithm is a self-starting fixed-step integration algorithm. It is a fourth-order method with truncation error $\varepsilon \sim (\Delta t)^5$ having a broad stability region of $-2.73 < \Delta t \, \lambda < 0$. It has an accurate but slow calculation procedure. Namely, to find Y_{n+1},

$$\begin{aligned} q_1 &= \dot{Y}_n \\ Y_1 &= Y_n + \frac{\Delta t}{2} q_1 \\ Y_2 &= Y_1 + \Delta t \, (1 - \sqrt{\tfrac{1}{2}})(\dot{Y}_1 - q_1) \\ q_2 &= q_1 + (1 - \sqrt{\tfrac{1}{2}})(2\dot{Y}_1 - 3q_1) \\ Y_3 &= Y_2 + \Delta t \, (1 + \sqrt{\tfrac{1}{2}})(\dot{Y}_2 - q_2) \end{aligned} \tag{10.18}$$

$$q_3 = q_2 + (1 + \sqrt{\tfrac{1}{2}})(2\dot{Y}_2 - 3q_2)$$

$$Y_{n+1} = Y_3 + \frac{\Delta t}{6}(\dot{Y}_3 - 2q_3) \tag{10.19}$$

Adams-Moulton variable-step algorithm: Index value is 8. The Adams-Moulton predictor-corrector algorithm offers such a wide range of stability and good accuracy that it is the default integration algorithm selected for CSSL-IV. It is a fourth-order variable-step integration method, with a stability region given by $-3.0 < \Delta t\, \lambda < 0$, and is not self-starting. The predictor is

$$P_{n+1} = Y_n + \frac{\Delta t}{24}(55\dot{Y}_n - 59\dot{Y}_{n-1} + 37\dot{Y}_{n-2} - 9\dot{Y}_{n-3}) \tag{10.20}$$

The corrector is

$$C_{n+1} = Y_{n+1} = Y_n + \frac{\Delta t}{24}(9\dot{P}_{n+1} + 19\dot{Y}_{n-1} - 5\dot{Y}_{n-1} + \dot{Y}_{n-2}) \tag{10.21}$$

The error is

$$e = -\tfrac{19}{270}(C_{n+1} - P_{n+1}) = D_{n+1} \tag{10.22}$$

and the step size is reduced by a factor of 2 if $|e| > $ XERROR or $|e| > $ MERROR $*$ (C_{n+1}). P_{n+1} and C_{n+1} are then recalculated. The calculation proceeds until the error is within limits or until MINTERVAL has been reached.

Milne algorithm: Index value is 9. The Milne predictor-corrector method is a fourth-order variable-step integration method that is not self-starting. It has a poor stability region of $-0.8 < \Delta t\, \lambda < -0.3$. The predictor is

$$P_{n+1} = Y_{n-3} + \frac{4\Delta t}{3}(2\dot{Y}_n - \dot{Y}_{n-1} + 2\dot{Y}_{n-2}) \tag{10.23}$$

The corrector is

$$C_{n+1} = Y_{n+1} = Y_{n-1} + \frac{\Delta t}{3}(\dot{P}_{n+1} + 4\dot{Y}_n + \dot{Y}_{n-1}) \tag{10.24}$$

The error is

$$e = -\tfrac{1}{29}(C_{n+1} - P_{n+1}) \tag{10.25}$$

and error control proceeds as in the Adams-Moulton variable-step integration algorithm.

The reader can conclude that for occasional problems the default Adams-Moulton variable-step integration algorithm is perhaps best (for problems where no discontinuities appear). For production runs, though, small experiments must first be carried out to establish which of the integration algorithms given is the fastest while providing acceptable accuracy.

Two special CSSL-IV simulation operators that pertain just to integration are LIMINT and MODINT. MODINT is discussed in Table 10.4.

LIMINT, Y = LIMINT (X, IC, LL, UL), limit integration. A BOUND operator, though limiting output, will not suppress integration. Integration continues unimpeded, and its response exceeds the established boundary for the output. This can cause problems if the sign of the derivative input to the integrator reverses; i.e., the function goes off the boundary. Because the integrator has continued to function, it cannot respond instantaneously to this change. The LIMINT command, however, establishes a direct lower limit (LL) and upper limit (UL) for the output of the integrator. The input X has initial conditions IC at $t = 0$.

E. CSSL-IV MACRO Structure

Though CSSL-IV provides 39 simulation operators, it cannot possibly cover all types of applications. The user may have to define a new simulation operator in subprogram form that is specifically tailored to his or her needs. To write a subprogram that will fit with the rest of the CSSL-IV system, the user may employ a PROCEDURAL directive, write a common Fortran subroutine and append it to the CSSL-IV program, or prepare a MACRO subroutine. Unlike the other two, a MACRO subroutine is sorted by the translator and can appear in the DERIVATIVE section of the program. MACRO subroutines are placed in the program where called for, when the program is first translated. Since a MACRO subroutine is written into the program, control need never be transferred, temporarily, to an external monitor, as it would have to be for a Fortran subroutine. System operating time is reduced in this manner. Table 10.5 offers a brief description of the CSSL-IV MACRO directives available. Since it is not the intention of this book to delve further than necessary into the art of simulation language programming, the interested reader is referred to the Nilsen Associates manual for more detail.

A MACRO is invoked in two ways: The stand-alone invocation and the algebraic invocation. The stand-alone invocation has the form

$$\text{NAME V1, V2, } \ldots \text{, VN.}$$

Name is the name of the MACRO. V1,..., VN is the argument list. The algebraic invocation has the form

$$\text{V1, V2, } \ldots \text{, VN} = \text{name (I1, I2, } \ldots \text{, IM)}.$$

Name is the name of the defined MACRO. V1, ..., VN is the output list corresponding to P(1), ..., P(N). The input list is I1, ..., IM and corresponds to P(N + 1), ..., P(N + M).

$$Y = \text{name (I1, I2, } \ldots \text{, IM)}.$$

This is the same invocation as above, but there is only one output variable required.

TABLE 10.5 CSSL-IV MACRO Directives

MACRO ASSIGN N   Assigns the total number of arguments in the MACRO invocaton argument list to the dummy variable N. N is the dimension of the vector P of arguments in the calling statement. See MACRO MACRO. It is modified by the INCREMENT, DECREMENT, DIVIDE, and MULTIPLY directives.

MACRO LXX ... CONTINUE   Provides a branch point for the MACRO GOTO and MACRO IF statements.

MACRO DECREMENT i The current value of the assigned variable N is decreased by the integer value i.

MACRO DIVIDE i The current value of the assigned variable N is divided by the integer value i. The result is the new value of N.

MACRO END   This statement ends the MACRO block.

MACRO EXIT   In combination with an IF MACRO, this directive allows for early termination of the MACRO subroutine.

MACRO GOTO LXX   Unconditional branching in the MACRO subroutine to a MACRO CONTINUE statement having statement labels LXX (X can be any integer).

MACRO IF $e_1 = e_2$ LXX   If the variables e_1 and e_2 are equal, the program will go to the MACRO CONTINUE statement with statement label LXX.

MACRO INCREMENT i Increases the current value of the assigned variable N by i.

MACRO MACRO name P, D1, D2, D3   This directive defines the start of a MACRO block. P is an array representing all the variable names in the argument list of the calling statement. D1, D2, and D3 are arrays containing the first, second, and third dimension of each variable of the argument list, respectively.

MACRO MULTIPLY i Multiplies the current value of the assigned variable N by i, and then replaces it by the new value.

MACRO PMACRO P, D1, D2, D3   This directive defines the start of a nonsortable procedural MACRO subroutine. It is similar to MACRO MACRO.

MACRO PRINT, alphanumeric string. MACRO write command.

MACRO RELABLE L1, L2, ...   With multiple calls to a MACRO subroutine, all labels assigned to Fortran statements (not MACRO directives) must be relabeled.

MACRO REDEFINE V1, V2, ...   As above, all Fortran variables in the argument list should be redefined at each entry to the subroutine. It is not necessary to redefine MACRO arguments.

MACRO STANDVAL P(i) = value   Provides default values for the arguments P(i) of the defined MACRO function that are left blank when the MACRO is invoked.

Sample Problem 10.7
Write a CSSLV4 MACRO subroutine that will simulate Coulomb friction. See Sample Problem 10.5.

Analysis: The MACRO subroutine is shown in Figure 10.25.   ■

```
PROGRAM MACEX1
MACRO MACRO COULOM P
MACRO RELABEL LA,LB
PROCEDURAL (P(1)=P(2),P(3))
P(1)=0.
IF (P(3).GE.0.) GO TO LA
P(1)=-P(2)
GO TO LB
LA..IF (P(3).EQ.0.) GO TO LB
P(1)=P(2)
LB..CONTINUE
END
MACRO END
INITIAL
CONSTANT M=0.5,K=1000.,UK=0.3,XD0=0.,X0=.015,...
TFIN=.3125,G=9.81
XCM=100.*X0
XDCM=100.*XD0
F=UK*M*G
END
DYNAMIC
CINTERVAL CI= .01171875
DERIVATIVE FXMACR
ALGORITHM IA=5,JA=5
COMMENT  MACRO FUNCTION COULOM WILL BE CALLED NEXT
FR=COULOM(F,XD)
XD=INTEG(XDD,XD0)
X=INTEG(XD,X0)
XDD=-(K*X+FR)/M
END
XCM=100.*X
XDCM=100.*XD
PREPAR XCM,XDCM
IF(T.GT.TFIN) GO TO L1
END
TERMINAL
L1..CONTINUE
END
END
```

Figure 10.25

PROBLEMS

*10.1 Write a CSSLV4 computer program for problem 2.18.

*10.2 Write a CSSLV4 computer program for problem 2.19.

*10.3 Write a CSSLV4 computer program for problem 2.20.

*10.4 Write a CSSLV4 computer program for problem 2.21.

*10.5 Write a CSSLV4 computer program for problem 2.22.

*10.6 Write a CSSLV4 computer program for problem 3.5.

*10.7 Write a CSSLV4 computer program for problem 3.7.

*10.8 Write a CSSLV4 computer program for problem 3.18.

*10.9 Write a CSSLV4 computer program for problem 3.20.

*10.10 Write a CSSLV4 computer program for problem 3.29.

*10.11 Write a CSSLV4 computer program for problem 7.6.

*10.12 Write a CSSLV4 computer program for problem 7.7.

*10.13 Write a CSSLV4 computer program for problem 7.37.

*10.14 Write a CSSLV4 computer program for problem 7.38.

*10.15 Write a CSSLV4 computer program for problem 7.39.

*10.16 Write a CSSLV4 computer program for problem 7.40.

*10.17 Write a CSSLV4 computer program for problem 8.6.

*10.18 Write a CSSLV4 computer program for problem 8.16.

*10.19 Write a CSSLV4 computer program for problem 8.28.

*10.20 Write a CSSLV4 computer program for problem 8.34.

*10.21 Write a CSSLV4 computer program for problem 9.40.

*10.22 Write a CSSLV4 computer program for problem 9.42.

*10.23 For the two-stage rocket of problem 4.25, write a CSSLV4 program that will calculate the highest altitude attained. For the calculation, assume a slow burning time so that a varying gravity force affects all phases of the motion. Consider also the atmospheric drag that retards the rocket. As in problem 4.25, the first-stage empty shell is discarded immediately, then the second stage ignites. The equation of motion for the rocket is

$$m\dot{v} = \dot{m}_i v_e - mg - D_i \quad \text{(a)}$$

where
$$\dot{m}_i v_e \sim I_{sp} b_i g \quad N \quad \text{(b)}$$

$$D_i = \tfrac{1}{2}\rho C_i S_i v^2 \quad N \quad \text{(c)}$$

$$\rho = 1.225 e^{-z/7250} \text{ kg/m}^3 \quad \text{(d)}$$

400 ANALYTICAL MECHANICS

$$C_1 S_1 = 30 \text{ m}^2$$
$$C_2 S_2 = 20 \text{ m}^2$$
$$g = 9.806 \left(\frac{6370 \times 10^3}{6370 \times 10^3 + z} \right)^2 \text{ m/s}^2 \tag{e}$$
$$b_1 = 200 \text{ kg/s} = \dot{m}_i$$
$$b_2 = 20 \text{ kg/s} = \dot{m}_2$$

and the burning times are

$$t_1 = 20 \text{ s}, \qquad t_2 = 40 \text{ s}$$

Also,

$$v_e = I_s g = 1960 \text{ m/s}$$

Output is required every 2.0 s. The student can determine after several runs what the final times must be for the calculaton.

*10.24 Shown below to the left is a branched torsional system driven by a motor. The total inertial load referred to the drive shaft is known to be $J = 180$ g·cm², the gear ratio is $N = 4$. Assuming the shaft to be rigid, write a CSSLV4 program to solve for the motion of the load. The two-phase, ac motor torque-speed characteristic curve is given below to the right. The load starts from rest. Use the Tabular Function and a communication interval of 0.004 s, and terminate the program at 0.18s.

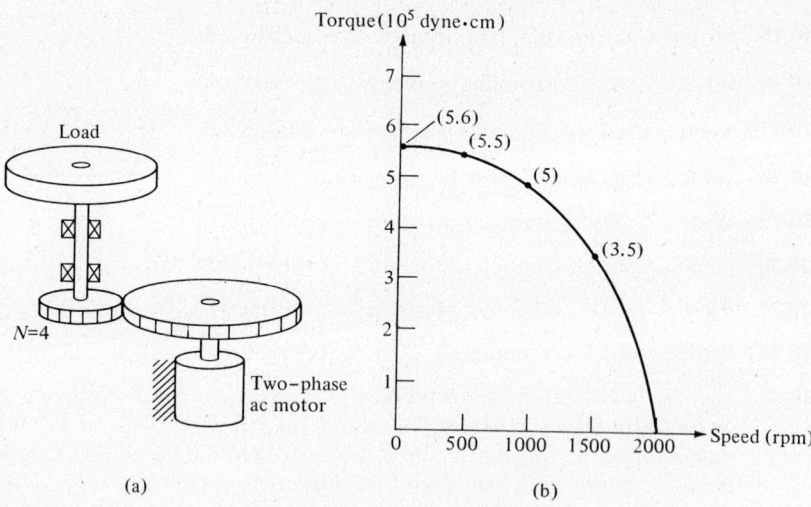

(a)   (b)

Appendix A

SHORT SUMMARY OF ELEMENTARY DYNAMICS

Particle Dynamics	Systems of Particles	Rigid Body Dynamics

Kinematics

Particle Dynamics:

$$\mathbf{v} \equiv \frac{d\mathbf{r}}{dt} \qquad \mathbf{a} \equiv \frac{d\mathbf{v}}{dt}$$

$$\mathbf{v} = \dot{x}\mathbf{i} + \dot{y}\mathbf{j} + \dot{z}\mathbf{k}, \quad \mathbf{v} = v\mathbf{i}_t$$
$$\mathbf{a} = \ddot{x}\mathbf{i} + \ddot{y}\mathbf{j} + \ddot{z}\mathbf{k}$$
$$= (\ddot{r} - r\dot{\theta}^2)\mathbf{i}_r + (r\ddot{\theta} + 2\dot{r}\dot{\theta})\mathbf{i}_\theta$$
$$= \dot{v}\mathbf{i}_t + \frac{v^2}{\rho}\mathbf{i}_n$$

Coordinate transformations

Systems of Particles:

$$\mathbf{R}_{c.m.} \equiv \frac{\Sigma m_i \mathbf{r}_i}{\Sigma m_i}$$

$\Sigma m_i \boldsymbol{\rho}_i = 0$, about the center of mass

Rigid Body Dynamics:

$$\mathbf{v}_B = \mathbf{v}_A + \mathbf{v}_{B/A} \qquad \mathbf{a}_B = \mathbf{a}_A + \mathbf{a}_{B/A}$$

$$\mathbf{v}_{B/A} = \boldsymbol{\omega}_{AB} \times \mathbf{r}_{B/A}$$
$$\mathbf{a}_{B/A} = \boldsymbol{\omega}_{AB} \times (\boldsymbol{\omega}_{AB} \times \mathbf{v}_{B/A})$$
$$\qquad + \dot{\boldsymbol{\omega}}_{AB} \times \mathbf{r}_{B/A}$$

or

$$\mathbf{v}_{B/A} = [\mathbf{v}_{B/A}]_{\text{rot}} + \boldsymbol{\Omega} \times \mathbf{r}_{B/A}$$
$$\mathbf{a}_{B/A} = [\mathbf{a}_{B/A}]_{\text{rot}} + 2\boldsymbol{\Omega} \times [\mathbf{v}_{B/A}]_{\text{rot}}$$
$$\qquad + \boldsymbol{\Omega} \times (\boldsymbol{\Omega} \times \mathbf{r}_{B/A}) + \dot{\boldsymbol{\Omega}} \times \mathbf{r}_{B/A}$$

Angular acceleration

Equations of Motion

Particle Dynamics:

$$\mathbf{F} = m\mathbf{a}$$
$$\mathbf{M}_O = \dot{\mathbf{H}}_O$$
$$\mathbf{H}_O = \mathbf{r} \times m\mathbf{v}$$

Systems of Particles:

$$\mathbf{F}_{\text{ext}} = M\mathbf{a}_{c.m.}$$
$$\mathbf{M}_O = \dot{\mathbf{H}}_O$$
$$\mathbf{M}_{c.m.} = \dot{\mathbf{H}}_{c.m.}$$

$$\mathbf{H}_O = \sum_{i=1}^n \mathbf{r}_i \times m_i \mathbf{v}_i$$

$$\mathbf{H}_{c.m.} = \sum_{i=1}^n \boldsymbol{\rho}_i \times m_i \dot{\boldsymbol{\rho}}_i$$

Rigid Body Dynamics:

$$\mathbf{F}_{\text{ext}} = M\mathbf{a}_{c.m.}$$
$$\mathbf{M}_O = \dot{\mathbf{H}}_O$$
$$\mathbf{M}_{c.m.} = \dot{\mathbf{H}}_{c.m.}$$

$$\{H_O\} = [I_O]\{\omega\}$$

$$\{H_{c.m.}\} = [\hat{I}]\{\omega\}$$

Parallel-axis theorem, principal moments of inertia, similarity transformations

For plane motion and principal axis z
$$F_x = M(a_{c.m.})_x, \qquad F_y = M(a_{c.m.})_y$$
$$(M_O)_z = I_{zz}\dot{\omega}_z, \quad (M_{c.m.})_z = \hat{I}_{zz}\dot{\omega}_z$$

Energy Principles

Particle Dynamics:

$$W_{1\to 2} = T_2 - T_1$$
$$T = \frac{mv^2}{2}$$
$$E = T_1 + V_1 = T_2 + V_2$$

Conservative system, potential energies

Systems of Particles:

$$W_{1\to 2} = T_2 - T_1$$
$$T = \sum_{i=1}^n \frac{m_i v_i^2}{2} = \frac{mv_{c.m.}^2}{2} + \frac{m_i \dot{\rho}_i^2}{2}$$

Rigid Body Dynamics:

$$W_{1\to 2} = T_2 - T_1$$
$$T = \frac{Mv_{c.m.}^2}{2} + \tfrac{1}{2}\{\omega\}^t[\hat{I}]\{\omega\}$$

$$T_O = \tfrac{1}{2}\{\omega\}^t[I_O]\{\omega\}$$

For plane motion

$$T = \frac{Mv_{c.m.}^2}{2} + \frac{I_{zz}\omega_z^2}{2}, \qquad T_O = \frac{(I_O)_{zz}\omega_z^2}{2}$$

Particle Dynamics	Systems of Particles	Rigid Body Dynamics
Momentum Principles		

Particle Dynamics	Systems of Particles	Rigid Body Dynamics
$\int_1^2 \mathbf{F}\,dt = \mathbf{p}_2 - \mathbf{p}_1$	$\int_1^2 \mathbf{F}_{ext}\,dt = \mathbf{P}_2 - \mathbf{P}_1$	$\int_1^2 \mathbf{F}_{ext}\,dt = \mathbf{P}_2 - \mathbf{P}_1$
$\mathbf{p} = m\mathbf{v}$	$\mathbf{P} = M\mathbf{v}_{c.m.} = \sum_{i=1}^{n} m_i \mathbf{v}_i$	$\mathbf{P} = M\mathbf{v}_{c.m.}$
$\int_1^2 \mathbf{M}_O\,dt = (\mathbf{H}_O)_2 - (\mathbf{H}_O)_1$	$\int_1^2 \mathbf{M}_O\,dt = (\mathbf{H}_O)_2 - (\mathbf{H}_O)_1$	$\int_1^2 \mathbf{M}_O\,dt = (\mathbf{H}_O)_2 - (\mathbf{H}_O)_1$
Central-force motion	$\int_1^2 \mathbf{M}_{c.m.}\,dt = (\mathbf{H}_{c.m.})_2 - (\mathbf{H}_{c.m.})_1$	$\int_1^2 \mathbf{M}_{c.m.}\,dt = (\mathbf{H}_{c.m.})_2 - (\mathbf{H}_{c.m.})_1$
	Impact	*For plane motion and principal axis z*
		$\int_1^2 F_x\,dt = (Mv_{c.m.})_{x,2} - (Mv_{c.m.})_{x,1}$
		$\int_1^2 F_y\,dt = (Mv_{c.m.})_{y,2} - (Mv_{c.m.})_{y,1}$
		$\int_1^2 M_O\,dt = (I_{zz}\omega_z)_2 - (I_{zz}\omega_z)_1$
		$\int_1^2 M_{c.m.}\,dt = (\hat{I}_{zz}\omega_z)_2 - (\hat{I}_{zz}\omega_z)_1$

Appendix B

MASS MOMENTS OF INERTIA OF COMMON SHAPES

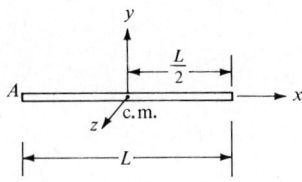

$$\hat{I}_{xx} \cong 0$$
$$\hat{I}_{yy} = \hat{I}_{zz} = \frac{mL^2}{12}$$
$$I_{AA} = \frac{mL^2}{3}$$

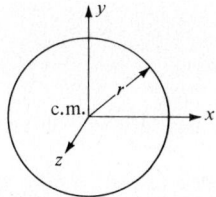

Thin disk

$$\hat{I}_{xx} = \hat{I}_{yy} = \frac{mr^2}{4}$$
$$\hat{I}_{zz} = 2\hat{I}_{xx}$$

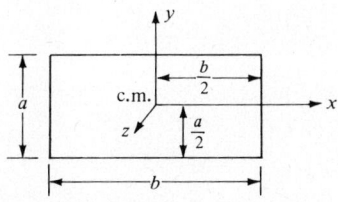

Thin rectangular plate

$$\hat{I}_{xx} = \frac{ma^2}{12}$$
$$\hat{I}_{yy} = \frac{mb^2}{12}$$
$$\hat{I}_{zz} = \frac{m}{12}(a^2 + b^2)$$

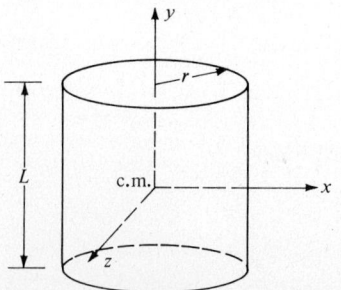

Right circular cylinder

$$\hat{I}_{xx} = \hat{I}_{zz} = \frac{m}{12}(3r^2 + L^2)$$
$$\hat{I}_{yy} = \frac{mr^2}{2}$$

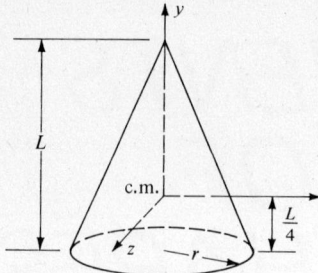

$$\hat{I}_{xx} = \hat{I}_{zz} = \frac{3m}{80}(4r^2 + L^2)$$

$$\hat{I}_{yy} = \frac{3}{10}mr^2$$

$$m = \frac{\rho\pi r^2 L}{3}$$

Right circular cone

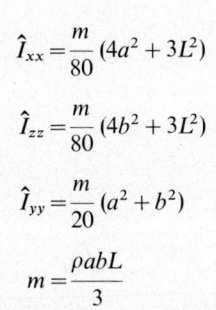

$$\hat{I}_{xx} = \frac{m}{80}(4a^2 + 3L^2)$$

$$\hat{I}_{zz} = \frac{m}{80}(4b^2 + 3L^2)$$

$$\hat{I}_{yy} = \frac{m}{20}(a^2 + b^2)$$

$$m = \frac{\rho abL}{3}$$

Right rectangular pyramid

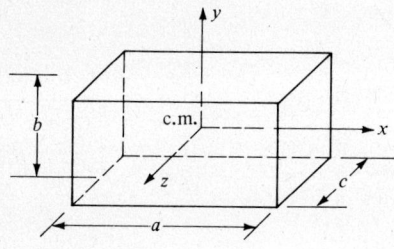

$$\hat{I}_{xx} = \frac{m}{12}(b^2 + c^2)$$

$$\hat{I}_{yy} = \frac{m}{12}(a^2 + c^2)$$

$$\hat{I}_{zz} = \frac{m}{12}(a^2 + b^2)$$

Rectangular prism

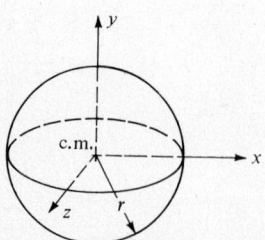

$$\hat{I}_{xx} = \hat{I}_{yy} = \hat{I}_{zz} = \tfrac{2}{5}mr^2$$

$$m = \tfrac{4}{3}\rho\pi r^3$$

Sphere

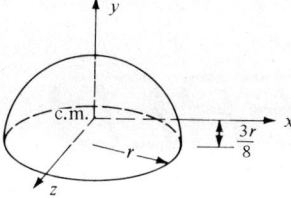

Half-sphere

$$\hat{I}_{xx} = \hat{I}_{zz} = \frac{83}{320} mr^2$$
$$\hat{I}_{yy} = \tfrac{2}{5} mr^2$$

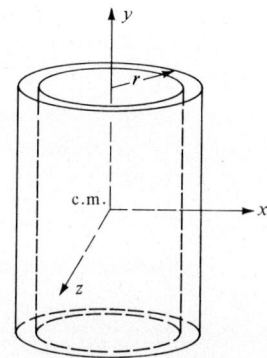

Thin circular cylindrical shell

$$\hat{I}_{xx} = \hat{I}_{zz} = \frac{m}{12}(6r^2 + L^2)$$
$$\hat{I}_{yy} = mr^2$$

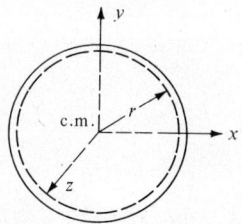

Thin spherical shell

$$\hat{I}_{xx} = \hat{I}_{yy} = \hat{I}_{zz} = \tfrac{2}{3} mr^2$$

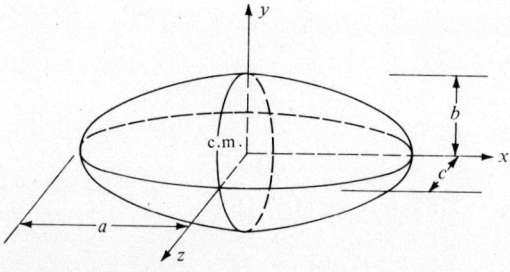

Ellipsoid

$$\hat{I}_{xx} = \frac{m}{5}(b^2 + c^2)$$
$$\hat{I}_{yy} = \frac{m}{5}(a^2 + c^2)$$
$$\hat{I}_{zz} = \frac{m}{5}(a^2 + b^2)$$
$$m = \tfrac{4}{3}\rho\pi abc$$

ANSWERS TO EVEN-NUMBERED PROBLEMS

Chapter 2

2.2 $t = \dfrac{c}{m} \ln\left(\dfrac{cv_0}{\mu mg} + 1\right)$

2.4 $y = \dfrac{-x^2 g^2 + v_0^4}{2v_0^2 g}$

2.6 $T = \dfrac{2\pi}{(g/L + k/9m)^{1/2}}$

2.8 $t = \dfrac{2\pi c_\theta}{\sqrt{3k_\theta}}$

2.10 $\delta = 2\pi\sqrt{3}$

2.12 $A = 15.3$ mm, $\phi = 116.9°$, M.F. $= 2.0$

2.14 M.F. $= \dfrac{1}{\{[1 - (\omega/\omega_n)^2]^2 + [2(c/c_c)(\omega/\omega_n)]^2\}^{1/2}}$

where $k = k_1 + k_2$, $c_c = 2m\omega_n$, and $\omega_n^2 = k/m$.

2.16 $\left|\dfrac{z_m}{y_0}\right| = \dfrac{(\omega/\omega_n)^2}{\{[1 - (\omega/\omega_n)^2]^2 + [2(c/c_c)(\omega/\omega_n)]^2\}^{1/2}}$, $\phi = \tan^{-1}\dfrac{\omega c}{k - m\omega^2}$

2.24 $\rho = \dfrac{v_0^2 \cos^2 \gamma}{g}$

2.26 $a_n = 1.13 L \omega^2$

2.28 $\rho = \dfrac{r^3 v_0^2}{bGM}$

2.30 $\mathbf{v}_P = 3R\dot{\phi}(\{\cos(\phi) + \cos(3\phi)\}\mathbf{i} + \{\sin(\phi) - \sin(3\phi)\}\mathbf{j})$

Chapter 3

3.2 $\dot{\theta} = (2g/L)^{1/2}$, energy

ANSWERS TO EVEN-NUMBERED PROBLEMS

3.4 $m(\dot{x} + L\dot{\theta}\cos\theta) = C$, linear momentum in the x direction

3.6 $m(\ddot{x} - L\dot{\theta}^2\sin\theta + L\ddot{\theta}\cos\theta) + k(x - L_0) = 0$

3.8 $\delta = \dfrac{md\omega^2}{k\sqrt{2}}$

3.10 $-k(r - r_0) + W\cos\theta = \dfrac{W}{g}(\ddot{r} - r\dot{\theta}^2)$

$-W\sin\theta = \dfrac{W}{g}(r\ddot{\theta} + 2\dot{r}\dot{\theta})$

3.12 $-k(r - r_0) = m(\ddot{r} - r\omega^2)$, $N_\theta = m2\dot{r}\omega$, $T = rN_\theta = 2mr\dot{r}\omega$

3.14 $r_{max} = \tfrac{9}{4}r_0$

3.16 $m(-R\dot{\phi}^2 - \tfrac{3}{4}R\dot{\theta}^2) = -0.54kR + N_r + \dfrac{mg}{2}$

$m\left(\dfrac{\sqrt{3}}{2}R\ddot{\theta} - R\dot{\phi}\dot{\theta}\right) = 0$

$m\left(R\ddot{\phi} + \dfrac{\sqrt{3}}{4}R\dot{\theta}^2\right) = -0.187kR + \dfrac{\sqrt{3}}{2}mg$

3.20 $m(\ddot{z} - L\ddot{\phi}\sin\phi - L\dot{\phi}^2\cos\phi) + k(z - L_0) = mg$

$H_0 = m(L^2\sin^2\phi)\dot{\theta}$, a constant

$m(L\ddot{\phi} - \ddot{z}\sin\phi - L\sin\phi\cos\phi\dot{\theta}^2) = -mg\sin\phi$

3.22 $V = -\tfrac{1}{2}\ln(x^2 - y^2) + C$, $x^2 > y^2$

3.24 $\omega = \left(\dfrac{6g}{R}\right)^{1/2}$

3.26 $d = -\dfrac{v_0^2}{4g} + \dfrac{1}{2}\left[\left(\dfrac{v_0^2}{2g}\right)^2 + 4R^2\right]^{1/2}$

3.28 $m(\ddot{r} - r\dot{\theta}^2) = \dfrac{-GMm}{r^2}$, $m(r\ddot{\theta} + 2\dot{r}\dot{\theta}) = 0$

$mr^2\dot{\theta} = $ constant, angular momentum

$\dfrac{m}{2}(\dot{r}^2 + r^2\dot{\theta}^2) - \dfrac{GMm}{r} = $ constant, energy

3.30 $v_{launch} \simeq 3000$ m/s

Chapter 4

4.2 $\mathbf{H}_0 = -m(9\mathbf{i} + 3\mathbf{k})$, $\mathbf{H}_{c.m.} = -m(6\mathbf{i} + 3\mathbf{j} + 3\mathbf{k})$, $T = 23.5m$

4.4 (a) $\mathbf{R}_{c.m.} = \frac{2}{3}L\mathbf{j}$; (b) $\mathbf{p} = mv_0\mathbf{i}$; (c) $\mathbf{H}_{c.m.} = \frac{-mLv_0}{3}\mathbf{k}$; (d) $T_{rot} = \frac{mv_0^2}{12}$

4.6 $f = \frac{-GM}{r^2}, r > a;\ f = \frac{-4}{3}\pi\rho r G, r < a$

4.8 $\delta_{max} = \frac{+\mu m_1 g}{k} - \frac{1}{2}\left[\left(\frac{2\mu m_1 g}{k}\right)^2 + 4\left(\frac{m_1 m_2}{m_1 + m_2}\right)\frac{v_0^2}{k}\right]^{1/2}$

4.10 $T = \frac{mL^2}{2}\dot\phi^2 + \frac{m}{2}[L^2(\dot\theta + \dot\phi)^2 + L^2\dot\phi^2 + 2L^2\dot\phi(\dot\theta + \dot\phi)\cos\theta]$

$V = -mgL[2\cos\phi + \cos(\phi + \theta)]$

4.12 $\hat{F} = 2m(2gL)^{1/2}$

4.14 $\lim_{n\to\infty} x = \frac{v_0^2 \sin 2\gamma}{g(1-\varepsilon)}$

4.16 $\mathbf{v}_A = [\{-(1-\varepsilon)0.14 + 0.47\}\mathbf{i} + \{(1-\varepsilon)0.08 + 0.82\}\mathbf{j}]v_0$

4.18 $\dot\theta = \frac{v_0\sqrt{2(1+\varepsilon)}}{7L}$

4.20 $P = v_0\gamma(v_0 + gt)$

4.22 $v_{final} = \sqrt{gL}$

4.24 $T \simeq \frac{dm}{dt}(v_e - v_a)$

Chapter 5

5.2 $\mathbf{v}_B = (R\phi)\dot\phi(\cos\phi\mathbf{i} + \sin\phi\mathbf{j})$

$\mathbf{a}_B = (R\dot\phi^2\cos\phi - R\phi\dot\phi^2\sin\phi)\mathbf{i} + (R\dot\phi^2\sin\phi + R\phi\dot\phi^2\cos\phi)\mathbf{j}$

5.4 $\mathbf{v}_m = L\dot\theta\cos\theta\mathbf{i} + (\dot z - L\dot\theta\sin\theta)\mathbf{k}$

$\mathbf{a}_m = (L\ddot\theta\cos\theta - L\dot\theta^2\sin\theta)\mathbf{i} + (\ddot z - L\ddot\theta\sin\theta - L\dot\theta^2\cos\theta)\mathbf{k}$

5.6 $\mathbf{v}_m = \dot x\mathbf{i} + 4R\dot\phi(\cos\phi\mathbf{i} + \sin\phi\mathbf{j})$

$\mathbf{a}_m = \ddot x\mathbf{i} + 4R(\ddot\phi\cos\phi\mathbf{i} - \dot\phi^2\sin\phi)\mathbf{i} + 4R(\ddot\phi\sin\phi + \dot\phi^2\cos\phi)\mathbf{j}$

5.8 $\mathbf{v}_B = [L\dot\theta\cos\theta + L(\dot\theta + \dot\phi)\cos(\theta + \phi)]\mathbf{i} + (L\dot\theta\sin\theta + L(\dot\theta + \dot\phi)\sin(\theta + \phi))\mathbf{j}$

$\mathbf{a}_B = [L\ddot\theta\cos\theta + L(\ddot\theta + \ddot\phi)\cos(\theta + \phi) - L\dot\theta^2\sin\theta - L(\dot\theta + \dot\phi)^2\sin(\theta + \phi)]\mathbf{i}$
$+ [L\ddot\theta\sin\theta + L(\ddot\theta + \ddot\phi)\sin(\theta + \phi) + L\dot\theta^2\cos\theta + L(\dot\theta + \dot\phi)^2\cos(\theta + \phi)]\mathbf{j}$

5.10 $\mathbf{v}_B = 0,\ \mathbf{a}_B = -a\omega_0^2\left(\frac{b+a}{b}\right)\mathbf{i},\ \theta = 0°$

$\mathbf{v}_B = -a\omega_0 \mathbf{i}, \; \mathbf{a}_B = \dfrac{a^2\omega_0^2}{(b^2-a^2)^{1/2}} \mathbf{i}, \; \theta = 90°$

5.12 $\omega_{BC} = \dfrac{\omega_0}{2} \mathbf{k}, \; \alpha_{AB} = \dfrac{-\omega_0^2}{4} \mathbf{k}, \; \alpha_{BC} = 0$

5.14 $\mathbf{v}_B = -r\omega_0(0.231\mathbf{i} + 0.932\mathbf{j})$
$\mathbf{a}_B = r\omega_0^2(0.03\mathbf{i} - 0.64\mathbf{j})$

5.16 $\mathbf{v}_m = \dot{r}(\cos\alpha\,\mathbf{i} + \sin\alpha\,\mathbf{j}) + r(\omega\sin\alpha - \dot{\phi}\cos\alpha)\mathbf{k}$
$\mathbf{a}_m = (\ddot{r}\cos\alpha - r\dot{\phi}^2\cos\alpha + 2r\dot{\phi}\omega\sin\alpha)\mathbf{i} + (\ddot{r}\sin\alpha - r\omega^2\sin\alpha)\mathbf{j}$
$+ (-r\ddot{\phi}\cos\alpha + 2\dot{r}\omega\sin\alpha - 2\dot{r}\dot{\phi}\cos\alpha)\mathbf{k}$

5.18 $\mathbf{v}_B = -(L(\dot{\phi}+\dot{\theta})\sin\phi)\mathbf{i}' + [L\dot{\theta} + L(\dot{\phi}+\dot{\theta})\cos\phi]\mathbf{j}'$
$-\{L\dot{\psi}[\sin\theta + \sin(\phi+\theta)]\}\mathbf{k}'$
$\mathbf{a}_B = [-L\dot{\theta}^2 - L\dot{\psi}^2\sin^2\theta - L\ddot{\theta}\sin\phi - L\dot{\phi}^2\cos\phi - 2L\dot{\phi}\dot{\theta}\cos\phi$
$- L\dot{\psi}^2\sin\theta\sin(\phi+\theta) - L\dot{\theta}^2\cos\phi - L\ddot{\theta}\sin\phi]\mathbf{i}'$
$+ [L\ddot{\theta} - L\dot{\psi}^2\sin\theta\cos\theta + L\ddot{\phi}\cos\phi - L\dot{\phi}^2\sin\phi - 2L\dot{\phi}\dot{\theta}\sin\phi$
$- L\dot{\theta}^2\sin\phi - L\dot{\psi}^2\cos\theta\sin(\phi+\theta) + L\ddot{\theta}\cos\phi]\mathbf{j}' + [-2L\dot{\psi}\dot{\theta}\cos\theta$
$- L\ddot{\psi}\sin\theta - 2L\dot{\phi}\dot{\psi}\cos(\phi+\theta) - 2L\dot{\theta}\dot{\psi}\cos(\phi+\theta) - L\ddot{\psi}\sin(\phi+\theta)]\mathbf{k}'$

5.20 $\mathbf{v}_B = R\dot{\theta}(\cos\theta\,\mathbf{i} - \sin\theta\,\mathbf{j}) - R\dot{\phi}\sin\theta\,\mathbf{k}$
$\mathbf{a}_B = (R\ddot{\theta}\cos\theta - R\dot{\theta}^2\sin\theta - R\dot{\phi}^2\sin\theta)\mathbf{i} + (-R\ddot{\theta}\sin\theta - R\dot{\theta}^2\cos\theta)\mathbf{j}$
$- (R\ddot{\phi}\sin\theta + 2R\dot{\theta}\dot{\phi}\cos\theta)\mathbf{k}$

5.22 $\mathbf{v}_{B/O} = -b\dot{\phi}\cos\theta\,\mathbf{i} + (-b\dot{\theta}\sin\theta + L\dot{\phi})\mathbf{j} + b\dot{\theta}\cos\theta\,\mathbf{k}$
$\mathbf{a}_{B/O} = (-L\dot{\phi}^2 + 2b\dot{\theta}\dot{\phi}\sin\theta - b\ddot{\phi}\cos\theta)\mathbf{i} + (L\ddot{\phi} - b\dot{\phi}^2\cos\theta - b\ddot{\theta}\sin\theta$
$- b\dot{\theta}^2\cos\theta)\mathbf{j} + (b\ddot{\theta}\cos\theta - b\dot{\theta}^2\sin\theta)\mathbf{k}$

5.24 $\mathbf{v}_B = \dot{r}\mathbf{i}_r + r\dot{\theta}\mathbf{i}_\theta - r\dot{\phi}\cos\theta\,\mathbf{k}$
$\mathbf{a}_B = (\ddot{r} - r\dot{\theta}^2 - r\dot{\phi}^2\cos^2\theta)\mathbf{i}_r + (r\ddot{\theta} + 2\dot{r}\dot{\theta} + r\dot{\phi}^2\cos\theta\sin\theta)\mathbf{i}_\theta$
$+ (-r\ddot{\phi}\cos\theta + 2r\dot{\phi}\dot{\theta}\sin\theta - 2\dot{r}\dot{\phi}\cos\theta)\mathbf{k}$

5.26 $\mathbf{v}_B = 3r\dot{\phi}\mathbf{i}; \; \mathbf{a}_B = (\tfrac{9}{2}r\dot{\phi}^2)\mathbf{j} + \dfrac{\sqrt{3}}{2}r\dot{\phi}^2\mathbf{k}$

5.28 $\dot{\phi} = \dfrac{r}{L}\dot{\theta}_1\sin\beta; \; \dot{\theta}_2 = \dot{\theta}_1\left(\cos\beta - \dfrac{c}{L}\sin\beta\right); \; \dot{\theta}_3 = \dot{\theta}_1\left(\cos\beta + \dfrac{c}{L}\sin\beta\right)$
$\mathbf{v}_{c.m.} = \left(\dfrac{dr}{L}\dot{\theta}_1\sin\beta\right)\mathbf{i} + (r\dot{\theta}_1\cos\beta)\mathbf{j}$

5.30 $\alpha_{disk} = -\dot{\phi}\dot{\theta}\mathbf{i} + \dot{\psi}\dot{\phi}\mathbf{j} + (\ddot{\phi} - \dot{\psi}\dot{\theta})\mathbf{k}$

$\mathbf{a}_E = (-L\ddot{\phi} + 2r\dot{\psi}\dot{\phi})\mathbf{i} + [-L\dot{\phi}^2 + (d+r)2\dot{\phi}\dot{\theta}]\mathbf{j} - [(d+r)\dot{\theta}^2 + r\dot{\psi}^2]\mathbf{k}$

5.32 $\mathbf{a}_B = -[r(\dot{\phi}^2 + \dot{\psi}^2) + 2r\dot{\psi}\dot{\phi}\sin\theta]\mathbf{i}$

5.36 $T \sim 9.799$ N

5.38 $y = \dfrac{-4v_0^3}{3g^2}(\omega_e \cos\lambda)$, to the west

Chapter 6

6.2 $\hat{z} = \dfrac{28r}{9\pi}$

$\hat{I}_{xx} = 1.5193 m'r^2$

$\hat{I}_{yy} = m'\left(\dfrac{L^2}{12} + 0.2693 r^2\right)$

$\hat{I}_{zz} = m'\left(\dfrac{L^2}{12} + 1.25 r^2\right)$

$\hat{I}_{xy} = \hat{I}_{xz} = \hat{I}_{yz} = 0$

6.4 $I_1 = 340$ kg·m², $e_{11} = 1$, $e_{12} = \tfrac{1}{3}$

$I_2 = 140$ kg·m², $e_{21} = 1$, $e_{22} = -3$

6.6 $I_{xx} = m\left(\dfrac{b^2 + 2L^2}{6}\right)$, $I_{yy} = m\left(\dfrac{a^2 + 2L^2}{6}\right)$, $I_{zz} = m\left(\dfrac{a^2 + b^2}{6}\right)$

$I_{xy} = \dfrac{-m}{12}(ba)$, $I_{xz} = \dfrac{maL}{6}$, $I_{yz} = \dfrac{mbL}{6}$

6.10 $I_1 = 7.33$ kg·m², $e_{11} = 1$, $e_{12} = e_{13} = 0$

$I_2 = 12.37$ kg·m², $e_{21} = 0$, $e_{22} = 1$, $e_{23} = 9.71$

$I_3 = 5.91$ kg·m², $e_{31} = 0$, $e_{32} = 1$, $e_{33} = -0.103$

6.12 $N = \dfrac{3W}{2}$; $f = \dfrac{\sqrt{3}W}{7}$; $\dot{\omega}_z = \dfrac{5\sqrt{3}}{14}\dfrac{g}{R}$

6.14 $\mathbf{a}_A = \dfrac{-2P}{mL}\mathbf{i}$

6.16 $\hat{I} = mR^2\left(\dfrac{g\sin\alpha t_1^2}{2x_1} - 1\right)$

6.18 $H = \dfrac{m_1}{2}L\omega^2 + \dfrac{3}{4}m_2 L\omega^2 - m_2\ddot{r}$

ANSWERS TO EVEN-NUMBERED PROBLEMS   411

$$T = \frac{3m_2}{2} L \dot{r} \omega$$

6.20 $\dot{\omega}_z = \frac{P}{m} \frac{3L}{6R^2 - L^2}$

6.22 $T = 2\pi \left[\frac{(R-r)(2m_1 + 9m_2)}{3g(m_1 + 2m_2)} \right]^{1/2}$

6.24 $\frac{D_0}{D_1} = 4$

6.26 $C_y = \frac{mg}{3}, \frac{2}{7} mg$

6.28 $v_{c.m.} = 0.499 \sqrt{ga}$

6.30 $\mathbf{v}_B = - \left[\frac{12gy(4a^2 - y^2)}{16a^2 - 3y^2} \right]^{1/2} \mathbf{j}$

6.32 $v_0 > 1.7 \sqrt{ga}$

6.34 $x = \tfrac{3}{8} a$

Chapter 7

7.2 $M_x = \dfrac{2\omega^2 m r_1 h}{3\pi}$

7.4 $A_y = \dfrac{-m\omega^2 b}{12}, \; B_y = \dfrac{-m\omega^2 b}{4}$

7.6 $\dot{\theta} = 20.31$ rad/sec

7.8 $f_C = \dfrac{1}{6} \dfrac{m_1 v^2}{r} = -f_D; \; N_D = N_C = \dfrac{2m + m_1}{2} g$

7.10 $\sigma = \dfrac{4(2\pi J \omega n / 60 + mgb)}{\pi r^3}$

7.12 $M_1 = \dfrac{J \Omega v}{r}$

7.14 $M_0 = -\left(\dfrac{4m_1 + 12m_2}{12}\right) a^2 \Omega^2 \cos\alpha \sin\alpha + m_2 ga \sin\alpha$

7.18 $B_1 = \dfrac{-J\Omega\omega}{2a} = -A_1$

7.20 $M_0 = \dfrac{\Omega^2 h R m}{20(R^2 + h^2)} (18h^2 + 3R^2) + \dfrac{3mgh^2}{4(R^2 + h^2)^{1/2}}$

7.22 $M_1 = \dfrac{mR^2}{2}\dot{\psi}\dot{\phi}$, $M_2 = \dfrac{-mR^2}{2}\dot{\psi}\dot{\theta}$, $M_3 = \dfrac{-mR^2}{2}\dot{\theta}\dot{\phi}$

7.24 $M_1 = \dfrac{mR^2}{2}(\dot{\psi} + \dot{\phi}\cos\theta)(\dot{\phi}\sin\theta) - \dfrac{mR^2}{4}\dot{\phi}^2\sin\theta\cos\theta$

$M_2 = \dfrac{-mR^2}{2}\dot{\theta}\dot{\psi}$; $M_3 = \dfrac{-mR^2}{2}\dot{\phi}\dot{\theta}\sin\theta$

7.26 $T = \dfrac{m_1 r^2}{4}(\dot{\psi} + \dot{\phi}\cos 60°)^2 + \left(\dfrac{m_1 r^2}{8}\right)(\dot{\phi}\sin 60°)^2 + \dfrac{mR^2}{4}\dot{\phi}^2$

7.28 $T = \tfrac{1}{12}[m_1 a^2(\dot{\theta}^2 + \dot{\phi}^2\sin^2\theta)] + \tfrac{1}{36}[m_1 b(b\dot{\phi}^2 + a\dot{\theta}\dot{\phi}\cos\theta)]$

7.30 $\omega_T = 1.732\dot{\psi}_0$; $H_T = 1.732 mR^2 \dot{\psi}_0$; $H_3 = mR^2 \dot{\psi}_0$

7.32 $\omega_T = \dfrac{\dot{\phi}_0}{2}$, $\gamma = 45°$

7.34 $4Rm_1 g \sin\theta = I_1\ddot{\theta} - I_1\dot{\phi}^2\sin\theta\cos\theta + I_3(-\dot{\psi} + \dot{\phi}\cos\theta)(\dot{\phi}\sin\theta)$

$M_2 = I_1(\ddot{\phi}\sin\theta + 2\dot{\phi}\dot{\theta}\cos\theta) - I_3\dot{\theta}(-\dot{\psi} + \dot{\phi}\cos\theta)$

$M_3 = I_3(-\ddot{\psi} + \ddot{\phi}\cos\theta - \dot{\theta}\dot{\phi}\sin\theta) = 0$

7.36 $\dot{\psi}^2 \geqslant \dfrac{(I_1 - I_3)(16 Rm_1 g \cos\theta_0)}{I_3^2}$

$\dot{\phi}_{\text{slow}} = \left(\dfrac{-\dot{\psi}}{63}\left\{1 - \left[1 - \dfrac{63(16)g\cos\theta_0}{R\dot{\psi}^2}\right]^{1/2}\right\}\right)\dfrac{1}{\cos\theta_0}$;

$\dot{\phi} = \dfrac{-8g}{\dot{\psi}R}$; stable for all spins for $\theta = 180°$

Chapter 8

8.2 $F = \left(\dfrac{9m_1}{4} + m_2\right)\ddot{x}_2 + \dfrac{9k}{4}x_2 + \dfrac{9c}{4}\dot{x}$

8.4 $(m_1 + m_2)\ddot{x}_1 + m_2\ddot{x}_2\cos\alpha + k_1(x_1 - L_1) = 0$

$m_2(\ddot{x}_2 + \ddot{x}_1\cos\alpha) + k_2(x_2 - L_2) - m_2 g\sin\alpha = 0$

8.6 $\dot{\theta} = 7.85117$ rad/sec

8.8 $mL^2\ddot{\theta} - mA\omega^2 L\cos\theta\cos\omega t - mgL\sin\theta = 0$

8.10 $2mL^2\ddot{\phi}(1 + \cos\theta) - 2mL^2\dot{\phi}\dot{\theta}\sin\theta + mL^2(\ddot{\theta}\cos\theta - \dot{\theta}^2\sin\theta)$

$+ mL^2(\ddot{\theta} + \ddot{\phi}) + mgL[2\sin\phi + \sin(\phi + \theta)] = 0$

$mL^2(\ddot{\theta} + \ddot{\phi}) + mL^2\ddot{\phi}\cos\theta + mL^2\dot{\phi}^2\sin\theta + mgL\sin(\phi + \theta) = 0$

8.12 $m(\ddot{r} - r\sin^2\beta\dot{\phi}^2) + 2kr + mg\sin\beta\cos\phi = 0$

ANSWERS TO EVEN-NUMBERED PROBLEMS

$$I\ddot{\phi} + mr^2 \sin^2 \beta \ddot{\phi} + 2mr\dot{r}\dot{\phi} \sin^2 \beta - mgr \sin \beta \sin \phi = T$$

8.14 $p_\phi = \left(2m_1 + \dfrac{m}{24}\right) L^2 \dot{\phi}$, ϕ is cyclic

$(m + 2m_1)\ddot{z} + kz - (2m_1 + m)g = 0$

8.16 $\phi = 56.729°, (90°)$

8.18 $p_\phi = (J + mR^2 \cos^2 \theta)\dot{\phi}$, a constant

$mR^2 \ddot{\theta} + mR^2 \sin \theta \cos \theta \dot{\phi}^2 + k[(R^2 + d^2 - 2Rd \sin \theta)^{1/2} - L_0]$

$\cdot \dfrac{-Rd \cos \theta}{(R^2 + d^2 - 2Rd \sin \theta)^{1/2}} - mgR \cos \theta = 0$

$N = \dfrac{k[(R^2 + d^2 - 2Rd \sin \theta)^{1/2} - L_0](R - d \sin \theta)}{(R^2 + d^2 - 2Rd \sin \theta)^{1/2}} - mg \sin \theta$

$- mR\dot{\theta}^2 - mR \cos^2 \theta \dot{\phi}^2$

8.20 $\dot{\theta} = \dfrac{-6}{7} \dfrac{\hat{F}}{mL}$

8.22 $v_{c.m.} = v_0 \cos \alpha; \ \dot{\theta} = \dfrac{v_0}{R}$

8.24 (a) $p_\phi = \left(MR^2 + \dfrac{3J}{4}\right) \dfrac{\dot{\phi}}{2} + \dfrac{J}{2}\left(\omega + \dfrac{\dot{\phi}}{2}\right)$, a constant

(b) $p_\phi = \left(mR^2 + \dfrac{5J}{4}\right) \dfrac{\dot{\phi}}{2} + \dfrac{L}{2}\dot{\theta}$, a constant

$p_\theta = J\left(\dot{\theta} + \dfrac{\dot{\phi}}{2}\right)$, a constant

8.26 $\dfrac{mL^2\ddot{\theta}}{3} - \dfrac{3p_\phi^2 \cos \theta}{mL^2 \sin^3 \theta} + \dfrac{mgL}{2} \sin \theta = 0$

8.28 $\theta = 56.1076°$

8.30 $N = \dfrac{65}{16}m_1 R\Omega^2 \cos \alpha \sin \alpha + \dfrac{m_1 R}{2}\Omega^2 \cos^2 \alpha + m_1 g \cos \alpha$

8.32 $A\ddot{\theta} + \dfrac{(p_\phi - p_\psi \cos \theta)p_\psi \sin \theta}{A \sin^2 \theta} - \dfrac{(p_\phi - p_\psi \cos \theta)^2 \cos \theta}{A \sin^3 \theta} = -k_\theta\left(\theta - \dfrac{\pi}{2}\right)$

8.36 $\Delta v = 0.293(v_{initial})$

Chapter 9

9.2 $\sigma = 47.8$ MPa

9.4 $(\alpha m + m)\ddot{z} - ma\ddot{\phi}\sin\phi - ma\dot{\phi}^2\cos\phi + k(z - L_0) - (\alpha m + m)g = 0$

$m(a^2\ddot{\phi} - a\ddot{z}\sin\phi) + mga\sin\phi = 0$

9.6 $\frac{7}{2}m\ddot{x} + m(R-r)[\ddot{\psi}(\cos\psi + \frac{1}{2}) - \dot{\psi}^2\sin\psi] = 0$

$\frac{3}{2}m(R-r)^2\ddot{\psi} + m(R-r)\ddot{x}(\cos\psi + \frac{1}{2}) + mg(R-r)\sin\psi = 0$

9.8 $\frac{3}{2}mR^2\ddot{\phi} + m_1(R^2 + r^2 - 2Rr\cos\phi)\ddot{\phi} + m_1 R\sin\phi(-\ddot{r} + r\dot{\phi}^2)$

$+ 2m_1\dot{r}\dot{\phi}(r - R\cos\phi) + m_1 gr\sin\phi = 0$

$m_1(\ddot{r} - R\ddot{\phi}\sin\phi - r\dot{\phi}^2) + k(r - L_0) - m_1 g\cos\phi = 0$

9.10 $mL^2(2\ddot{\phi} + \ddot{\theta}) + 2mgL\phi = 0$

$mL^2(\ddot{\theta} + \ddot{\phi}) + mgL\theta = 0$

9.12 $(m + m_1)\ddot{x} + m_1 L\ddot{\theta} + kx = 0$

$m_1 L^2\ddot{\theta} + m_1 L\ddot{x} + m_1 gL\theta = 0$

9.14 $\omega_{1,2} = \sqrt{3}\omega_0, \sqrt{3}\dfrac{\omega_0}{2}$

9.16 $\omega_{1,2} = 1.985\omega_0, 1.07\omega_0$

9.18 $\omega_{1,2} = \left(\dfrac{g}{L}\right)^{1/2}, \left[\dfrac{31}{45}\left(\dfrac{g}{L}\right)\right]^{1/2}$

9.20 $\omega_1^2 = \dfrac{2+\sqrt{2}}{4}\left(\dfrac{g}{R}\right), \dfrac{R\Phi}{X} = -\dfrac{1}{2(\sqrt{2}+1)}$

$\omega_2^2 = \dfrac{2-\sqrt{2}}{4}\left(\dfrac{g}{R}\right), \dfrac{R\Phi}{X} = \dfrac{1}{2(\sqrt{2}-1)}$

9.22 $\omega_1^2 = \dfrac{3g}{a}, \dfrac{X}{a\Theta} = -1$

$\omega_2^2 = \dfrac{3g}{5a}, \dfrac{X}{a\Theta} = \dfrac{1}{3}$

9.24 $\omega_{1,2}^2 = \dfrac{\omega_0^2}{2}, \dfrac{3\omega_0^2}{2}$

$\{u_1\} = \begin{Bmatrix} 0.775 \\ 0.632 \end{Bmatrix}, \{u_2\} = \begin{Bmatrix} -0.775 \\ 0.632 \end{Bmatrix}$

9.26 $\omega_{1,2}^2 = \omega_0^2(2 \pm \sqrt{2}), \omega_0^2 = \dfrac{k_2}{J_2}$

$\{u_1\} = \begin{Bmatrix} 1 \\ 1 + \sqrt{2} \end{Bmatrix}, \{u_2\} = \begin{Bmatrix} 1 \\ 1 - \sqrt{2} \end{Bmatrix}$

9.28 $\omega_1 = \left(\dfrac{g}{L} + \dfrac{2kd^2}{mL^2}\right)^{1/2}, \omega_2 = \left(\dfrac{g}{L}\right)^{1/2}$

$\theta_1 = \dfrac{\theta_0}{2}(\cos\omega_1 t + \cos\omega_2 t)$

$\theta_2 = \dfrac{\theta_0}{2}(\cos\omega_1 t - \cos\omega_2 t)$

9.30 $x = \cos\sqrt{\dfrac{3g}{a}}\,t + \cos\sqrt{\dfrac{3g}{5a}}\,t$

$\theta = -\dfrac{1}{a}\cos\sqrt{\dfrac{3g}{a}}\,t + \dfrac{3}{a}\cos\sqrt{\dfrac{3g}{5a}}\,t$

9.32 $x_1 = \dfrac{(\omega_0^2 - \omega^2)F_0 \sin\omega t}{m[(\omega_0^2 - \omega^2)(3\omega_0^2 - \omega^2) - 2\omega_0^4]}, \omega_0^2 = \dfrac{k}{m}$

$x_2 = \dfrac{\omega_0^2 F_0 \sin\omega t}{m[(\omega_0^2 - \omega^2)(3\omega_0^2 - \omega^2) - 2\omega_0^4]}$

9.34 $\{u_1\} = \begin{Bmatrix} 2 \\ 1-\sqrt{3} \\ 1 \end{Bmatrix}$ for $\hat{\omega}_1^2 = (2+\sqrt{3})\omega_0^2$, with $\omega_0^2 = \dfrac{k}{m}$

$\{u_2\} = \begin{Bmatrix} 2 \\ 1+\sqrt{3} \\ 1 \end{Bmatrix}$ for $\hat{\omega}_2^2 = (2-\sqrt{3})\omega_0^2$

9.36 $\begin{Bmatrix} x_1 \\ x_2 \end{Bmatrix} = \begin{bmatrix} \dfrac{-(\sqrt{33}+3)}{6} & \dfrac{(\sqrt{33}-3)}{6} \\ 1 & 1 \end{bmatrix} \begin{Bmatrix} \dfrac{\hat{F}\sin(1.57\omega_0 t)}{6.56m\omega_0} \\ \dfrac{1.033\hat{F}\sin(0.74\omega_0 t)}{m\omega_0} \end{Bmatrix}$

9.38 $J_2 = \dfrac{k_2}{\omega^2}$

9.46 $(\omega_z')^2 \geq \dfrac{4mgl I_1}{I_3^2}$, for stable motion

9.48 $(k - m_2\omega^2) >$, stable

LIST OF SYMBOLS

a	Magnitude of the acceleration
a, b	Semimajor axis, semiminor axis of an ellipse
a_0	Initial acceleration
a_x	Acceleration component in the x direction
$a_{x,0}$	Initial acceleration in the x direction
$(a_p)_x$	Acceleration of the point P in the x direction
$(a_p)_n$	Component of acceleration of point P in the normal direction
$(a_p)_{x,0}$	Initial acceleration of P in the x direction
$(a_{c.m.})_x$	Acceleration of the center of mass in the x direction
$(a_{c.m.})_{x,2}$	Acceleration of the center of mass at time 2 in the x direction
$\dot{A}$	Areal velocity
$\mathbf{a}, \mathbf{a}_p$	Acceleration, acceleration of point P
$\mathbf{a}_i$	Acceleration of particle i
$\mathbf{a}_{P/B}$	Acceleration of point P relative to a frame at B that translates
$\mathbf{a}_{c.m.}$	Acceleration of the center of mass
$\mathbf{a}_{\text{Coriolis}}$	Coriolis acceleration
$(\mathbf{a})_n$	Normal acceleration
$(\mathbf{a}_{P/B})$	Normal acceleration of point P relative to a frame at B that translates
$[\mathbf{a}_{D/O}]^n_{\text{rot}}$	Normal acceleration of point D relative to a frame at O that rotates
c	Damping coefficient
c_c	Critical damping
C	Capacitance
$\mathbf{C}_i$	Constraint force acting on particle i
D	Rayleigh dissipation function
ds	Arc length
$d\mathbf{r}$	Real displacement
e_1	Voltage level at 1
$e(t)$	Varying voltage
e_{11}, e_{12}, e_{13}	Components of the unit vector for the first principal axis direction along axes 1, 2, 3
E	Energy
$\mathbf{e}_1$	First principal axis direction unit vector
f	Natural frequency
$f(t)$	Function of time
$F(s)$	Laplace transform of $f(t)$
$\mathbf{F}$	Force
$\mathbf{F}_i$	Sum of external forces acting on particle i
$\mathbf{F}_{\text{ext}}$	Sum of external forces
$\mathbf{F}_i^{(T)}$	Total force acting on particle i
$\mathbf{f}_{ij}$	Internal force acting on particle i exerted by particle j and directed toward particle j

LIST OF SYMBOLS 417

g	Local acceleration of gravity
G	Universal gravity constant
h	Angular momentum per unit mass
H	The hamiltonian
H_ψ	Angular momentum about the ψ axis
$(H_{c.m.})_x$	The component in the x direction of the angular momentum of a body about its center of mass
$(H_O)_x$	The component in the x direction of the angular momentum of a body about fixed point O
$(H_O)_{x,2}$	The component of angular momentum about point O acting along the x direction at time 2
$\mathbf{H}_{c.m.}$	Angular momentum about the center of mass
$\mathbf{H}_O$	Angular momentum of the body about fixed point O
$\mathbf{H}_{Ai}$	Angular momentum of particle i about point A
i	Current
$I_{i.c.}$	Mass moment of inertia of the body referenced to the instantaneous center
I_1, I_2, I_3	Principal mass moments of inertia
I_{xx}	Mass moment of inertia about axis x
I_{xy}	Mass product of inertia for the xy plane
$\hat{I}_{zz}$	Mass moment of inertia of the body referenced to the center of mass about axis z
$(I_C)_{zz}$	Mass moment of inertia of the body referenced to point C about axis z
$\mathbf{i}, \mathbf{j}, \mathbf{k}$	Unit vectors for cartesian coordinate axes
$\mathbf{I}, \mathbf{J}, \mathbf{K}$	Unit vectors for inertial or fixed coordinate axes
$\mathbf{i}_m$	Unit vector along arbitrary direction m
k	Spring constant
k_B	Mass radius of gyration of the body about point B
K_i	Element i of the generalized stiffness matrix
k_{jl}	Element jl of the stiffness matrix
l	Length
L	Length, inductance, the lagrangian
l_{xY}	Direction cosine between axis x and axis Y
m	Number of degrees of freedom
m, M	Mass
m_i	Mass of particle i
m'_1, m'_2	Increased, reduced mass
M_i	Element i of the generalized mass matrix
m_{jl}	Element jl of the mass matrix
M.F.	Magnification factor
$\mathbf{M}_{i.c.}$	The moment of forces about the instantaneous center
M_1, M_2, M_3	Components of the moment along the principal axes
$\mathbf{M}_{c.m.}$	Sum of the moments of external forces about the center of mass
$\mathbf{M}_O$	Sum of the moments of external forces about a fixed origin O
$\mathbf{M}_{Oi}$	Total moment of the forces acting on particle i about the fixed origin O
0	Subscript, initial condition
O	Point O, a fixed origin

LIST OF SYMBOLS

P	Power
p_j	Generalized linear momentum
p	Linear momentum of a particle
P	Total linear momentum of a system of particles
q	Electric charge
q_i	Generalized coordinate i
Q_j	Generalized force associated with the coordinate q_j
R	The routhian, resistance, radius of a circle, the horizontal component of the position vector in the cylindrical coordinate system
r	Magnitude of the position vector
$\dot{r}, \ddot{r}$	Change and rate of change of the length of the position vector with respect to time
r, **r**$_P$	Absolute position vector, absolute position vector to point P
r$_i$	Absolute position vector to mass i
r$_{P/B}$	Relative position vector from point B to point P
R$_{c.m.}$	Absolute position vector to the center of mass
R$_{c.g.}$	Absolute position vector to the center of gravity
S_i	Scale factor i
t	Time
T	Period, kinetic energy
$u, \dot{u}$	Relative speed, relative acceleration measured in a rotating frame
u_{21}	The first normal mode vector's components along the q_2 direction
v	Speed
v_c	Speed in a circular orbit
v_{esc}	Magnitude of the escape velocity
$\dot{v}$	Acceleration
v_0	Initial speed
v_x	Speed in the x direction
$v_{x,0}$	Initial speed in the x direction
$v_{B,1}$	The speed of particle B at time 1
$(v_P)_y$	Component of the velocity of point P in the y direction
$(v_{c.m.})_y$	Component of the velocity of the center of mass in the y direction
$(v_P)_{y,0}$	Component of the initial velocity of P in the y direction
$(v_{c.m.})_{y,2}$	Component of the velocity of the center of mass in the y direction at time 2
V	Potential energy
v, **v**$_p$	Velocity, velocity of point p
v$_i$	Velocity of point i
v$_{P/B}$	Velocity of point P relative to a frame at B that translates
v$_{c.m.}$	Velocity of the center of mass
$[\mathbf{v}_{D/O}]_{rot}$	Velocity of point D relative to a rotating frame at O
W	Weight
$W_{1 \to 2}$	Work done on a particle in its motion from point 1 to point 2
x_s	Static solution
x_c, x_p	Complementary, particular solution for x
x, y, z	Cartesian coordinates
x_0, y_0, z_0	Initial position
$\dot{x}, \ddot{x}$	Velocity, acceleration component in the x direction

LIST OF SYMBOLS

X, Y, Z	Inertial or fixed coordinates
z_i	State variable i
α	Angular acceleration
β	Heading angle
δ_{ij}	Kroneker delta function
$\delta \mathbf{r}$	Virtual displacement
$\Delta \mathbf{r}$	Change in $\mathbf{r}$
ε	Coefficient of restitution, eccentricity
ζ	Damping ratio
θ, ϕ, ψ	Eulerian angles
$\dot{\theta}, \dot{\phi}, \dot{\psi}$	Angular speeds
$\ddot{\theta}, \ddot{\phi}, \ddot{\psi}$	Angular accelerations
λ	Eigenvalue, latitude angle
λ_s	Lagrange multiplier
μ	Coefficient of friction
μ_k	Coefficient of kinetic friction
ρ	Radius of curvature
$\boldsymbol{\rho}_i$	Relative position of particle i referenced to the center of mass
ϕ	Precession angle, phase angle
ω	Forcing frequency
ω_n	Circular frequency
ω_d	Damped natural frequency
$\omega_1, \omega_2, \omega_3$	Component of the angular velocity along principal axes directions
$\boldsymbol{\omega}$	Angular velocity
$\dot{\boldsymbol{\omega}}$	Angular acceleration
$\boldsymbol{\omega}_{AP}$	Angular velocity of member AP
$\boldsymbol{\omega}_e$	Angular velocity of the earth
$\boldsymbol{\Omega}$	Angular velocity, coordinate axes angular velocity
∇	Gradient operator
L^{-1}	Inverse Laplace transform
$\{\ \}$	Column matrix
$[\ ]$	Matrix
$[\ ]^t$	Transposed matrix
$[\ ]^{-1}$	Inverse matrix
Det $[\ ]$	Determinant of a matrix
Tr $[\ ]$	Trace of a matrix
$[\ ]_{rot}$	Vector derivative measured in a rotating reference frame
$[k]$	Stiffness matrix
$[l]$	Direction cosine matrix
$[m]$	Mass matrix
$[A]$	Rotation matrix, system matrix
$[B]$	Input matrix
$[1]$	Identity matrix
$[K]$	Generalized stiffness matrix
$[M]$	Generalized mass matrix
$\{u_i\}$	Normal mode vector i
$\{u\}^i$	Trial vector i
$[U]$	Modal matrix
$[\phi]$	Transition matrix

INDEX

Acceleration
  absolute, 4, 10
  angular, 102–108
  in cartesian coordinates, 11–15
  in cylindrical coordinates, 29–30
  of gravity, 5, 13
  in normal and tangential
    coordinates, 25–26
  in radial and transverse
    coordinates, 27–28
  relative
    with respect to a rotating
      frame, 132–143
    with respect to a translating
      frame, 31–33
  in spherical coordinates, 30–31
Analog computer, 42, 367–373
Angular acceleration, 102–108
Angular momentum (moment of
    linear momentum)
  of a particle, 54
  of a rigid body
    about an arbitrary point, 181–182
    about the center of mass, 158–159,
      205
    about a fixed point, 158–159, 205
  of a system of particles
    about the center of mass, 80
    about a fixed point, 80
  time rate of change observed in a
    rotating reference frame of, 206
Angular velocity, 102–108
Aphelion, 290
Areal velocity, 288

Beat phenomenon, 362
Bernoulli brothers, 300
Body cone, 224–225
Body-fixed principal axes, 207
Brachistrochrone problem, 300–303

Calculus of variations, 259, 295, 301
Canonical momentum, 277

Center
  of curvature, 25
  of gravity, 76
  of mass, 76, 159
  of percussion, 199
Central-force motion, 67–70, 288
Centrodes
  body, 112, 149
  space, 112, 149
Characteristic equation, 18, 169
Chasles' theorem, 108–110
CINTERVAL specification, 375, 393
Circular frequency, 16
Circular orbit, 71, 291
Coefficient
  of kinetic friction, 61, 162, 184
  of restitution, 90
  of static friction, 162–163, 182–184
Configuration space, 295
Conic sections, 289–290
Conjugate momentum, 277
Conservation
  of angular momentum, 67, 91
  of linear momentum, 64, 84
  of mechanical energy, 58, 82
Conservative forces, 56–59
Constraints
  equations of, 100, 259
  holonomic, 261
  nonholonomic, 261, 282, 299
  nonintegrable, 261
  rheonomic, 262
  scleronomic, 262
  unilateral, 263
Continuous system simulation
    languages, 373
Coriolis, G., 133
Coriolis acceleration, 133, 137–138, 145
CSMP, 374, 379–380
CSSL-IV, or CSSL, version 4, 374
  directivies
    input/output, 387
    Macro, 397

422 INDEX

[CSSL-IV]
   run-time, 386
   translator, 385
  simulation operators, 388–391
Cuspidal motion, 234
Cylic coordinates, 277

D'Alembert, J., 269
D'Alembert's principle, 269
Degrees of freedom, 100
DERIVATIVE section, 375, 385
Derivative of a vector, 104–105
  referenced to a rotating frame, 108,
    121–126, 132–133
Digital computer simulation, 373
Dynamic coupling, 327
DYNAMIC section, 375, 385

Eccentricity, 289
Eigenvalue, 167–169, 173, 177–178, 323,
  337
Eigenvector, 167–169, 173, 177–178, 323,
  337
Energy
  conservation, 58, 82, 184
  kinetic
    of a particle, 56
    of a rigid body, 184, 213
    rotational, 82, 184
    of a system of particles, 81–82
    translational, 82, 184
  mechanical, 57
  potential, 57–64
Equations of motion
  for a particle, 53–54
  for a rigid body, 157–158, 205–207
  for a system of particles, 78–81
Escape velocity, 74, 291
Euler, L., 76
Eulerian angles, 107, 218–221
Euler-Lagrange equations, 298
Euler's equations
  for body-fixed principal axes, 207
  modified, 216–217
Euler's method (*see* Numerical
    integration, Euler's method)
Euler's theorem, 101, 174, 177–181
Euler's trapezoidal method (*see*
    Numerical integration, modified
    Euler method)

Flexibility matrix, 322, 342
Foucault, J., 156, 238
Foucault pendulum, 156
Free vectors, 110
Freudenstein's equation, 120–121

Gauss-Jordan method, 342
General motion of a rigid body, 108, 110,
  205–207
Generalized constraint force, 283
Generalized coordinates, 259
  independent, 260, 269–270
Generalized forces, 264–266
  conservative, 266
  of constraint, 283
Generalized impulse, 314
Generalized mass, 324
  matrix, 324
Generalized momentum, 258
Generalized stiffness, 324
  matrix, 324
Gravitation
  Newton's law of, 4
  universal constant of, 4
Gravitational
  field strength, 94
  potential (*see* Energy, potential; *see*
    also Potential functions)
  potential per unit mass, 94
Gravity acceleration of, 5, 13
Gyroscope, 155–156, 238
Gyroscopic instruments
  directional gyro, 241, 245
  drift
    apparent, 242
    random, 242
  floated integrating gyro, 240
  gyrocompass, 245
  inertial navigation, 243–245
  rate gyro, 239
  single degree of freedom gyro, 239
  stable platform, 243
  two degrees of freedom gyro, 240
  vertical gyro, 241
Gyroscopic mill, 318

Hamilton, W. R., 258
Hamiltonian, 259, 303
Hamilton's equations, 259, 308

Hamilton's principle, 258, 295
　for nonconservative systems, 299
HDR directive, 377, 386, 387
Herpolhode, 223, 225
Hohmann transfer orbit, 75
Householder's method, 337
Hybrid computation, 373

Impact
　direct central, 89
　elastic, 90
　force, 89
　oblique, 89, 190–193
　oblique central, 89
　plastic, 90
Impulse
　angular, 66, 91, 189
　linear, 64, 84, 189
Increased mass, 288
Independent coordinates, 100, 259
Inertia
　ellipsoid, 170, 222
　mass moments of, 159
　mass products of, 159
　principal axes of, or
　principal directions of, 160–161, 167–169, 207
　principal moments of, 160–161, 167–169, 207
　transformations
　　coordinate rotation (similarity transformation), 165–166
　　coordinate translation (parallel-axis theorem), 164–165
Inertial reference frame, 7
INITIAL section, 375, 385
INPUT directive, 376, 386
Input matrix, 333
Instantaneous axis of rotation, 101, 104, 112
Instantaneous center of rotation, 112–116, 118
INTEG command, 374, 388
Intermediate modes, numerical methods for, 345–352
Internal forces, 78–79, 81
Invariable plane, 222–223
Inverse of a matrix, 342–344

Jacobi's method, 173–176, 337

Kaiser glocke, 361
Kepler, J., 288
Kepler's laws of planetary motion, 288, 290, 292
Kinetic energy (*see* Energy, kinetic)
Kroneker delta function, 345

LABEL directive, 379, 386, 387
Lagrange, J., 258
Lagrange multipliers, 262, 282–283
Lagrange's equations, 258, 271
Lagrangian, 258, 270
Line of nodes, 107, 218–219
Linear momentum
　moment of (*see* Angular momentum)
　of a particle, 64
　of a rigid body, 189
　of a system of particles, 84

MACRO directives, 396–398
Magnification factor, 17
Matrix inversion, 342–344
Matrix iteration, 338
Mechanical efficiency, 56
Modal matrix, 324
Modal sweeping, 345–352
Mode shape, 323
Mode vector, 328
Modified Hamilton's principle, 307
Moment
　of a force
　　about an arbitrary point, 182
　　about the center of mass, 79
　　about a fixed point, 54, 79
　of linear momentum (*see* Angular momentum)
Momentum
　angular (*see* Angular momentum)
　linear (*see* Linear momentum)
　canonical, 277
　conjugate, 277
　generalized, 258, 277

Natural frequency, 16
Newton, I., 3
Newton-Raphson method, 44–47, 120–121
Newton's laws, 3
Normal coordinates, 327–332
Normal mode vector, 323

Normal modes, 322–327
Numerical integration
  Euler's method, 33–37, 280–281, 393
  forward difference formula, 38
  modified Euler method (Euler's trapezoidal method), 37–40, 393
  predictor-corrector methods, 38, 394–395
  real-time integration, 42, 367, 373
  round-off error, 34
  Runge-Kutta fourth-order method, 40–44, 279, 296, 394–395
  self-starting methods, 33
  stiff systems, 34
  truncation error, 34
Nutation, 107, 233

Orthogonal matrix, 129
Orthogonal transformations, 129
Orthogonality
  conditions, 129
  of eigenvectors, 169–170
  of normal modes of vibration 324
OUTPUT directive, 377, 386

Parallel-axis theorem, 164–165
Perihelion, 290
Period
  of orbital motion, 75, 291–292
  of simple harmonic motion, 16
Perturbed steady motion, 352–355
Phase space, 303, 333
Plane motion of a rigid body
  equations of motion in, 161–163
  kinematics of, 110–121
  kinetic energy in, 184–185
  momentums in, 189–190
PLOT directive, 379, 386
Poinsot, L., 222
Poinsot ellipsoid, 222–223, 225
Polhode, 223, 225
Position vector
  absolute, 10, 31
  relative, 31
Potential energy (*see* Energy, potential; *see also* Potential functions)
Potential functions
  constant gravitational force, 58, 187
  varying gravitational force, 58
  linear restoring force, 58–59

Power, 56, 186
Power method, 338
Precession
  direct, 226–227
  rate of, 107, 215, 235
  retrograde, 226–227
  steady, 156, 226–227, 234–235
Principal axes of inertia (*see* Inertia, principal axes of)
Principal moments of inertia (*see* Inertia, principal moments of)
Principle
  D'Alembert's, 269
  Hamilton's, 258, 295
  of conservation of mechanical energy, 58, 82, 184
  of impulse and momentum
    for a particle, 64–70
    for a rigid body, 189–193
    for a system of particles, 83–93
  of work and kinetic energy
    for a particle, 56
    for a rigid body, 184–188
    for a system of particles, 81–83
  of virtual work for applied forces, 264
PROCEDURAL directive, 384, 385

Radius
  of curvature, 26
  or gyration, 159
Rayleigh dissipation function, 271
Real-time integration, 42, 367, 373
Reduced mass, 288
Relative acceleration (*see* Acceleration, relative)
Relative motion
  with respect to the rotating earth, 124, 143–147
  with respect to a rotating frame, 124–126, 132–143
  with respect to a translating frame, 31–33
Relative velocity (*see* Velocity, relative)
Repeated roots, 170, 327
Rigid body mode, 323
Rocket
  burning rate, 87
  burnout mass, 87
  mass ratio, 99
  multistage, 99

[Rocket]
  specific impulse 88
  thrust (static), 88
  two stage, 98, 399–400
Rotation
  about a fixed axis, 102–103
  about a fixed point, 101–108
  with three degrees of rotational
    freedom, 106–108
  with two degrees of rotational
    freedom, 104–106
Routh, E. J., 258
Routh-Hurwitz criteria, 352
Routhian, 258, 284–285
Runge-Kutta (*see* Numerical integration,
  Runge-Kutta fourth-order method)

Scotch-yoke mechanism, 135–136, 196
Self-starting methods, 33
Simple harmonic motion, 16–25
Sleeping top, 236–237
Space cone, 224–225
Spectral matrix, 335
Spin rate, 107, 227, 233, 235
Spinning top, 107, 230–237
Stability
  of perturbed steady motion, 352–355
  Routh-Hurwitz criteria for, 352
State space, 333
State variables, 333
Static coupling, 327
Steady precession (*see* Precession,
  steady)
Stiffness matrix, 322, 342
System matrix, 333

Tensor, second-order, 165–166
TERMINAL section, 375, 385
TERMT command, 375, 385
TFIN specification, 375
Time of orbital travel, 292
Trace of a matrix, 167, 178
Transformations
  coordinate, 126–128
  of inertia (*see* Inertia, transformation
  orthogonal (*see* Orthogonal
    transformations)
  similarity, 165–167, 177
  vector, 126–132
Transition matrix, 334

Translation
  of axes (*see* Parallel-axis theorem)
  of a rigid body, 101
Two-body problem, 286–295

Units
  absolute, 4
  American, 5
  gravitational, 6
  SI (Système International d'Unités), 4

Van der Pol's equation, 364
Variable mass system, 86–89
Vector transformations (*see*
  Transformations, vector)
Velocity,
  absolute, 10
  angular, 102–108
  in cartesian coordinates, 11–15
  in cylindrical coordinates, 29–30
  in radial and transverse coordinates,
    27–28
  relative
    with respect to a rotating frame,
      124–126
    with respect to a translating frame,
      31–33
  in spherical coordinates, 30
  tangential, 10
Vibration,
  absorber, 332
  accelerometer, 49
  circular frequency, 16
  critically damped, 18–19
  damped
    forced single degree of freedom, 19–25
    forced multidegrees of freedom,
      333–337
    free, 18–19
  undamped
    forced, 16–18
    free, 16
  logarithmic decrement, 48
  magnification factor, 17
  natural frequency, 16
  period, 16
  seismometer, 49
Virtual displacement, 263
Virtual work, 263

Vis viva integral, 291

Whitworth quick-return mechanism,
　　136–138
Work
　external, 62, 81, 186

[Work]
　of a force, 55
　internal, 62, 81
　rotational, 82, 186
　translational, 82, 186